Advances in
Nuclear Science
and Technology

VOLUME 11

Advances in Nuclear Science and Technology

Series Editors

Ernest J. Henley
University of Houston, Houston, Texas, U. S. A.

Jeffery Lewins
University of London, London, England

Martin Becker
Rensselaer Polytechnic Institute, Troy, New York, U.S.A.

Editorial Board

A Continuation Order Plan is available for this series. A continuation order will bring delivery of each new volume immediately upon publication. Volumes are billed only upon actual shipment. For further information please contact the publisher.

Advances in Nuclear Science and Technology

VOLUME 11

Edited by

Ernest J. Henley
University of Houston
Houston, Texas, U.S.A.

Jeffery Lewins
University of London
London, England

and

Martin Becker
Rensselaer Polytechnic Institute
Troy, New York, U.S.A.

SPRINGER SCIENCE+BUSINESS MEDIA, LLC

The Library of Congress cataloged the first volume of this title as follows:

Advances in nuclear science and technology. v. 1 —
   1962—
New York, Academic Press.
   v. Illus., diagrs. 24 cm. annual.
   Editors: 1962—   E. J. Henley and H. Kouts.
   1. Nuclear engineering—Yearbooks. 2. Nuclear physics—Yearbooks.
   I. Henley, Ernest J., ed. II. Kouts, Herbert J., 1919-   ed.
   TK9001 .A3   621.48058   62-13039

Library of Congress Catalog Card Number 62-13039
ISBN 978-1-4613-2864-3   ISBN 978-1-4613-2862-9 (eBook)
DOI 10.1007/978-1-4613-2862-9

© 1979 Springer Science+Business Media New York
Originally published by Plenum Press, New York in 1979

Preface

The present volume in our annual review series reviews a wide range of developments, giving a broad interpretation to the "technology" of our title. Starting at the beginning, Science, we have the review of basic nuclear physics data of Walker and Weaver for reactor kinetics, particularly, therefore, delayed neutron data. In the search for better and better accuracy, it is being realized that this involves the closest scrutiny of fundamental data, given to us here from the Birmingham school. Associated with this review of data is the review from Italy by Professor Pacilio and his co-workers of the theory of reactor kinetics in the stochastic form, and a valuable compilation of the theory underlying a wide range of practical techniques.

Tending more to technology come the papers by Jervis, reviewing the application of digital computers to the control of large nuclear power stations as developed in both the United Kingdom and Canada, Pickman's review of the design of fuels for heavy water reactors, and the account by Ishikawa and Inabe of the new Japanese Research Reactor Program, itself initially directed largely to fuel element studies.

The balance of the volume is made up of more philosophical contributions to the practicalities of nuclear power. Much has been said about plutonium (indeed, much that did not deserve to be said) in the last two years; Puechl's review of its role in relation to breeding and recycling is a welcome, sane, summary of the problems and advantages. Similarly, the paper by McNelly on the usefulness of Nuclear Power Parks comes at an opportune time in the development of national programs that seek to retain and exploit the advantages of nuclear power in acceptable ways. All those who are engaged in such applications have a responsibility to see that the education and training of professional scientists and engineers has the highest technical and ethical content, so it is not inappropriate to see the review of computer-assisted learning for nuclear engineers by Smith included in this volume as well.

The year of the Carter initiative has passed, and it is not for us to give premature judgment. Clearly, much that has been aired needed saying, if only to expose it to critical examination. Not all the initiatives that came from the U.S.A. appear permanent, either within the U.S.A. itself, or by adoption elsewhere in the world. Clearly, the major European countries led by France and Germany see the valuable benefits that nuclear programs can bring to countries with little indigenous coal left and no indigenous oil; the United Kingdom has no pressing reason for an early decision (for which she is no doubt thankful) but the present review of the Japanese research reactor program given here must point up the difficult economic circumstances of that country and the strong pressure there will be to seek the benefits of nuclear power.

Some of the issues concerning fast breeders and the recycling of plutonium have been dealt with here within this and previous volumes of the series; other issues raise broader questions of public policy than can be analyzed here. Yet it is important to remember that sound analysis, judgment, and selection of a course of action in any nation requires acknowledgment of physical and technical facts, rather than prejudices and speculation. We may hope that in this volume, therefore, we and our authors contribute to such a balance.

                                        E. J. Henley

                                        J. Lewins

                                        M. Becker

Contents

Fuel for the SGHWR
    D. O. Pickman, J. H. Gittus,
    and K. M. Rose

The Nuclear Safety Research Reactor (NSRR)
        in Japan
    M. Ishikawa and T. Inabe

Contents of Volume 10

Volumes 1 - 9 of *Advances in Nuclear Science and Technology*
were published by Academic Press, New York.

NUCLEAR PHYSICS DATA FOR REACTOR KINETICS

J. Walker and D. R. Weaver

Department of Physics and Radiation Centre
University of Birmingham, Birmingham, England

I. PRINCIPLES OF REACTOR KINETICS AND DATA REQUIREMENTS

The type of information required for reactor kinetics calculations can be seen by inspection of a neutron balance equation, even the following simple one without independent (external) sources of neutrons.

$$\frac{dN(t)}{dt} = \frac{\rho - \beta_{eff}}{\Lambda} N(t) + \sum_{i=1}^{I} \lambda_i C_i(t) \tag{1}$$

where $N(t)$ is the number of neutrons, Λ is their reproduction time (closely, the neutron lifetime), ρ is the reactivity of the system, β_{eff} is a factor (usually assumed to be constant but which actually changes with time to some extent) which allows for the fraction of neutrons being delayed and for the difference in importance between them and the prompt neutrons, and $\sum \lambda_i C_i(t)$ gives the neutrons from the decay of precursors formed earlier. The summation in the last term is carried out over the $\underline{I}$ (usually 6) groups, each with a characteristic decay constant, that are used to give a mathematical representation of the decay of the delayed neutrons after a fission burst.

The last term makes immediately obvious the need for data on delayed-neutron precursors that are well known as a special group of nuclides in the whole range of fission products, but they are not the only ones of importance to reactor kinetics; neutron absorbers such as ^{135}Xe may have a marked effect on ρ, and therefore have to be included. Contained in ρ are also the effects of changes in the heavy elements in the reactor fuels. These changes and the growth of some of the fission product absorbers influence reactor behavior on a much longer time scale than do delayed neutrons or the short-lived absorbers, but we have included them in this review.

The impact of nuclear properties can be seen in more detail from the multigroup form of the neutron balance equation in which the whole spectrum of neutrons is divided into a number of energy groups, with a balance equation such as equation (2) applying to each; information on the delayed-neutron precursors is provided by equation (3).

$$
\begin{aligned}
\frac{1}{v^g} \frac{\partial \phi^g(\underline{r},t)}{\partial t} =\ & \nabla \cdot D^g(\underline{r},t) \nabla \phi^g(\underline{r},t) \\
& - \Sigma_a^g(\underline{r},t) \phi^g(\underline{r},t) \\
& + \sum_{g'=g+1}^{G} \Sigma_s^{g' \to g}(\underline{r},t) \phi^{g'}(\underline{r},t) \\
& + \sum_{m=1}^{M} (1-\beta_m) \chi_m^g \sum_{g'=1}^{G} \nu_m^{g'} \Sigma_{f_m}^{g'}(\underline{r},t) \phi^{g'}(\underline{r},t) \\
& + \sum_{m=1}^{M} \sum_{i=1}^{I} \chi_m^{i,g} \lambda_m^i c_m^i(\underline{r},t) \quad\quad (2)
\end{aligned}
$$

$$
\frac{dc_m^i(\underline{r},t)}{dt} = \beta_m^i \sum_{g'=1}^{G} \nu_m^{g'} \Sigma_{f_m}^{g'}(\underline{r},t) \phi^{g'}(\underline{r},t) - \lambda_m^i c_m^i(\underline{r},t)
$$

$$
(3)
$$

In these equations, m represents one of M fissioning nuclides, g represents one of G energy groups, and i represents one of I delayed-neutron groups, and:

$\phi^g(\underline{r}, t\,\underline{\ell})$ is the flux in energy group g at the point $\underline{r}$ and at time t

D^g is the group diffusion constant

Σ_a^g is the cross-section for removal from group g, and includes absorption and scattering

$\Sigma_s^{g' \to g}$ is the cross-section for scattering from group g' to group g (down scattering assumed)

$\Sigma_{f_m}^{g'}$ is the fission cross-section of nuclide m for neutrons in group g'

$\nu_m^{g'}$ is the number of neutrons for each of these fissions

$\beta_i (= \sum\limits_i^I \beta_m^i)$ is the fraction of neutrons emitted by fission of the nuclide m which are delayed (β_m^i is the delayed neutron fraction emitted from nuclide m into delayed neutron group i)

χ_m^g is the fraction of fission neutrons that enter group g

$\chi_m^{i,g}$ is the fraction of delayed neutrons that enter group g

c_m^i is the concentration of precursor i from fissioning nuclide m

The roles of the various cross-sections are clear from the equations, as is the need for information on the energies of delayed neutrons (and of the prompt ones, for that matter) as well as on their yields and decay constants. The actual

selection of the necessary nuclear data obviously becomes
very complex and extensive, and because of this, the main
objective of this paper has been more to direct readers to
the sources of data rather than to attempt to present the
data themselves. This is the reasoning behind the inclusion
of comments on the various computer libraries that are now
available; in addition to the libraries, the compilation of
neutron cross-sections by the Brookhaven National Laboratory
(BNL325)(1,2) and the CINDA bibliography(3) are important
sources of information. Actual data have been included when
necessary to illustrate main features, when they could be
kept within acceptable bounds.

It is important to realize that full reactor kinetics
calculations embrace much more than nuclear information
and, indeed, more than the properties of a reactor itself;
changes in coolant temperature, for example, depend not only
on flux changes in the reactor but also on the equipment
outside the reactor core. Changes in temperature within
the core alter densities, and consequently, macroscopic
cross-sections, as well as neutron spectra, and thus influence
reaction rates. The speed with which a temperature change
affects reactivity depends, of course, on where it occurs,
and on the materials and structure of a reactor. When addit-
ionally, it is realized that equations (2) and (3) must be
integrated over the whole of a reactor and thus also introduce its structural features, it becomes clear that kinetic
calculations in their entirety must be related to specific
systems. Nothing on these matters has been included in this
paper, which concentrates on the explicitly nuclear aspects
of the data for reactor kinetics.

II. EVALUATED NUCLEAR DATA LIBRARIES

Few, if any, kinetics calculations have the ability or
the necessity to use all the·data now available in the several libraries. They are usually performed with multigroup
data sets which, in the case of complex neutronics-hydraulics
calculations, may be limited to very few groups indeed for
the sake of computer storage. However, in order to generate
the flux-averaged cross-section sets, it is necessary to use
the more detailed information in the libraries. A frequent
situation that may arise is that the library data, which
are fairly generally available, are collapsed into group
format and then adjusted to fit the results of integral
measurements. Reactor manufacturers often regard

these adjusted group data sets as proprietary information
(Poncelet et al. (4)) so that these are not available outside the company. As in addition, the group data sets are
so many and various, it is not possible, as mentioned, to
contemplate a complete review of data that are the immediate
input to any kinetics program. It is more sensible to step
back from the group data to the basic libraries that are
more widely available and not limited to particular reactor
systems; Pearlstein has discussed the development of the
libraries in a previous paper in this series (5). The whole
question of whether the private data set used by a reactor
vendor to calculate core characteristics should be available
to a customer is discussed in the paper by Poncelet et al.
on "Proprietary, Standard and Government-supported Nuclear
Data Bases" (4). Table I is a list of the nuclides for which
neutron cross-sections are given in the five most frequently
used libraries.

A.   ENDF - Evaluated Nuclear Data File

       Probably the most extensive and widely-used file is the
ENDF/B file maintained by the National Neutron Cross-Section
Center (NNCSC) at the Brookhaven National Laboratory. It
has passed through several versions since 1968, the current
one being ENDF/B IV which was released in 1974. Descriptions
of data formats and procedures can be found in Garber et al.
(6); this paper, the latest of a series, brings together
the descriptions of neutron and photon formats. Two different evaluated data libraries are kept at the NNCSC; the
ENDF/A library consists of evaluations of data that may not
be complete in that only limited energies and cross-sections
are covered; also, more than one evaluation of the same
cross-section for the same nuclide may appear in the A library. The B library, on the other hand, is complete, and
a single evaluation is presented for each cross-section of
a particular material. The task of selecting the data sets
to be contained in the B library is undertaken by the Cross-
section Evaluation Working Group, a body composed of repre-
sentatives of the U. S. national laboratories, the reactor
vendors and other interested parties. From time to time
they may recommend a change in the data set which represents
the cross-section for a particular material; the recommen-
dation may arise from users indicating the need for an improved representation of a cross-section, or it may arise
from new and significant experimental results, either from

TABLE I

LIST OF NUCLIDES FOR WHICH NEUTRON
CROSS—SECTIONS ARE AVAILABLE IN LIBRARIES

		ENDF/B IV	ENDL	UKNDL	KEDAK 3	SOKRATOR
1	H	1, 2, 3	1, 2, 3	1 in H_2O 2 in D_2O 3	1, 1 in H_2 1 in H_2O, 2	2, 3
2	He	3, 4	3, 4	3, 4	3, 4	3, 4
3	Li	6, 7	6, 7	6, 7		6, 7
4	Be	9	7, 9	9		9
5	B	10, 11	10, 11	10, 11		natural
6	C	12	12	natural	12	12
7	N	14	14	natural	natural	14
8	O	16	16	natural	16	16
9	F	natural	19	19		19
10	Ne					
11	Na	23	23	23	23	23
12	Mg	natural	natural, 24			24
13	Al	27	27	27	27	27
14	Si	natural	natural	natural		28
15	P		31			31
16	S	32	32			32
17	Cl	natural	natural	natural	natural 35, 37	

TABLE I. (cont'd)

		ENDF/B IV	ENDL	UKNDL	KEDAK 3	SOKRATOR
18	Ar		natural			
19	K	natural	natural	natural		39
20	Ca	natural	natural	natural		natural
21	Sc	45	45			
22	Ti	natural 46, 47, 48	natural	natural		48
23	V	natural	51	natural		51
24	Cr	natural	natural	natural	natural 50, 52 53, 54	52
25	Mn	55	55			natural
26	Fe	natural 54, 56, 58	natural 54, 58	natural[a]	natural 54, 56 57, 58	natural 56
27	Co	59	59			59
28	Ni	natural 58, 60	natural 58	natural	natural 58, 60, 61, 62, 64	59
29	Cu	natural 63, 65	natural 63	natural 63, 65		65
30	Zn		64			64
31	Ga		natural	natural		

TABLE I.   (cont'd)

		ENDF/B IV	ENDL	UKNDL	KEDAK 3	SOKRATOR
32	Ge	72, 73 74, 76				
33	As	75				
34	Se	76, 77, 78, 80, 82				
35	Br	78, 81				
36	Kr	78, 80, 82, 83, 84, 85, 86				
37	Rb	85, 86, 87				
38	Sr	86, 87, 88, 89, 90				
39	Y	89, 90, 91				89
40	Zr	90, 91, 92, 93, 94, 95, 96	natural 90	natural		91
41	Nb	93, 94, 95	93	93		93
42	Mo	natural 94, 95, 96, 97, 98, 99, 100	natural	natural	natural 92, 94, 95, 96, 97, 98, 100	96
43	Tc	99				

TABLE I.   (cont'd)

	ENDF/B IV	ENDL	UKNDL	KEDAK 3	SOKRATOR
44 Ru	99, 100, 101, 102, 103, 104, 105, 106				
45 Rh	103, 105				
46 Pd	104, 105, 106, 107, 108, 110				
47 Ag	107, 109, 111	107, 109	107, 109		
48 Cd	natural, 108, 110, 111, 112, 113, 114, 115m, 116	natural, 114	natural, 113	natural,	
49 In	113, 115	115			
50 Sn	115, 116, 117, 118, 119, 120, 122, 123, 124, 125, 126	natural			119

TABLE I.   (cont'd)

		ENDF/B IV	ENDL	UKNDL	KEDAK 3	SOKRATOR
51	Sb	121, 123, 124, 125, 126				
52	Te	122, 123, 124, 125, 126, 127m, 129m, 130 132				
53	I	127, 129, 130, 131, 135				
54	Xe	124, 126, 128, 129, 130, 131, 132, 133, 134, 135, 136	135			
55	Cs	133, 134, 135, 136, 137				
56	Ba	134, 135, 136, 137, 138, 140	138			

TABLE I. (cont'd)

		ENDF/B IV	ENDL	UKNDL	KEDAK 3	SOKRATOR
57	La	139, 140				
58	Ce	140, 141,				
		142, 143,				
		144				
59	Pr	141, 142,				
		143				
60	Nd	142, 143,				
		144, 145,				
		146, 147,				
		148, 150				
61	Pm	147, 148,				
		148m, 149,				
		151				
62	Sm	147, 148,				
		149, 150,				
		151, 152,				
		153, 154				
63	Eu	151, 152,	natural	151, 153		
		153, 154,				
		155, 156,				
		157				

J. WALKER AND D. R. WEAVER

TABLE I.   (cont'd)

		ENDF/B IV	ENDL	UKNDL	KEDAK 3	SOKRATOR
64	Gd	natural, 154, 155, 156, 157, 158, 160	natural			
65	Tb	159, 160				
66	Dy	160, 161, 162, 163, 164				
67	Ho	165	165			
68	Er	166, 167				
69	Tm					
70	Yb					
71	Lu	175, 176				
72	Hf					
73	Ta	181, 182	181	natural		181
74	W	182, 183, 184, 186	natural, 186	natural		natural
75	Re	185, 187	185, 187			
76	Os					
77	Ir		191, 193			
78	Pt		natural			
79	Au	197	197	197		

TABLE I.   (cont'd)

		ENDF/B IV	ENDL	UKNDL	KEDAK 3	SOKRATOR
80	Hg					
81	Tl					
82	Pb	natural	natural	natural		207
83	Bi					209
84	Po					
85	At					
86	Rn					
87	Fr					
88	Ra					
89	Ac					
90	Th	232	232	232		232
91	Pa	233		233		
92	U	233, 234, 235, 236, 238	233, 234 235, 236, 237, 238 239, 240	233, 234 235, 236, 238, 239 239, 240	235, 238	235, 238
93	Np	237	237			
94	Pu	238, 239, 240, 241 242	238, 239, 240, 241 242, 243	238, 239[a], 240[a], 241[a], 242	238, 239, 240, 241, 242	239
95	Am	241, 243	241, 242, 243	241, 243		

J. WALKER AND D. R. WEAVER

TABLE I.   (cont'd)

	ENDF/B IV	ENDL	UKNDL	KEDAK 3	SOKRATOR
96 Cm	244	242, 243,	244		
		244, 245			
		246, 247,			
		248			
97 Bk		249			
98 Cf		249, 250,			
		251, 252			

a   Two evaluations are included in the file.

direct measurements of cross-sections or from integral experiments that show that the existing data are in error. The next ENDF/B library, version V, is not expected to appear before August, 1977 (Dunford (7)).

A useful document by Garber (8) gives brief information on the evaluation of each data file in greater detail than in the textual information of the file itself, but in less detail than in the original evaluation report. Similarly, Magurno has given information on the ENDF/B IV dosimetry file (9) which contains 36 reactions in 26 nuclides, and a paper on the ENDF/B fission product decay data is in press (7).

A recent development in the data file is the inclusion of uncertainty information; Peelle (10) and Perey (11) discussed the provision of covariance files in ENDF/B at the 1975 Washington conference on Nuclear Cross-Sections and Technology. This information is invaluable in sensitivity calculations (e.g., Weisbin et al. (12)) and more uncertainty data can be expected as the file progresses.

B.   ENDL - The Livermore Evaluated Nuclear Data Library

This evaluation was begun in 1958 and now contains 84 nuclides; for 72 of these, the data are completely specified in the energy range* 10^{-4} eV to 20 MeV and include cross-sections, energy distribution for secondary particles, and angular distribution. For the other 12 isotopes, only one or two reactions are covered. Data on the production of gamma rays and on cross-sections for reactions with them are also included, as well as charged-particle cross-sections for a few light nuclides. The current description by Howerton et al. (13) notes the guideline on which the library is based: that the least detail is used to present the experimental information. Some of the evaluations are based on theoretical calculations and some experimental data are not included if there is no requirement for them in neutron and photon calculations in physical systems. A 'fission product' material is included that simulates to some degree the average behavior of all fission products.

---

* 1 eV = 0.160206 aJ.

There are two versions of ENDL extant simultaneously, a difference compared with ENDF/B. The first is the 'floor' version which is maintained unchanged for about a year, whereas the 'current' version is updated continuously. Translation of the ENDL 76 version into ENDF/B format is explained in Howerton (14).

C.   UKNDL - United Kingdom Nuclear Data Library

The last major reediting of the UKNDL was performed in 1973 and is documented by Pope (15). It contains 64 files for 60 materials, with more than one file for Fe, ^{239}Pu and ^{241}Pu; those for plutonium represent separate evaluations at Winfrith and Saclay, but both are in the UKNDL format. Tabulations of thermal neutron cross-sections, resonance integrals and reaction cross-sections averaged over a fission spectrum are given by Pope and Story (16). A separate file, less complete in that it contains only a few cross-sections for each material, is the Fission Product Library; it contains 524 files that have been derived from a variety of sources and a description is given by Pope (17).

D.   KEDAK - Karlsruhe Evaluated Data File

KEDAK-3 was introduced in October, 1975, and is documented by Goel and Krieg (18). A useful graphical presentation of the data has been published for the nonfissile materials (Goel)(19) and it is intended to produce a second section for fissile materials. Programs for management and retrieval of data from KEDAK are described by Krieg (20).

E.   SOKRATOR

This Russian file resembles the UKNDL format in many ways; a comparison of the two libraries has been made by Ribon (21). The contents of the library have been transmitted to CCDN Saclay, and a list of data held is given in the CCDN newsletters 74.1 and 76.1; the relevant information for Table I was obtained from the newsletters (22). Recently, a note on changes and additions in the format of the SOKRATOR evaluated library has been produced by Nikolaev (23).

F.   A Few Comments on the Libraries

It is important to understand the way in which a par-

ticular evaluated library is conceived because this will
have influenced the data it contains. One example of a
divergence of opinion between evaluators is the question of
whether or not to adjust evaluations based on microscopic
data so that agreement with integral experiments is obtained;
Karam(24) gives an indication of the literature on this subject. It is not the role of this review to speak for or
against such adjustments, but rather to indicate that the
user of library data should be clear about their origins.

The recent meeting in Trieste on "Nuclear Theory in
Neutron Nuclear Data Evaluation" (25) summarized the current
accuracy of data on cross-sections. For ^{235}U, ^{238}U and
^{239}Pu, it noted that the 2200 m/s cross-sections are known
to better than 1%, but despite decades of work, the fission
cross-sections for fast neutrons are known only to about 5%,
whereas 1% is desirable for reactor calculations; for other
nuclides, the situation is not even as good. The effects of
data uncertainties on reactor design have been reviewed by
McKnight et al. (26) and Paik (27).

III. FISSION-PRODUCT NUCLEAR DATA

It is perhaps within the field of fission-product data
that the greatest changes have occurred over the last few
years in the evaluations available for kinetics calculations.
While the basic cross-section data in the libraries (ENDF/B,
UKNDL, etc.) have been refined, there has not been the large-
scale reorganization that can be seen for fission products.
ENDF/B has gained a fission-product file and the UK Chemical
Nuclear Data Committee is engaged on a new file of decay
schemes that will be in ENDF/B format. Perhaps an indication
of the quantity of information involved is the size of the
report of the Panel on Fission-Product Nuclear Data, held
in Bologna in November, 1973 (28); it extends to three volumes and totals over 1,000 pages. Unfortunately for the
present review, there will not be another meeting of the
panel until September, 1977, (Cuninghame)(29) which is after
the closing date for this particular report.

A.   Definitions

For the sake of clarity, we shall define the terms that
it will be necessary to use in this section. When a nucleus
undergoes fission, it divides most frequently into two

fission fragments (binary fission). In about 10% to 20% of
fissions, a neutron may be emitted at the time of scission
(Nair et al. (30)) and in less than 1% of fissions a light
charged particle is emitted (ternary fission); very rarely
are three similar-sized fragments formed. In times of the
order of 10^{-14}s and 10^{-11}s, respectively, the fission fragments emit neutrons and gamma rays and leave the primary
fission products; these then undergo beta decay to form
secondary fission products that continue decaying down a
chain until stability is attained. The term prompt neutrons
is used for those neutrons emitted following fission and before beta decay of the primary fission products; it is not
usual to distinguish between scission neutrons and those
emitted from fission fragments. The independent or direct
fission yield is the probability of formation of a particular
primary fission product; that is, before beta decay, but
after any prompt neutron emission. Most frequently, independent yields are quoted as fractions of the chain yield.
Cumulative yield is the probability of formation of a particular fission product after the beta decay of all its precursors. It is thus the sum of its own independent yield
and the independent yields of its precursors including, if
necessary, a contribution from delayed-neutron emission.
Like independent yields, cumulative yields are often quoted
as fractions of the chain yield. A chain yield is the probability of formation of the stable fission product at the
end of a beta-decay chain, and is therefore equal to the
cumulative yield of that stable nuclide. In some circumstances, it is necessary to define a total chain yield; in
the mass chain 136, for example, ^{136}Xe is stable and therefore does not decay to ^{136}Cs. However, ^{136}Cs has a direct
yield of approximately 1% in the fission of ^{233}U and ^{239}Pu
(Walker (31)). Thus, the chain yield of ^{136}Xe does not represent the chain yield of the whole mass 136 chain, and in
this situation, the total chain yield is defined to include
the additional independent yields of those nuclides beyond
a stable mid-chain nuclide. In the 136 chain, ^{136}Cs is said
to be 'shielded' as it can have no yield from beta decay of
the chain.

        For the purposes of this review, the data relating to
delayed-neutron emission have been treated in a separate
section because of their particular importance to reactor
kinetics. In many respects, however, it is not sensible to
separate delayed-neutron precursors from other fission pro-

ducts and there will be implicit reference to them in the sections on yields and decay data.

B.   The Bologna Panel on Fission Product Nuclear Data

Prior to this IAEA meeting, 18 review papers were circulated, and it was particularly requested that these should contain users' requirements with priorities, and that the papers should include information from laboratories around the world and thus be international in scope. Of particular relevance to kinetics are: Valente (32), who gives a list of compilations and evaluations of fission-product nuclear data and associated computer codes; Tyror (33) and Devillers (34) on the importance of fission products to the physics and engineering aspects, respectively, of reactor design; Ribon and Krebs (35) on fission-product neutron cross-sections; Walker (31) and Cuninghame (36) on yields from thermal and fast fission, respectively; Rudstam (37) on decay data; and Amiel (38) on delayed-neutron data. A useful review of fission product systematics and their use in predicting unknown data was presented by Musgrove et al. (39).

Having had the reviews prior to the meeting, the panel was able to concentrate on producing a number of observations and recommendations. The publication of newsletters on current progress in fission-product nuclear data was recommended and this resulted in a series of reports, "Progress in Fission Product Nuclear Data," published by the International Nuclear Data Committee (40,41). These reports also include a section on discrepancies that have become apparent in the data. An area that particularly exercised the Bologna panel was the difficulty that evaluators had in interpreting measurements on fission products, and in assigning importance to particular data because of lack of information in experimental reports. A note for measurers has been prepared to indicate the detail required during evaluation. This may be obtained from G. Lammer of the Nuclear Data Section of the IAEA, Vienna.

C.   Yields

One of the recommendations of the Bologna panel was that several evaluations should continue around the world. This was felt to be better than just one evaluation by an international group because individual evaluators have particular methods of analysis and opinions on data, which en-

courages constructive comparisons; for example, Walker (42)
places particular emphasis on mass spectrometric results,
whereas Crouch (43) does not. However, the panel did suggest
that there should be an internationally agreed data base for
evaluations.

Recently published evaluation are those by Crouch (44-
46), Walker (42), Meek and Rider (47), Devillers et al. (48)
and Lammer and Eder (49). In the Bologna paper (31), Walker
also gave a review of the methods used to measure fission
yields and the procedures employed by evaluators, while
Cuninghame in the companion paper (36) included a useful section on ternary fission yields. The recent ENDF/B IV Fission
Product File made particular use of the evaluations by Meek
and Rider and by Walker.

Despite the large quantity of experimental information,
it is not possible to construct the whole yield curve from
experimental results, even for ^{235}U; the less common nuclides
such as ^{233}U or ^{241}Pu have little data available. To fill
the gaps, use is made of the observed systematics of fission-
product formation; the work by Musgrove et al. (39) on this
topic, already mentioned, includes an analysis of the ^{235}U
yield curve as the sum of five Gaussian peaks and is plotted
in Figure 1. Obviously, certain conservation laws must be
obeyed in the formation of fission products; the number of
nucleons is constant, as is the number of protons before
beta decay; the sum of all the chain yields must equal two
and the two peaks must each contribute an equal proportion
to the total yield. (The division between the light and
heavy peaks should be taken at $(A - \tilde{\nu}_{sym})/2$, where A is the
mass of the compound nucleus which undergoes fission and
$\tilde{\nu}_{sym}$ is the average number of neutrons emitted in symmetrical
fission). Once systematics have been employed to fill in
missing data, the results can be normalized to obey the
conservation laws. The evaluations of Walker (42) and Meek
and Rider (47) have been normalized so that each peak sums
to 100%, and Crouch has applied a method of simultaneously
fitting to all the conservation laws (44,45) in his latest
evaluation (46).

To calculate independent yields from mass distribution,
it is necessary to know the distribution of yield as a function of charge for each isobar. As might be expected, this·
information is scarce, and in many cases where there are no
data, assumptions on the distribution have to be made.

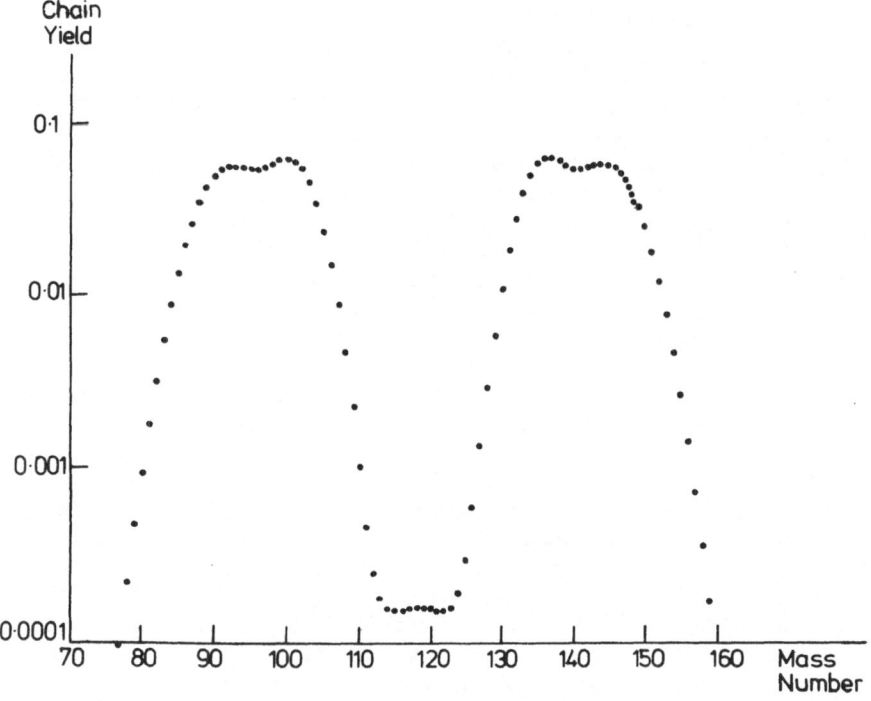

Figure 1

^{235}U Mass Yield Curve
(from analysis by Musgrove et al.)

        The most common form assumed is that the charges are
distributed normally, with a standard deviation σ, about a
mean Z_p, the most probable charge for a given mass chain;
following Wahl et al. (50,51) a value of 0.56 ± 0.06 for σ
has often been used when actual information was not available. However, with the new mass separators such as Lohen-
grin at Grenoble becoming available, more experimental evidence is being published. For example, Siegert et al. (52)
have shown that in thermal neutron fission of ^{235}U, the
width of the charge distribution in the light mass peak
varies sinusoidally as a function of the chain mass number,

and exhibits minima when the mean charge, Z_p, for each mass
number is even. A recent tabulation of Z_p as a function of
mass number will be found in the paper by Nethaway (53). A
second feature of the charge distribution that is still not
well known for all nuclides is the odd-even effect. This is
the observation that those nuclides in a particular isobar
with an even number of protons have an enhanced yield relative to those with an odd number. The effect can be quite
marked; for thermal fission of ^{233}U and ^{235}U, Amiel and
Feldstein (54) give an enhancement for even proton numbers
of 22 ± 7% over the average distribution for both even and
odd, but only 8 ± 4% in the fast fission of ^{235}U. It is
thought that an equivalent neutron odd-even effect, which
may be present in the initial fission-fragment distribution,
is lost after the emission of prompt neutrons.

Maeck (55) has indicated the need for care when refer-
ring to 'fast' neutron fission because the term can signify
spectra over a large range of energies; this has been confirmed by recent measurements with monoenergetic neutrons by
Cuninghame et al. (56) which have shown that yields in the
wings and valley of the yield distribution rise monotonically
with energy in the range 0.13 - 1.7 MeV. Also, it is known
(e.g., Van Assche et al. (57)) that the ratios of peak-to-
valley yields vary significantly from resonance to resonance
at lower energies.

There is no intention to give a complete review here of
recent yield measurements, but information concerning current
techniques may be obtained from references (58-73).

D.   Decay Data

Complementary to the yield data for the calculation of
the quantities of fission products present in a reactor at
any given time is the knowledge of their decay parameters.
The topic was reviewed at the Bologna meeting by Rüdstam (37).
Compilations of data are maintained; Blachot (74) has a series
that contains half-lives, Q values and branching ratios for
decay modes, energies and intensities of β^- and β^+ particles,
isometric transitions and gamma rays. Tobias (75) has a similar file that contains diagrammatic displays of the beta-
decay chains for mass numbers 72-161 and 166. Work is in
progress on a new Chemical Nuclear Data File in the United
Kingdom, which will be discussed more fully in the section

on data libraries. The ENDF/B Fission Product File also contains decay data, a description of which will be found in the paper by Reich et al. (76), and a convenient summary of the file in tabular form has been published by England and Schenter (77). Valente's paper at the Bologna meeting (32) lists evaluations, in particular subsections such as half-lives or gamma-ray spectra. Again, it is not intended to review actual measurements, but some idea of the data being published will be obtained from references (71) and (78-89).

E.   Neutron Cross-sections

For kinetics calculations, the absorption cross-sections of certain fission products are of prime importance; the xenon poisoning of a thermal reactor is a well-known example. The Bologna review was performed by Ribon and Krebs (35) and they note that while the libraries contain some 800 fission products, only about 150 have cross-sections that must be accurately known.

It is common to divide the fission products into 'saturating' or 'nonsaturating'. The former group consists of those products that decay rapidly or that have high absorption cross-sections, leading to the condition $(\sigma_c\phi+\lambda)t \gg 1$ being satisfied; σ_c is the microscopic capture cross-section, ϕ the neutron flux, λ the decay constant and t the time that the reactor has been at power. If the condition is satisfied, the ratio of the neutron capture rate in the fission product to the total absorption rate in the reactor becomes independent of time, and if $\sigma_c\phi \gg \lambda$, then this ratio is also independent of σ_c. For example, the effect is seen in ^{135}Xe, ^{147}Pm and ^{149}Sm in resonances in the thermal region. Because there are no equivalent regions of high cross-section, saturation does not occur at energies of interest to fast reactors. Nonsaturating fission products are those in which the condition on cross-section, flux, decay constant and time is not obeyed, and in these cases, the ratio of neutrons absorbed in the fission product increases with time. A more complete discussion of the effect will be found in Tyror (33).

Evaluations of cross-sections of fission products tend to occur in a more piecemeal fashion than evaluation of their yields. The current evaluation being used in ENDF/B will be found in the summary documentation by Garber (8), and England and Schenter (77) note that 181 nuclides have complete cross-sections from 10^{-5} eV to 20 MeV in the fission product file.

The UKNDL contains fission product cross-sections from
several evaluations, some of which include the same nuclides.
Pope compared these evaluations (17) by considering the
thermal neutron cross-sections averaged over a Maxwellian
spectrum, epithermal resonance integrals over a 1/E (energy)
flux from 0.55 eV to 2 MeV, and cross-sections averaged over
a fission spectrum. An evaluation of 28 fission products
was reported by Igarasi et al. (90) at the Washington conference on Nuclear Cross-Sections and Technology 1975, and
Gruppelaar (91) referred to the calculational models used in
estimating fission product cross-sections at the Trieste
consultants' meeting on the use of nuclear theory in neutron
nuclear data evaluation. Recent measurements of cross-
sections will be found in references (67) and (92-96), and
the CINDA bibliography (3) is invaluable for references on
particular nuclides.

F.   Major Libraries of Fission Product Data

ENDF/B IV Fission Product File

This file, the work of about 30 people over 2-1/2 years,
expanded from 55 nuclides to 824 the data on cross-sections,
decay parameters, and yields. The main motivation was to
satisfy the needs of calculations on loss-of-coolant accidents, but as can be seen from the Bologna meeting, fission
product data are required for many other aspects of reactor
work, and the file certainly will find wide uses.

UK Chemical Nuclear Data File

Work on this new file has been started, and a library
of alpha decay schemes has been completed already by Rogers
(97), albeit in non-ENDF format. The main aim is the production of a decay-scheme file in ENDF/B format for fission
products that will include nuclides produced by activation
of stable fission products. Some work is in progress on decay data for activation products of structural materials and
actinides, but other nuclides have lower priority. As to
yields, Crouch is continuing his series of evaluations; the
results of adjusting the evaluated chain and independent
yields so that conservation of physical laws is obtained (44,
45) have been converted to ENDF/B format by James (98), and
as already mentioned, a new evaluation of chain and independent yields for thermal, fast and 14 MeV fission has been
submitted for publication (46).

Until the work of the team on the Chemical Nuclear Data
File is sufficiently advanced, the earlier work of Tobias
(75) is available; so is the ENDF/B file. In addition, there
are cross-sections in the Fission Product File of the UK Nuclear Data Library.

French File of Fission Product Properties

This file, containing yield and decay parameters, was
last described by Devillers (48), but it is understood that
new documentation is in preparation (99).

G. General Comments

The quantity of data now available for fission products
allows calculations of decay heat to be made by summation
of the contributions of individual fission products; with
this approach, Devillers et al. (100) found that experimental results fell within the uncertainty (approximately
15%) of their calculations. Of those reactor properties
that depend on the dynamics of reactor operation and that
relate to fission products in particular, perhaps decay heat
has the greatest amount of recently published information
(100-109), but other features such as burnup (55,110), fuel
handling (111,112), nuclear safeguards (113) and environmental
effects (114,115) are prominent. Fission-product data are
also important in accident and safety analyses. As with
cross-section data, sensitivity studies are becoming more
common; papers by Schmittroth (108) and Ilberg et al. (116)
consider the effects of uncertainties in data on decay heat
and ρ calculations, respectively.

IV. DATA ON DELAYED NEUTRONS

Delayed-neutron emitters have the obvious distinction
compared with other fission products that they contribute
source terms to the neutron balance equation. Beta decay
of a fission product may lead to a state above the binding
energy of a neutron in the daughter nuclide which subsequently and rapidly emits a neutron. An excellent description
of the mechanism of delayed-neutron emission has been given
by Tomlinson (117). Since his paper was written, more details on the modes of decay have become available; Figure 2
illustrates the latest position. As we are concerned here
with the data and not the effect of the delayed neutrons

in the kinetics equations, we shall not dwell further on
the stabilizing effects they have on the dynamics of reactors. However, perhaps it is necessary to draw attention
to papers such as those by Hetrick (*118*) and Smets (*119*) that
will show that delayed neutrons are not always a stabilizing influence, contrary to the commonly-held view.

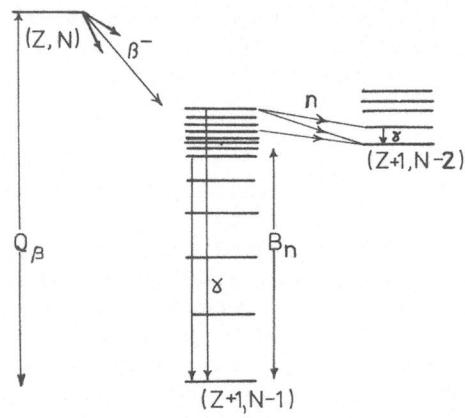

Figure 2

Delayed Neutron Emission

The parameters of delayed-neutron emission that are
important to reactor kinetics are the half-life of the
precursor, the neutron yield per fission and the spectrum
of the neutrons emitted. Although it is now possible to
identify many fission products as delayed-neutron precursors (for example, Tomlinson (*120*) mentions 57), delayed
neutrons usually have been included in kinetics calculations by the use of group data. The neutron yield curve
as a function of time after fission is analyzed as the sum
of a small number, say five or six, exponential decays,
with appropriate yields assigned to each group. The work
by Keepin et al. (*121-123*) has been used most frequently;
there, six groups were fitted because this number gave the
best fit to the experiments performed on the reactor
GODIVA. With many more than six delayed-neutron precursors
being known, it is obvious that the six groups have no par-

ticular physical meaning, but their use adequately describes
the yield of delayed neutrons as a function of time. With
the longer-lived precursors, it is possible to assign particular precursors to the groups: Group I with a decay constant of 0.0127 s^{-1} can be associated with ^{87}Br, and Group
II largely with ^{88}Br and ^{127}I. It is fortunate for the
person performing kinetics calculations that as small a number
as six groups is sufficient, since the usual method of analysis requires the simultaneous solution of one equation for
each delayed group, and one for prompt neutrons. In those
thermal reactors where only one uranium isotope is responsible for the fisssions, this six-group system is reasonable
enough, but with long fuel irradiations some modification
of the constants may be desirable to account for the formation of other fissile materials. In the case of heavy-
water moderated reactors, extra groups may be included to
allow for photo-neutrons. In a fast reactor, where many
nuclides can be of significance, the use of six groups for
each nuclide would lead to large numbers of simultaneous
equations. One way of avoiding this is to use weighted
averages of the relative group yields and decay constants
(Tuttle (124)). An alternative is to reanalyze the delayed-
neutron yield curves of the fissioning nuclides using a
fixed set of decay constants; this method has been described
by Cahalan and Ott (125), while Tuttle (124) discusses the
abilities of these and other methods to cope with particular
situations.

Four reviews and evaluations of particular note that
have appeared over the last few years are those of Tomlinson (126), Amiel (38), Cox (127), and Tuttle (124). Each has
a slightly different emphasis and purpose: Tomlinson gave
particular stress to the data on identified delayed-neutron
precursors; Amiel was providing a review for the Bologna
meeting on Fission Product Nuclear Data; Cox made an evaluation for ENDF/B IV, with particular emphasis on absolute
(total) yields and group data; Tuttle gave considerable
attention to absolute yields.

A.   Absolute Yields

Tuttle (124) gave a brief description of 30 measurements
of the yield of delayed neutrons per fission, dating from
the 1947 report of Wilson (128) to that of Cox (127) in 1974;
the summary is a particularly useful collection of information on the various experiments. Later corrections and

adjustments to results are noted as are the adjustments made
by Tuttle himself to correct for systematic effects not known
at the time of the experiments, to convert to yields those
results that reported other related quantities, and finally,
to bring all the quoted errors into a consistent form so that
one standard deviation is given. Short-lived groups, missed
because of a delay between irradiation and counting, provide
an example of the first point; the multiplication of delayed-
neutron fractions by $\bar{\nu}$ to give yields is an example of the
second. In evaluating the yields, Tuttle calculated a
weighted average for each of the nuclides ^{232}Th, ^{233}U, ^{235}U,
^{238}U, ^{239}Pu, ^{240}Pu, ^{241}Pu, ^{242}Pu. The weight of each data
point depended not only on its quoted standard deviation but
also on whether it involved some other standard value; for
instance, many values are quoted relative to ^{235}U, and consequently, systematic and random errors in the standard can
affect the deduced results. The evaluation produced delayed-
neutron yields for thermal fission, fast fission, and combined values where all results were used, irrespective of
the energy of the stimulating neutron.

Cox concluded in his evaluation (127) that the absolute
yield of delayed neutrons is constant for fission with fast
neutrons up to about 4 MeV. Above this energy, the data
indicate a fall in yield that might be associated with the
onset of second-chance fission. Cox also concluded that the
absolute yield for thermal neutrons tended to be lower than
for fast neutrons in the region up to 4 MeV. This result is
borne out in the evaluation by Tuttle where ^{235}U, ^{239}Pu and
^{241}Pu show a rise of about 5% in yield between thermal and
fast fission; in ^{233}U, the rise is nearer 10%. However, if
one determines the expected absolute yield by the summation
of the effects of known delayed-neutron precursors, one obtains the inverse result; fast neutron yields are calculated
to be less than those for thermal neutron fission. We have
performed this calculation for thermal and fast fission on
^{235}U, using the fission product yield data of Crouch (45) and
the delayed-neutron emission probabilities of individual precursors given by Tomlinson (120) and with later data from
Asghar (129,130) injected where appropriate. The results are
given in Table II; clearly, it is very desirable for the
errors in the summation method to be reduced in order to
check more carefully the energy dependence of yield.

TABLE II

SUMMATION CALCULATION OF ABSOLUTE DELAYED-NEUTRON YIELD
FOR THERMAL AND FAST-NEUTRON FISSION OF ^{235}U

	Delayed neutrons/fission	
	This work	Tuttle[124]
Thermal	$0.0165 \pm 10\%$[a]	$0.01654 \pm 2.5\%$
Fast	$0.0157 \pm 10\%$	$0.01714 \pm 1.3\%$

(a)   The error assigned to the calculated yield is derived entirely
from the errors on the delayed-neutron emission probabilities
as there is no means of assessing the errors on the cumulative
yields derived from Crouch[45].

As indicated by Cox (127) there seem to be rather large
variations in the published yields for ^{238}U; the results of
Masters et al. (131) and Clifford (132) are in agreement to
about 3%, as are the results of Keepin (123) and Cox (127);
but the two groups lie about 15% apart. However, a recent
measurement by Besant et al. (133), based on an improved
form of the experimental arrangement used by Clifford, gives
a value in agreement with Keepin's. Meadow's work (134) on
^{238}U is based on the method of Masters but cannot be used to
test his high value because only relative data are given.
Table III collects together recent information on absolute
yields.

B.   Group Yields

As mentioned earlier, group yields have maintained a
prominent position in reactor kinetics owing to the simplification they afford to calculations; it is worth noting that
Rudstam (135) feels their continued use is unjustified when

TABLE III

ABSOLUTE YIELDS
(Delayed Neutrons/Fission)

	Tuttle[124]			Cox[127]	Besant et al[133]
	Thermal	Fast	Combined	All neutrons < 4 MeV	Fast fission
^{232}Th		0.0547 ± 2.2%	0.0545 ± 2.1%	0.0527 ± 7.6%	
^{233}U	0.00664 ± 2.8%	0.00729 ± 2.7%	0.00698 ± 1.9%	0.0074 ± 5.4%	
^{234}U		0.0106 ± 11.3%	0.0106 ± 11.3%		
^{235}U	0.01654 ± 2.5%	0.01714 ± 1.3%	0.01697 ± 1.2%	0.01668 ± 4.2%	0.0164 ± 3.6%
^{236}U		0.0231 ± 11.3%	0.0231 ± 11.3%		
^{238}U		0.04510 ± 1.3%	0.04508 ± 1.3%	0.0460 ± 5.4%	0.0432 ± 4.0%
^{238}Pu		0.00456 ± 11.3%	0.00456 ± 5.4%		
^{239}Pu	0.00624 ± 3.9%	0.00664 ± 2.0%	0.00655 ± 1.8%	0.00645 ± 6.2%	0.00598 ± 3.7%
^{240}Pu		0.0096 ± 11.5%	0.0096 ± 11.5%	0.0090 ± 10%	
^{241}Pu	0.0156 ± 10.0%	0.0163 ± 10.0%	0.0160 ± 10.0%	0.0157 ± 9.6%	
^{242}Pu		0.0228 ± 11.0%	0.0228 ± 11.0%		

the properties of 25 delayed-neutron precursors are sufficiently well known to account for about 87% of the delayed-
neutrons from ^{235}U. However, bearing in mind that some
spatial or neutronic/hydraulic kinetics calculations still
approximate with only one delayed-neutron group (Buckner and
Stewart (136); Hendricks and Henry (137)) for computational
economy, it seems very likely that the six-group formalism
will continue to be of value for some time.

As Cox points out, it is important to realize that the
values of the group yields and decay constants form a set,
and that comparison of individual values is not particularly
meaningful. In his review, he reconstructs the decay curves
from the parameters published in various papers and compares
them graphically by displaying the ratio of the yields at
particular decay times with those given by Keepin's results.
In this way, Cox is able to compare experimental results where
the number of groups analyzed is not just six as in Keepin's
work. At thermal neutron energies and up to about 4 MeV,
Cox shows that there is good agreement between Keepin (123)
and Cox and Whiting (138), except for small differences at
decay times greater than 100 seconds, when the yield is small
anyway. The other data that he shows are those of Maksyutenko (139-141) which differ from Keepin's values, particularly for ^{238}U where differences up to 15% are seen. For
energies between 5 and 13 MeV, only the data of Maksyutenko
are shown by Cox, and these exhibit in ^{238}U at 7.2 and 7.5
MeV a ratio of over 2 1/2 times to Keepin's value for decay
times between 10 and 50 seconds, whereas, the results at
7.1 and 7.8 MeV show, at most, an 80% deviation from Keepin's.
Cox notes that the absolute measured yields by Krick and
Evans (142) show no sign of this structure, but their data do
not extend beyond 6.9 MeV, and hence do not cover the region
of the structure seen by Maksyutenko. At 14 to 15 MeV, there
is reasonable agreement between most experimenters for ^{235}U
and ^{238}U; in ^{233}U and ^{239}Pu, measurements by East et al. (143)
show similar time dependence for the two nuclides, but differences of up to 20% from Keepin's fast-fission decay curves.

Recent measurements by Besant et al. (133) on ^{235}U,
^{238}U and ^{239}Pu have been compared by us in the same manner
as Cox; for ^{235}U and ^{239}Pu, the measurements differ from
Keepin's by less than 5% up to 100s, while for ^{238}U, the
difference is less than 7%. However, if the errors quoted
on the parameters are treated as independent (an unlikely

possibility, but the only one open to us in the absence of
further information) the differences in the ratios are found
to be not statistically significant.

Tomlinson, Cox and Tuttle all recommend the use of
Keepin's set of fractional group yields and decay constants;
Tomlinson has quoted absolute group yields by multiplying
the fractional yields by his own evaluations of absolute
yields. Tuttle has added a data set for ^{242}Pu and the data
he quotes are given in Table IV. Also included in the table
are results from the work by Cahalan and Ott (125); we have
already mentioned this work in which the yield curves of
several nuclides are analyzed using a fixed set of decay
constants for all the nuclides. We have converted their absolute group yields into relative group yields so that comparisons may be made and the underlining of the decay constants given in the table for ^{239}Pu indicates that this is
the set to be used with the data of Cahalan and Ott.

Before leaving the group yield data, it is worth drawing attention to the note by Tomlinson concerning apparent
delayed-neutron groups of low intensity but very long half-
life; e.g., 3, 12 and 15 minutes. These, he states, are due
to high energy gamma rays from fission products causing
photo-fission in fuel materials. Moreover, in some neutron
detectors, hard gamma rays can be mistaken for neutrons and
thus provide another source of apparent delayed neutrons.

C. Spectra

Experimental developments in the last few years have
meant that much more information is now available on delayed-
neutron spectra than in the late 1960's when only the data
of Batchelor and Hyder (144), Bonner et al. (145), and Burgy
et al. (146) were available. Of particular note has been the
development of ^{3}He spectrometers of very good resolution
(the group at Mainz now quote 19 keV f.w.h.m. for 1 MeV
neutrons (147)) and proton recoil spectrometers. The use of
these spectrometers with on-line mass separators has permitted the analysis of the spectra from particular delayed-
neutron precursors. This will be dealt with in a separate
section. We shall concentrate here on measurements relating
to individual fissioning nuclides and their delayed-neutron
groups.

TABLE IV

RELATIVE GROUP YIELDS AND DECAY SONSTANTS
FOR DELAYED—NEUTRON EMISSION

		Tuttle[124]		Cahalan and Ott[125] a
		Fractional group yields	Decay constants (s^{-1})	Fractional group yields
^{232}Th	1	0.034 ± 8.8%	0.0124 ± 2.4%	0.029 ± 8.4%
	2	0.150 ± 4.7%	0.0334 ± 4.8%	0.156 ± 5.9%
	3	0.155 ± 20%	0.121 ± 5.8%	0.170 ± 12%
	4	0.446 ± 4.9%	0.321 ± 5.0%	0.435 ± 4.8%
	5	0.172 ± 11%	1.21 ± 11%	0.169 ± 6.3%
	6	0.043 ± 21%	3.29 ± 13%	0.041 ± 17%
^{233}U	1	0.086 ± 4.7%	0.0126 ± 4.8%	0.076 ± 5.7%
	2	0.274 ± 2.6%	0.0334 ± 6.3%	0.281 ± 6.1%
	3	0.227 ± 23%	0.131 ± 5.3%	0.250 ± 14%
	4	0.317 ± 5.0%	0.302 ± 12%	0.303 ± 6.1%
	5	0.073 ± 29%	1.27 ± 31%	0.067 ± 30%
	6	0.023 ± 43%	3.13 ± 32%	0.023 ± 100%
^{235}U	1	0.038 ± 11%	0.0127 ± 2.4%	0.036 ± 8.3%
	2	0.213 ± 3.2%	0.0317 ± 3.8%	0.221 ± 3.6%
	3	0.188 ± 13%	0.115 ± 3.5%	0.212 ± 6.9%
	4	0.407 ± 2.5%	0.311 ± 3.9%	0.381 ± 2.4%

Table IV.   (cont'd)

	5	0.128 ± 9.4%	1.40	± 8.6%	0.108 ± 7.8%
	6	0.026 ± 15%	3.87	± 14%	0.042 ± 7.1%
	1	0.013 ± 7.7%	0.0132	± 3.0%	0.012 ± 10%
	2	0.137 ± 2.2%	0.0321	± 2.8%	0.131 ± 5.0%
	3	0.162 ± 19%	0.139	± 5.0%	0.165 ± 14%
^{238}U	4	0.388 ± 4.6%	0.358	± 5.9%	0.371 ± 6.3%
	5	0.225 ± 8.4%	1.41	± 7.1%	0.203 ± 3.9%
	6	0.075 ± 9.3%	4.02	± 8.0%	0.118 ± 7.4%
	1	0.038 ± 11%	0.0129	± 2.3%	0.038 ± 8.3%
	2	0.280 ± 2.1%	0.0311	± 2.3%	0.279 ± 5.1%
	3	0.216 ± 13%	0.134	± 3.0%	0.216 ± 10%
^{239}Pu	4	0.328 ± 4.6%	0.331	± 5.4%	0.329 ± 5.8%
	5	0.103 ± 13%	1.26	± 13%	0.103 ± 11%
	6	0.035 ± 20%	3.21	± 12%	0.035 ± 14%
	1	0.028 ± 14%	0.0129	± 4.7%	0.032 ± 11%
	2	0.273 ± 2.1%	0.0313	± 2.2%	0.268 ± 6.8%
	3	0.192 ± 41%	0.135	± 12%	0.183 ± 28%
^{240}Pu	4	0.350 ± 8.6%	0.333	± 14%	0.354 ± 12%
	5	0.128 ± 21%	1.36	± 22%	0.120 ± 25%
	6	0.029 ± 31%	4.04	± 29%	0.044 ± 21%

Table IV.   (cont'd)

	1	0.010 ± 30%	0.0128 ± 1.6%	0.012 ± 21%
	2	0.229 ± 2.6%	0.0299 ± 2.0%	0.240 ± 2.7%
^{241}Pu	3	0.173 ± 14%	0.124 ± 10%	0.179 ± 16%
	4	0.390 ± 13%	0.352 ± 5.1%	0.347 ± 17%
	5	0.182 ± 10%	1.61 ± 9.3%	0.201 ± 14%
	6	0.016 ± 31%	3.47 ± 49%	0.021 ± 22%
	1	0.004 ± 25%	0.0128 ± 2.3%	0.023 ± 36%
	2	0.195 ± 16%	0.0314 ± 4.1%	0.164 ± 33%
^{242}Pu	3	0.161 ± 30%	0.128 ± 7.0%	0.169 ± 28%
	4	0.412 ± 37%	0.325 ± 6.2%	0.379 ± 29%
	5	0.218 ± 40%	1.35 ± 6.7%	0.174 ± 20%
	6	0.010 ± 30%	3.70 ± 12%	0.091 ± 41%

a.   The group decay constants to be used with the data of Cahalan and Ott are those of ^{239}Pu as listed by Tuttle.   For this reason they have been underlined in the table.   So that both Tuttle's data and that of Cahalan and Ott will show fractional group yields, we have divided the absolute group yields by the absolute yields quoted by Cahalan and Ott.

Table V shows the experiments reported on spectra from irradiations of thorium, uranium and plutonium. In general, the [3]He spectrometers have better resolution than the proton recoil ones in the energy range above 100 keV, but are not as suitable for lower energies because subtraction of the high-energy wing of the thermal peak can result in large uncertainties. This is borne out by comparison of the results of Sloan and Woodruff (151) and Eccleston and Woodruff (152), where considerable structure can be seen in the delayed neutron spectrum at energies less than 100 keV, with those of Shalev and Cuttler (148) and Evans and Krick (154), where it would be difficult to be certain of structure in the same region. Evans has informed us that his earlier data (153) have been modified slightly to account for a 38 mm lead shield. This corrected information is given in Figure 3.

Shalev has analyzed his experimental data from various irradiation cycles to give spectra for delayed neutron groups 1 and 2, while Saphier et al. (157) have deduced equilibrium spectra from the same data. Fieg (149) also gives group spectra for fission of ^{235}U by thermal and 14 MeV neutrons.

The importance of the spectrum of delayed neutrons is greater in fast reactors than in thermal ones, particularly under fault conditions (see Saphier et al. (157)) because the average energy of delayed neutrons is several hundred keV, which is much nearer the mean of the fast reactor spectrum than is the mean of the prompt neutron spectrum (about 2 MeV). Even in fast reactors, slow variations in reactivity are less sensitive than quick ones to the spectrum of delayed neutrons.

D.   Delayed Neutron Precursors

As already mentioned, Tomlinson (120) gives data (half-lives and emission probabilities) for 57 delayed neutron emitters produced in fission. Rudstam notes the identification of 16 more in a paper that is also a useful review of the use of on-line separators (158), and the half-lives for these precursors are given by Rudstam and Lund (159,160).

Experimental work on emission probabilities and half-lives has been recently reported also by the group at Grenoble (Asghar et al. (161,162); Ristori (163)), at Mainz by Kratz et al. (147,164), by Roeckl et al. (165), and by Reeder et al. (71), while Izak-Biran and Amiel (166) and Nir-

TABLE V

EXPERIMENTS REPORTING DELAYED-NEUTRON SPECTRA
FROM THORIUM, URANIUM AND PLUTONIUM

Author	Reference	Detector	Target	Neutron source
Shalev and Cuttler	148	^{3}He	^{232}Th, ^{233}U ^{235}U, ^{238}U, ^{239}Pu	Reactor core spectrum
Fieg	149	Proton recoil	^{235}U, ^{238}U, ^{239}Pu ^{235}U	14MeV thermal
Fieg and Lalovic	150	--------corrections to Fieg$^{(149)}$---------		
Sloan and Woodruff	151	Proton recoil	^{235}U	thermal
Eccleston and Woodruff	152	Proton recoil	^{232}Th, ^{233}U ^{235}U, ^{238}U ^{239}Pu	Be(p, n)
Evans and Krick	153, 154	^{3}He	^{235}U, ^{239}Pu ^{238}U	Li(p, n) 0.5MeV Li(p, n) 1.7MeV
Iwasaki et al.	155	T.O.F.	^{238}U	Photofission
Chulick et al.	156	T.O.F.	^{252}Cf	Spontaneous fission

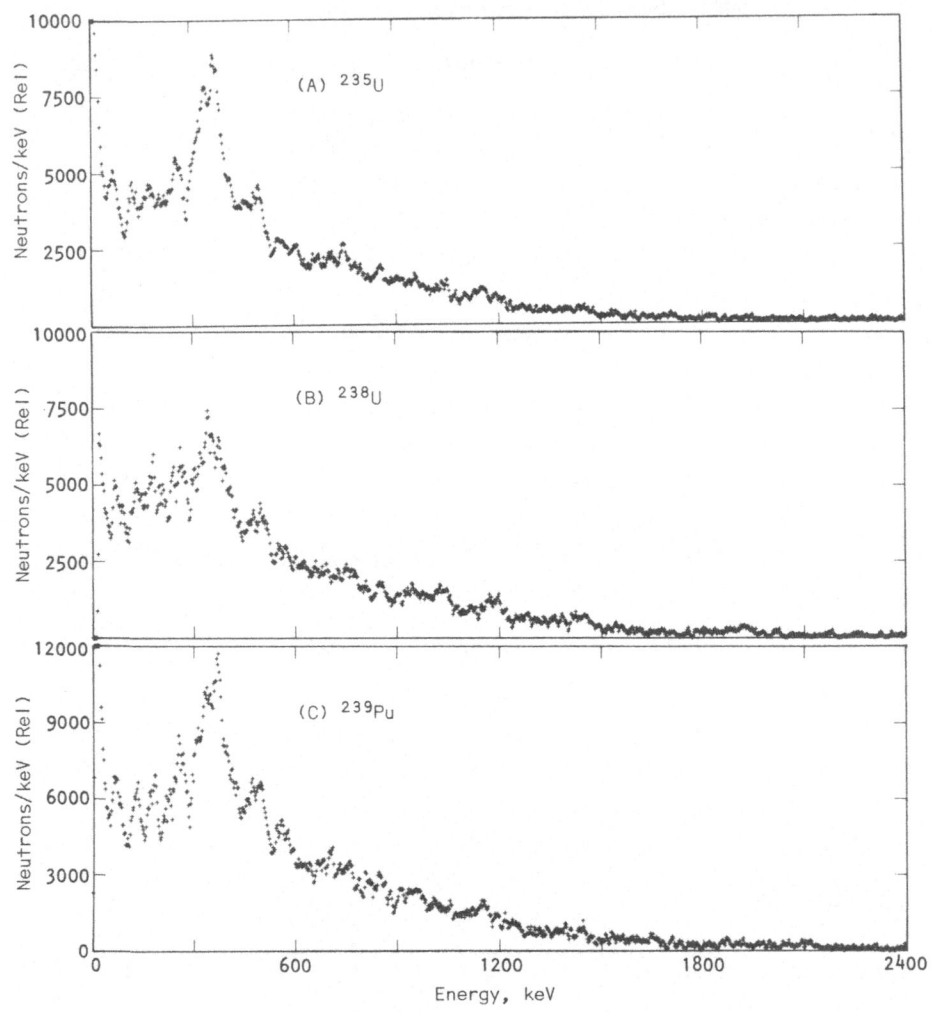

Figure 3

Delayed-Neutron Spectra of ^{235}U, ^{238}U and ^{239}Pu
From Fast-Neutron Fission (Evans and Krick)

El and Amiel (167) have made calculations of delayed-neutron emission probabilities based on fission-product systematics.

E. Neutron Spectra from Single Precursors

Table VI gives a list of measurements that have been made of the spectra emitted by particular precursors. Saphier (157) and Rudstam (135) have used the spectra for individual precursors, together with fission product yields and neutron emission probabilities to estimate the group or equilibrium spectra from the summation of the effects of the precursors. The existence of definite line shapes in the spectra has led to considerable analysis of the mechanism of delayed-neutron production. Experiments (Crawford et al. (179); Kratz et al. (81)) have shown that delayed-neutron emission can populate excited states as well as the ground state in the daughter nucleus. It is also worth noting that gamma rays have been seen from levels in delayed-neutron emitters above the neutron binding energy, and Tovedal (89) and Nuh (84) have shown there is competition between neutron and gamma emission. In the recent conference on Nuclei Far From Stability, Kratz et al. (147), Klapdor (180), and Shihab-Eldin et al. (181) presented experimental results and theoretical analyses of delayed-neutron emission that go a long way toward explaining the existence of the discrete lines.

F. A Note of Caution

Before leaving the section on delayed neutrons, it is worth noting that despite the rapid improvement in the knowledge of them, it is still not always possible to predict accurately their impact on certain experiments, as illustrated by Mihalczo's paper (182) where calculations of the effective delayed-neutron fraction in unreflected uranium spheres were different from experimental values by 11%. Similarly, Fischer (231) suggests that the well-known 'central worth' discrepancies in fast reactors may be attributable to erroneous effectiveness factors.

V. DATA ON HEAVY ELEMENTS

This topic may be divided into two groups, the first to deal with the nuclides of major interest in reactors (^{232}Th, ^{233}U, ^{235}U, ^{238}U, ^{239}Pu) and the second to cover those remaining.

TABLE VI

EXPERIMENTS REPORTING DELAYED—NEUTRON EMISSION
FROM SEPARATED PRECURSORS

Author	Reference	Precursor
Shalev and Rudstam	168	^{137}I
Shalev and Rudstam	169	^{85}As, ^{87}Br, ^{134}Sn, ^{135}Sb, ^{136}Te, ^{137}I
Rudstam and Shalev	170	^{89}Br, ^{91}Br, ^{93}Rb, ^{139}I, $^{141}(I+Cs)$, ^{143}Cs
Shalev and Rudstam	171	^{88}Br, ^{90}Br, ^{138}I, ^{140}I, $^{142}(Xe+Cs)$, ^{144}Cs
Chrysochoides et al.	172	^{87}Br, ^{88}Br
Franz et al.	173	^{85}As, ^{135}Sb
Kratz	147	^{85}As, ^{87}Br, ^{135}Sb, ^{137}I
Ray and Kenney	174	^{87}Br
Kellie et al.	175, 176	^{93}Rb, ^{94}Rb, ^{95}Rb, ^{96}Rb, ^{97}Rb
Saphier et al.	157, 177	^{94}Rb, ^{95}Rb (preliminary data)
Rudstam	178	$^{79}(Zn+Ga)$, ^{80}Ga, ^{81}Ga, ^{130}In (work in progress)

A.   The Major Fuel Nuclides

Because of their special place in reactor calculations, these five nuclides have been studied intensively, and there is continuous refinement of their evaluations; for example, Lemmel (*183*) reported on the third IAEA evaluation of 2200 m/s and 20°C Maxwellian-averaged neutron data for ^{233}U, ^{235}U, ^{239}Pu and ^{241}Pu at the 1975 Washington conference, and de Saussure et al. (*184*) have recently published a discussion on the shortcomings of the ENDF/BIV file for ^{238}U cross-sections in the region of the resolved resonances. In general, the 2200 m/s cross-sections are known to 1% or better, (*1*) and for ^{235}U, ^{238}U and ^{239}U, the variation with incident neutron energy of the fission cross-section is known to 3% to 5% in the fast neutron region (*25*). However, reactor physics calculations would benefit from a greater accuracy in fission cross-sections, and the work of Czirr and Sidhu (*185-188*) is one attempt to cover the whole energy range from thermal to 20 MeV in the measurement of the neutron-induced fission cross-section of ^{235}U to an accuracy of 1%.

Although not a member of the group of fuel nuclides, ^{252}Cf has a special relevance to their data because its $\bar{\nu}$ value is frequently used as a standard. Until recently, the experimental values of $\bar{\nu}$ have not been consistent to better than 1%, and this led to requests for an improvement to 0.25%. However, Lemmel's report (*183*) stated that the third IAEA evaluation had eliminated the discrepancies between experiments and quoted a value of 3.731 ± 0.008; furthermore, Reich (*189*) has noted that a new measurement (by J. R. Smith) is in progress.

Recent data for the major fuel isotopes will be found in references 190-202, but with these nuclides, any list we give will be incomplete by the time of publication. The reader is therefore directed to the CINDA bibliography (*3*) and reminded that information added to CINDA since its last published volume is available in the form of computer listings from the data centers.

B.   Other Transactinium Nuclides

The burnup of fuel nuclides and the production of higher actinides affects the operation and dynamics of a reactor, and conversely, the operating condition, power level and

refueling schedule can alter markedly the quantities of
these nuclides produced. Although the production of individual nuclides can be large, the problems of separating
them from the highly active waste products and the difficulties of measuring active samples have led to gaps or uncertainties in our knowledge of their data.

Transactinium nuclear data were discussed at an IAEA
group meeting held at Karlsruhe in November, 1975 (203); as
at the Bologna meeting on fission product data, this group
was provided with review papers that covered the status of
the data and the requirements of users. Recommendations
from the meeting fall into two parts: first, a number of
general observations on how the results of experiments and
evaluations of them might be coordinated internationally,
by means of an Actinide Newsletter and the exchange of data
through the IAEA; and second, detailed recommendations on
the three areas covered by the working groups: thermal
reactors, fast reactors, and waste management. For some
uses, the present data appear to be sufficient; for example, the thermal reactor group concluded that current
data were adequate for calculations of heat production in
discharged fuel (204). In other situations where no value
exists, an order-of-magnitude estimate of a cross-section
would be acceptable. Table VII combines the unsatisfied
requests for data compiled by the working groups on thermal
and fast reactors and indicates the accuracies required.
These requirements came from effects in the core such as
reactivity, power output, spontaneous fission rate, burnup
and shutdown heating (^{239}Np is the heavy nuclide of major
importance) and from problems relating to fuel out of the
core, including the radiation hazards from fresh and spent
fuel while in transit, storage and reprocessing, criticality, and the determination of isotopic content of fresh and
irradiated fuel.

Of the reviews of requirements for transactinium data,
the one by Barré and Bouchard (206) is concerned most closely
with the dynamics of reactor operation. If one excludes the
obvious effects of the fuel nuclides, the most significant
transactinium nuclides as far as reactivity is concerned are
^{236}U, ^{240}Pu and ^{241}Pu, which have an effect greater by an
order of magnitude than the next most significant group.
Mutual cancellation of reactivity effects occur in the nuclides present in low concentrations; Barré and Bouchard

give an example of a fast reactor where the reactivity contributions of ^{237}Np, ^{239}Np, ^{238}Pu, ^{241}Am, ^{242m}Am, ^{243}Am, ^{242}Cm combine to give a negligible effect. These nuclides are also very much less significant than ^{240}Pu and ^{241}Pu in calculations of the internal breeding gain in fast reactors. The way transactinium elements affect the dynamics of a reactor has its converse when one considers the way in which the operation of a reactor can alter the concentrations of these nuclides in discharged fuel. In the second part of their paper, Barré and Bouchard compare the accuracies required in the data for in-core and out-of-core calculations so that the more stringent request may be identified.

It is not possible to include succinctly the requests from the third working group which considered those topics not specifically related to the development and operation of thermal and fast reactors, and which, therefore, included matters such as the disposal of actinide waste by nuclear incineration and other methods. While some aspects of this work fall within our brief because they depend on the dynamics of reactor behavior, they are on the fringe, and we simply draw attention to the tabulation in the proceedings of the Karlsruhe panel (207). A description of methods in use for the management of actinide wastes (and fission product wastes) has been given by Olivier (112).

An indication of the large discrepancies that can appear in calculations has been given by Kuesters and Lalovic (208), who note that the ^{236}U yield in a 1000 MW(e) fast breeder reactor (LMFBR) predicted by the ORIGEN code (209) differs by a factor of 7 from that given by a more refined calculation where spatial effects are taken into account. To be fair, ^{236}U is the worst case in that study; other differences fall within a range of about ± 40%. These authors also brought out the point that self-shielding effects can be modified significantly by changes in fuel composition and hence, can influence burnup and the production of heavy elements. A coarse energy-group structure can also introduce discrepancies; predicted concentrations of uranium and plutonium isotopes in a boiling-water reactor differ by up to 30%, according to whether a two-group or a five-group calculation is used (208). Equally, the data files connected with particular codes can lead to differences of two or three times in the predicted production rates; Harte (210) has told us of a comparison of ORIGEN and HYACINTH (211) that

TABLE VII

SUMMARY OF UNSATISFIED DATA REQUESTS
For Thermal and Fast Reactors from the
Karlsruhe[a] Meeting (203)

Nuclide	Thermal reactor requests[b]		Fast reactor requests[c]				Half-life requests
	$\sigma_{n\gamma}$	$I_{n\gamma}$	$\sigma_{n\gamma}$	σ_{nf}	$\sigma_{n,2n}$	$\bar\nu$	
^{233}Pa	10(12)						
^{234}U		5(12)					
^{236}U	4(6)	4(6)					
^{238}U					20(-)		
^{237}Np			30(50)		50(-)	50(-)	
^{239}Np	100(-)[d]		20(-)	50(-)		50(-)	
^{236}Pu	100(-)[d]	100(-)[d]	50(-)	50(-)			
^{238}Pu			20(30)	7(10)		4(-)	0.5(1)
^{239}Pu					50(-)		0.5(1)
^{240}Pu		1(12)	5[e]	2			
^{241}Pu			8	1.5			1(3)
^{242}Pu			8	4			
^{241}Am			5(30)	15[f]		10(-)	
^{242m}Am	50(-)	50(-)	50(-)	15(50)		10(-)	

TABLE VII.   (cont'd)

^{243}Am	10(-)		25(-)
^{242}Cm	50(-)	25(-)	15(-)
^{244}Cm	50(-)		

Notes:

a   All accuracies are given in % with the achieved values in parentheses.

b   Thermal reactor requests are for (i) $\sigma_{n\gamma}$ = 2200 m/s or thermal reactor spectrum averaged cross section

(ii) $I_{n\gamma}$ = capture resonance integral; this was used by the working group as a means of parametrising the non-thermal requirement. In general, energy dependence of cross section across major resonance will be required.

c   Fast reactor requests are for (i) $\sigma_{n\gamma}$ = capture cross section 0.5 - 100 keV

(ii) σ_{nf} = fission cross section 1 keV - 2 MeV

(iii) $\sigma_{n, 2n}$ = (n, 2n) cross section above threshold

d   This is taken to mean that an approximate value is required.

e   The fast reactor working group did not survey fully requests for the higher isotopes of plutonium because they are sufficiently covered by requests in, for example, WRENDA[205].   The figures given in the table are values noted by the working group as being typical.

f   Highly dissimilar experimental values exist.

shows such an effect on the yields of ^{242}Cm and ^{244}Cm, but
it disappears when the data of one code are used for both
calculations. Clearly, uncertainties of two or three times
in the yields of nuclides that are prominent alpha-particle
emitters or that undergo spontaneous fission are of importance.

The papers that reviewed available data at the Karlsruhe meeting, and that are of particular relevance to this
survey are those by Benjamin (212), James (213), Moore (214),
Igarasi (215), and Yiftah et al. (216). Benjamin and Igarasi
produced extensive reviews, element by element, of data in
the thermal and fast regions, respectively; the latter included figures that show diagrammatically the requested and
achieved accuracies in the energy range 1 keV to 10 MeV.
Moore reviewed the information obtained from underground
nuclear explosions on cross-sections of the transactinium
elements and noted the usefulness of this technique for
samples that are radioactive or of small size (about 1 μg is
required for a fission cross-section measurement; about 30
mg for radiative capture). James considered the data available in the resonance region from linear accelerator measurements, and his paper contained a review of statistical methods
used to test for structure in experimental results. Yiftah,
et al. discussed the evaluations of transactinium elements;
they suggested that WRENDA (205) should include requests for
evaluations, so that it matches CINDA more closely. A review
of alpha-decay data was given by Baranov et al. (217), and
Reich performed a similar survey of information on beta decay, gamma decay and spontaneous fission (189). This last
paper mentioned that a file of 'Actinide Nuclear Data' is
being prepared for ENDF/B V, and a discussion of the format
of the decay data, which will be part of this file, was included as an appendix.

Because of the lack of experimental information, prediction of data for the transactinium elements from nuclear
theory has been important; Lynn presented a paper at Karlsruhe (218), and the whole subject of the use of nuclear
theory to evaluate neutron-nuclear data was discussed at the
IAEA meeting at Trieste in December, 1975 (25). The success
of theoretical predictions was mentioned by the Karlsruhe
working group on fast reactor data (219); they noted that the
reliability of such calculations is ± 25% for isotopes of U,
Np, Pu and Am, ± 30% for medium-weight isotopes of Cm and
decreases to ± 50% for other isotopes of Cm and for Bk and

Cf. It should be noted that the accuracies are better than
those requested for some of the nuclides in Table VII; however, the working group pointed out that this did not remove the need for experiments on these nuclides, at least
not for the purpose of checking theory. An indication of
the possible errors that might occur by too much reliance
on theoretical analysis is given by calculations for thorium
and protoactinium; σ_{ny} can be calculated to ± 25%, but the
predicted fission cross-sections appear to be too high by a
factor of two (219).

In addition to the papers already discussed, the Karlsruhe proceedings contain a number of evaluations and experimental results; and other recent work will be found in
references (94,191,197,220-228).

It may be surprising that the half-lives of two common
plutonium isotopes, ^{239}Pu and ^{241}Pu, still appear in a list
of unsatisfied requests. The ^{239}Pu request reflects the
accuracy required for the determination of the isotopic
composition of fuels; the WRENDA request for a 0.2% accuracy for use in calorimetric assay of plutonium is actually
more stringent. Results of a recent set of measurements on
^{239}Pu at Harwell are expected (229). For ^{241}Pu, there remain
considerable differences in published data; Wilkins surveyed
these up to the end of 1974, when he published the results
of his mass spectrometer measurements (228). He obtained a
value of 15.02 ± 0.10 years for the half-life, but observed
that this differed significantly from the result of Strohm
et al. (230) (14.355 ± 0.007 years) who used a calorimetric
method. Recently, McKean and Crouch (223) obtained results
by mass spectrometry (14.53 ± 0.12 years, 14.24 ± 0.12
years, 14.32 ± 0.10 years and 14.34 ± 0.13 years), which are
closer to Strohm et al. than Wilkins. Clearly, the experimental evidence has not yet resolved the differences in the
data.

ACKNOWLEDGMENTS

We are grateful to the many people who have provided
papers or assisted in the collection of information, and we
thank Mrs. P. M. Courtney for her work in the preparation of
the typescript.

REFERENCES

1. Mughabghab, S. F. and Garber, D. I., <u>Neutron Cross-
    Sections</u>, Vol. <u>1</u>, <u>Resonance Parameters</u>, Brookhaven
    National Laboratory Report <u>BNL 325,</u> third edition,
    June, 1973.

2. Garber, D. I. and Kinsey, R. R., <u>Neutron Cross-Sections</u>,
    Vol. <u>2</u>, <u>Curves,</u> Brookhaven National Laboratory Report,
    <u>BNL325</u>, third edition, January, 1976.

3. CINDA, An Index to the literature on Microscopic Neutron
    Data, published by IAEA, April 1976.

4. Poncelet, C. G., Ozer, O. and Harris, D. R., Proprietary,
    Standard and Government-supported Nuclear Data Bases,
    Los Alamos Scientific Laboratory Report <u>LA6023MS</u>, Aug-
    ust, 1975.

5. Pearlstein, S., <u>Advances Nuclear Science & Technology</u>,
    Vol. <u>8</u>, pp 115-139, 1975.

6. Garber, D., Dunford, C. and Pearlstein, S., Data Formats
    and Procedures for the Evaluated Nuclear Data File, ENDF,
    Brookhaven National Laboratory Report <u>BNL-NCS-50496</u>,
    October, 1975.

7. Dunford, C., Brookhaven National Laboratory, Upton,
    New York, private communication, August, 1976.

8. Garber, D., ENDF/B Summary Documentation, Brookhaven
    National Laboratory Report <u>BNL-17541</u>, October, 1975.

9. Magurno, B. A., ENDF/B IV Dosimetry File, Brookhaven
    National Laboratory Report <u>BNL-NCS-50446</u>, April, 1975.

10. Peelle, R. W., Proceedings of the Conference on Nuclear
    Cross-sections and Technology, Washington, D. C.,
    pp 173-176, March, 1975.

11. Perey, F. G., Proceedings of the Conference on Nuclear
    Cross-sections and Technology, Washington, D. C.,
    pp 842-847, March 1975.

12. Weisbin, C. R., Oblow, E. M., Ching, J., White, J. E.,
    Wright, R. Q. and Drischler, J., Ibid, pp 825-833.

13. Howerton, R. J., Cullen, D. E., Haight, R. C., MacGregor, M. H., Perkins, S. T. and Plechaty, E. F., "The LLL Evaluated Nuclear Data Library (ENDL): Evaluation Techniques, Reaction Index, and Descriptions of Individual Evaluations," Lawrence Livermore Laboratory Report UCRL-50400, Vol. 15, Part A, September, 1975.

14. Howerton, R. J., Ibid Part C, "Translation of ENDL Neutron-induced Interaction Data into the ENDF/B Format, April, 1976.

15. Pope, A. L., The Current edition of the Main Tape NDL-1 of the UK Nuclear Data Library, UKAEA Reactor Group Report AEEW-M 1208, March, 1973.

16. Pope, A. L. and Story, J. S., "Minigal Output from UK Nuclear Data Library - NDLI, Thermal Cross-section Resonance Integrals and Fission Spectrum Averages," UKAEA Reactor Group Report AEEW-M1191, April, 1973.

17. Pope, A. L., "Minigal Output from 'The Fission Products' Data Files of the UK Nuclear Data Library," UKAEA Reactor Group Report AEEW-M1266, April, 1974.

18. Goel, B. and Krieg, B., Status of the Nuclear Data Library KEDAK-3, Kernforschungszentrum Karlsruhe Report KFK2234, October, 1975.

19. Goel, B., Graphical Representation of the German Nuclear Data Library KEDAK, Part 1: Non-fissile Materials, Kernforschungszentrum Karlsruhe Report KFK2233, 1975.

20. Krieg, B., Handling and service programs for KEDAK, Part 1, "Management and Retrieval Programs," Kernsforschungszentrum Karlsruhe Report KFK1725, June, 1973.

21. Ribon, P., Evaluated-data Libraries: "A Comparative Study of the Proposed USSR Format and the UKAEA Format," Centre d'Etudes Nucléaires, Saclay, France Report SMPNF/851, April, 1971.

22. CCDN Newsletter, published by N.E.A. Neutron Data Compilation Centre, B. P. 9, 91190-Gif-sur-Yvette, France.

23. Nikolaev, M. N., Changes in and Additions to the Format
    of the SOKRATOR Evaluated Nuclear Data Library, Inter-
    national Nuclear Data Committee Report INDC(CCP)-69L,
    October, 1975.

24. Karam, R. A., Trans. American Nuclear Society 19, P. 389,
    1974.

25. Proceedings of a Consultants Meeting on the Use of Nuclear
    Theory in Neutron Nuclear Data Evaluation, Trieste, Dec-
    ember, 1975, IAEA Report IAEA-190, Vienna, 1976.

26. McKnight, R. D., Le Sage, L. G., Christenson, J. M.,
    Proceedings of the Conference on Nuclear Cross-sections
    and Technology, Washington, D. C., pp 385-388, March,1975.

27. Paik, N. C., Proceedings of the Conference on Nuclear
    Cross-sections and Technology, Washington, pp 39-44,
    March, 1975.

28. Proceedings of a Panel on Fission Product Nuclear Data,
    Bologna, November, 1973, IAEA Report IAEA-169, 1974.

29. Cuninghame, J. G., AERE, Harwell, England, private
    communication.

30. Nair, S., Gayther, D. B., Patrick, B. H., Bowey, E. M.,
    UK Nuclear Data Progress Report, UKNDC(76)P80, pp 18-20,
    D. B. Gayther, Editor, AERE, Harwell, England, August,
    1976.

31. Walker, W. H., Proceedings of a Panel on Fission Product
    Nuclear Data, Bologna, IAEA Report IAEA-169, Vol. 1,
    pp 285-352, 1974.

32. Valente, S., Proceedings of a Panel on Fission Product
    Nuclear Data, IAEA Report IAEA-169, Vol. 1, pp 9-27.
    1974.

33. Tyror, J. G., Ibid, pp 51-82.

34. Devillers, C., Ibid, pp 83-162.

35. Ribon, P. and Krebs, J., Ibid, pp 235-283.

36. Cuninghame, J. G., Ibid, pp 353-434.

37. Rudstam, G., Ibid, Vol. 2, pp 1-32.

38. Amiel, S., Ibid, pp 33-52.

39. Musgrove, A. R. de L., Cook, J. L., Trimble, G. D.,
    Ibid, pp 163-200.

40. Progress in Fission Product Nuclear Data, INDC(NDS)-
    70/G + P, November, 1975, available from IAEA Nuclear
    Data Section.

41. Ibid, 75/G + P, May, 1976.

42. Walker, W. H., "Fission Product Data for Thermal
    Reactors," Part II: Yields, Atomic Energy of Canada
    Limited Report AECL3037, April, 1973.

43. Crouch, E. A. C., "Fission Product Chain Yields from
    Experiments in Thermal Reactors," Proceedings of the
    Symposium on Applications of Nuclear Data in Science
    and Technology, Paris, IAEA-SM-170/94, Vol. 1,
    pp 393-458, 1973.

44. Crouch, E. A. C., "Chain and Independent Fission Product
    Yields Adjusted to Conform with Physical Conservation
    Laws," AERE Harwell Report AERE-R7785, February, 1975.

45. Ibid, Report AERE-R8152, January, 1976.

46. Crouch, E. A. C., "Chain and Independent Fission Product
    Yields for Thermal, Fast and 14 MeV Neutron-Induced
    Fission," submitted to At. Data Nuclear Data Tables,
    1976.

47. Meek, M. E. and Rider, B. F., "Compilation of Fission
    Product Yields," General Electric Company, Vallecitos
    Nuclear Center Report NEDO-12154-1, January, 1974.

48. Devillers, C., Blachot, J., Lott, M., Nimal, B.,
    N'Guyen Van Dat, Noel, J. P. and de Tourreil, R.,
    Proceedings of the Symposium on Applications of
    Nuclear Data in Science and Technology, Paris,
    IAEA-SM-170/63, Vol. 1, pp 477-503, March, 1973.

49.   Lammer, M., and Eder, O. J., Proceedings of the Sym-
       posium on Applications of Nuclear Data in Science
       and Technology, Paris, IAEA-SM-170/13, Vol. 1,
       pp 505-551, 1973.

50.   Wahl, A. C., Ferguson, R. L., Nethaway, D. R.,
       Troutner, D. E. and Wolfsberg, K., Physical Review,
       126, pp 1112-1127, 1962.

51.   Wahl, A. C., Norris, A. E., Rouse, R. A., Williams,
       J. C., Proceedings of the Second IAEA Symposium on
       the Physics and Chemistry of Fission, Vienna,
       IAEA-SM-122/116, pp 813-843, 1969.

52.   Siegert, G., Greif, J., Wollnik, H., Fiedler, G.,
       Decker, R., Asghar, M., Bailleul, G., Bocquet, J. P.,
       Gautheron, J. P., Schrader, H., Armbruster, P., and
       Ewald, H., Physical Review Letters, 34, pp. 1034-1036,
       1975.

53.   Nethaway, D. R., Tables of Values of Z_p, the Most
       Probable Charge in Fission, Lawrence Livermore
       Laboratory Report UCRL-51640, September, 1974.

54.   Amiel, S. and Feldstein, H., Physical Review, C 11,
       pp 845-858, 1975.

55.   Maeck, W. J., Proceedings of the Conference on
       Nuclear Cross-sections and Technology, Washington,
       pp 378-384, March, 1975.

56.   Cuninghame, J. G., Goodall, J. A. B. and Willis, H.
       H., J. Inorganic Nuclear Chemistry 36, pp 1453-1457,
       1974.

57.   Van Assche, P. H. M., Vandenput, G., Jacobs, L.
       Van den Cruyce, J. M. and Silverans, R., Proceedings
       of a Symposium on Nuclear Physics with Thermal and
       Resonance Energy Neutrons, Reactor Centrum Nederland,
       Petten, RCN-203, pp 95-99, 1973.

58.   Adams, D. E., James, W. D., Beck, J. N. and Kuroda,
       P. K., J. Inorganic Nuclear Chemistry 37, pp 419-424,
       1975.

59.   Blachot, J., Carraz, L. C., Cavallini, P., Chauvin,
        C., Ferrieu, A. and Moussa, A., J. Inorganic Nuclear
        Chemistry 36, pp 495-501, 1974.

60.   Blachot, J., Cavallini, P., Ferrieu, A. and Louis, R.,
        J. Radioanal. Chemistry 26, pp 107-125, 1975.

61.   Di Iorio, G., Wehring, B. W., Proceedings of the
        International Conference on the Interaction of
        Neutrons with Nuclei, Lowell, Massachusetts, Paper
        PB1/J6, July 1976.

62.   Eaker, R. W. and Choppin, G. R., J. Inorganic Nuclear
        Chemistry 38, pp 31-36, 1976.

63.   Eaker, R. W., Kandil, A. T. and Choppin, G. R., J.
        Inorganic Nuclear Chemistry 38, pp 969-973, 1976.

64.   Egger, C., von Gunten, H. R., Schmid, A. and Pruys,
        H. S., Radiochim. Acta 21, pp 200-202, 1974.

65.   Flynn, K. F., Nagy, S., Glendenin, L. E., Gindler,
        J. E. and Meadows, J. W., Trans. American Nuclear
        Society 22, pp 677-679, 1975.

66.   Flynn, K. F., Gindler, J. E., Glendenin, L. E. and
        Sjoblom, R. J., J. Inorganic Nuclear Chemistry 38,
        pp 661-664, 1976.

67.   Gäggeler, H., von Gunten, H. R. and Pruys, H. S.,
        J. Inorganic Nuclear Chemistry 38, pp 205-210, 1976.

68.   Lipinski, R. J. and Wehring, B. W., Proceedings of
        the International Conference on the Interactions of
        Neutrons with Nuclei, Lowell, Massachusetts, Paper
        PG1/J18, July, 1976.

69.   Rajagopalan, M., Pruys, H. S., Grütter, A., von
        Gunten, H. R., Hermes, E. A., Richmond, R., Rössler,
        E., Schmid, A. and Wydler, P., Nuclear Science Eng.
        58, pp 414-419, 1975.

70.   Rajagopalan, M., Pruys, H. S., Grütter, A., Hermes,
        E. A. and von Gunten, H. R., J. Inorganic Nuclear
        Chemistry 38, pp 351-352, 1976.

71.   Reeder, P. L., Wright, J. F. and Anderl, R. A., Pro-
       ceedings of the Conference on Nuclear Cross-sections
       and Technology, Washington, pp 401-404, March, 1975.

72.   Strom, P. O., Love, D. L., Greendale, A. E., Delucchi,
       A. A., Sam, D. and Ballou, N. E., Physical Review 144,
       pp 984-993, 1966.

73.   Tarpley, W. and Min, K. S., Proceedings of the Inter-
       national Conference on the Interactions of Neutrons
       with Nuclei, Lowell, Massachusetts, Paper PB1/J7,
       July, 1976.

74.   Blachot, J., Devillers, C., de Tourreil, R., Nimal, B.,
       Fiche, C. and Noel, J. P., Bibliothèque de Données
       Nucléaires Relatives aux Produits de Fission, Centre
       d'Etudes Nucléaires de Fontenay-aux-Roses Report
       CEA-N-1822, October, 1975.

75.   Tobias, A., Data for the Calculation of Gamma Radia-
       tion Spectra and Beta Heating from Fission Products
       (Revision 3), Central Electricity Generating Board
       Berkeley Nuclear Laboratories Report RD/B/M-2669,
       June, 1973.

76.   Reich, C. W., Helmer, R. G., Putnam, M. H., Radio-
       active-Nuclide Decay Data for ENDF/B, Aerojet Nuclear
       Co., Idaho Falls Report ANCR-1157, August, 1974.

77.   England, T. R. and Schenter, R. E., ENDF/B IV
       Fission Product Files: Summary of Major Nuclide
       Data, Los Alamos Scientific Laboratory Report
       LA-6116-MS, September, 1975.

78.   Ahrens, H., Patzelt, P. and Herrmann, G., J. Inorganic
       Nuclear Chemistry 38, pp 191-192, 1976.

79.   Bögl, W., Bäckmann, K. and Büttner, K., Radiochim.
       Acta 21, pp 33-40, 1974.

80.   Erten, H. N. and Blachot, J., Radiochim. Acta 21,
       pp 1-7, 1974.

81.   Kratz, J. V., Franz, H., Kaffrell, N. and Herrmann,
       G., Nuclear Physics A, 250, pp 13-37, 1975.

82.   Kratz, J. V. and Herrmann, G., J. Inorganic Nuclear
       Chemistry 32, pp 3713-3723, 1970.

83.   Chien-Chang Lin, J. Inorganic Nuclear Chemistry 38,
       pp 1409-1411, 1976.

84.   Nuh, F. M., Slaughter, D. R., Prussin, S. G., Kratz,
       K. L., Franz, H. and Herrmann, G., Physics Letters B
       53, pp 435-438, 1975.

85.   Nuh, F. M., Slaughter, D. R. and Prussin, S. G.,
       Nuclear Physics A 250, pp 1-12, 1975.

86.   Schussler, F., Cavallini, P., Carraz, L. C., Brissot,
       R., Crançon, J., Monnand, E., Ristori, C. and Moussa,
       A., Radiochim. Acta 21, pp 8-12, 1974.

87.   Schussler, H. D. and Herrmann, G., Radiochim. Acta,
       18, pp 123-133, 1972.

88.   Slaughter, D. R., Nuh, F. M. and Prussin, S. G.,
       J. Inorganic Nuclear Chemistry 38, pp 1753-1755, 1976.

89.   Tovedal, H. and Fogelberg, B., Nuclear Physics A 252,
       pp 253-259, 1975.

90.   Igarasi, S., Iijima, S., Kawai, M., Nakagawa, T.,
       Kikuchi, Y., Maki, K. and Matsunobu, H., Proceedings
       of the Conference on Nuclear Cross-sections and
       Technology, Washington, pp 320-323, March, 1975.

91   Gruppelaar, H., Proceedings of a Consultants Meeting
       on the Use of Nuclear Theory in Neutron Nuclear Data
       Evaluation, Trieste, December, 1975, IAEA Report
       IAEA-190, Vol. 2, pp 61-94, 1976.

92.   Bustraan, M., Proceedings of a Panel on Fission Pro-
       duct Nuclear Data, Bologna, November 1973, IAEA
       Report IAEA-169, Vol. 2, pp 53-113, 1974.

93.   Gruppelaar, H., Dragt, J. B., Janssen, A. J. and
       Dekker, J. W. M., Proceedings of the Conference on
       Nuclear Cross-sections and Technology, Washington,
       pp 165-168, March, 1975.

94.   Hockenbury, R. W., Knox, H. R. and Kaushal, N. N.,
         Proceedings of the Conference on Nuclear Cross-
         sections and Technology, Washington, pp 905-907,
         March, 1975.

95.   Priesmeyer, H. G. and Harz, V., Proceedings of the
         Conference on Nuclear Cross-sections and Technology,
         Washington, pp 744-747, March, 1975.

96.   Widder, J. F., Nuclear Science Engineering 60,
         pp 53-61, 1976.

97.   Rogers, F. J. G., Alpha-Decay Schemes, AERE Harwell
         Report AERE-R8005, to be published.

98.   UK Nuclear Data Progress Report for the period April,
         1975 to March, 1976, UKNDC(76)P80, D. B. Gayther,
         AERE, Harwell, Editor, page 44, August, 1976.

99.   Evardson, L., CCDN, Saclay, private communication,
         August, 1976.

100.   Devillers, C., Nimal, B., Fiche, C., Noël, J. P.,
         Blachot, J. and de Tourreil, R., Proceedings of
         the Conference on Nuclear Cross-sections and
         Technology, Washington, pp 29-38, March, 1975.

101.   England, T. R., Schenter, R. E. and Whittemore, N. L.,
         Gamma and Beta Decay Power following ^{235}U and ^{239}Pu
         Fission Bursts, Los Alamos Scientific Laboratory
         Report LA-6021-MS, July, 1975.

102.   England, T. R. and Stamatelatos, M. G., Trans. Ameri-
         can Nuclear Society, 23, pp 493-494, 1976.

103.   Gunst, S. B., Conway, D. E. and Connor, J. C.,
         Nuclear Science Engineering 56, pp 241-262, 1975.

104.   Gunst, S. B., Connor, J. C. and Conway, D. E.,
         Nuclear Science Engineering 58, pp 387-413, 1975.

105.   Lott, M., Proceedings of a Panel on Fission Product
         Nuclear Data, Bologna, November 1973, IAEA Report
         IAEA-169, Vol. 2, pp 115-162, 1974.

106. Morrison, G. W., Weisbin, C. R., Kee, C. W., Proceedings of the Conference on Nuclear Cross-sections and Technology, Washington, pp 455-458, March, 1975.

107. Schenter, R. E. and Schmittroth, F., Ibid, pp 21-28.

108. Schmittroth, F., Nuclear Science Engineering 59, pp 117-139, 1976.

109. Tobias, A., Annals of Nuclear Energy 2, pp 3-10, 1975.

110. Maeck, W. J., Proceedings of a Panel on Fission Product Nuclear Data, Bologna, November, 1973, IAEA Report IAEA-169, Vol. 1, pp 163-190, 1974.

111. Merz, E. and Laser, M., Proceedings of a Panel on Fission Product Nuclear Data, Bologna, November, 1973, IAEA Report IAEA-169, Vol. 1, pp 213-223, 1974.

112. Olivier, J. P., Advances Nuclear Science Technology, 8, pp 141-172, 1975.

113. Weitkamp, C., Proceedings of a Panel on Fission Product Nuclear Data, Bologna, November, 1973, IAEA Report IAEA-169, Vol. 1, pp 191-212, 1974.

114. Kühn, W. K. G. and Niemann, E. G., Ibid, pp 225-230.

115. Alpen, E. L., Ibid, pp 231-234.

116. Ilberg, D., Saphier, D. and Yiftah, S., Nuclear Science Engineering 58, pp 445-449, 1975.

117. Tomlinson, L., Nuclear Technology 14, pp 42-52, 1972.

118. Hetrick, D. L., Transactions American Nuclear Society 10, pp 248-249, 1967.

119. Smets, H. B., Nuclear Science Engineering 25, pp 236-241, 1966.

120. Tomlinson, L., At. Data Nuclear Data Tables 12, pp 179-194, 1973.

121.   Keepin, G. R., Wimett, T. F. and Zeigler, R. K.,
        Physical Review, 107, pp 1044-1049, 1957.

122.   Keepin, G. R., Wimett, T. F. and Zeigler, R. K.,
        J. Nuclear Energy 6, pp 1-21, 1957.

123.   Keepin, G. R., Physics of Nuclear Kinetics,
        Addison Wesley, Reading, Massachusetts, 1965.

124.   Tuttle, R. J., Nuclear Science Engineering 56,
        pp 37-71, 1975.

125.   Cahalan, J. E. and Ott, K. O., Nuclear Science Eng.
        50, pp 208-215, 1973.

126.   Tomlinson, L., "Delayed Neutrons from Fission,"
        AERE Harwell Report AERE-R6993, February, 1972.

127.   Cox, S. A., Delayed Neutron Data - Review and
        Evaluation, Argonne National Laboratory Report
        ANL/NDM-5, April, 1974.

128.   Wilson, R. R., Physical Review 71, Page 560, 1947.

129.   Asghar, M., Crançon, J., Gautheron, J. P. and
        Ristori, C., J. Inorganic Nuclear Chemistry 37,
        pp 1563-1567, 1975.

130.   Asghar, M., Crançon, J., Gautheron, J. P. and
        Ristori, C., J. Inorganic Nuclear Chemistry 37,
        pp 2229-2233, 1975.

131.   Masters, C. F., Thorpe, M. M. and Smith, D. B.,
        Nuclear Science Engineering 36, pp 202-208, 1969.

132.   Clifford, D. A., as reported by M. H. McTaggart,
        in Fast Integral Assembly Newsletter, No. 19-22,
        March, 1972.

133.   Besant, C. B., Challen, P. J. McTaggart, M. H.,
        Tavoularidis, P., and Williams, J. G., "Absolute
        Yields and Group Constants of Delayed Neutrons in the
        Fast Fission of ^{235}U, ^{238}U and ^{239}Pu," J. British
        Nuclear Energy Society, 16, pp 161-176, 1977.

134. Meadows, J. W., "Delayed Neutron Yield of ^{238}U and ^{241}Pu," Argonne National Laboratory Report ANL/NDM18 November, 1975.

135. Rudstam, G., "Characterization of Delayed-Neutron Spectra," Swedish Research Councils' Laboratory Studsvik Report LF-70, 1976.

136. Buckner, M. R. and Stewart, J. W., Nuclear Science Engineering 59, pp 289-297, 1976.

137. Hendricks, J. S. and Henry, A. F., "Finite Difference Solution of the Time-dependent Neutron Group Diffusion Equations," MIT Report MITNE-176, August, 1975.

138. Cox, S. A. and Dowling Whiting, E. E., Argonne National Laboratory Report ANL-7610, January, 1970.

139. Maksyutenko, B. P., Atomnya Energiya 7, pp 474-475, 1959.

140. Maksyutenko, B. P., J. Exp. Theoretical Physics, 35, pp 815-816, 1958.

141. Maksyutenko, B. P., Soviet J. Nuclear Physics 6(1), Page 16, 1968.

142. Krick, M. S. and Evans, A. E., Nuclear Science Eng. 47, pp 311-318, 1972.

143. East, L. V., Augustson, R. H. and Menlove, H. O., Nuclear Assay Research Group, Program Status Report, Los Alamos Scientific Laboratory Report LA-4605-MS, pp 15-17, January, 1971.

144. Batchelor, R. and McK. Hyder, H. R., J. Nuclear Energy 3, pp 7-17, 1956.

145. Bonner, T. W., Bame, S. J., Jr., and Evans, J. E., Physical Review, 101, pp 1514-1515, 1956.

146. Burgy, M., Pardue, L. A., Willard, H. B. and Wollan, E. O., "Energies of the Delayed Neutrons from ^{235}U Fission Products," War Department, Corps of Engineers, Manhattan District, Oak Ridge Report MDCC-16, June, 1946.

147.   Kratz, K. L., Rudolph, W., Ohm, H., Franz, H.
        Ristori, C., Zendel, M., Herrmann, G., Nuh, F. M.,
        Slaughter, D. R., Shihab-Eldin, A. A. and Prussin,
        S. G., Proceedings of the Third International
        Conference on Nuclei Far From Stability, Cargèse,
        CERN Report CERN76-13, pp 304-310, May, 1976.

148.   Shalev, S. and Cuttler, J. M., Nuclear Science Eng.
        51, pp 52-66, 1973.

149.   Fieg, G., J. Nuclear Energy 26, pp 585-592, 1972.

150.   Fieg, G. and Lalovic, M., Nuclear Science Eng. 58,
        pp 260-261, 1975.

151.   Sloan, W. R. and Woodruff, G. L., Nuclear Science
        Eng. 55, pp 28-40, 1974.

152.   Eccleston, G. W. and Woodruff, G. L., "Measured
        Near-Equilibrium Delayed Neutron Spectra Produced
        by Fast-Neutron-Induced Fission of ^{232}Th, ^{233}U,
        ^{235}U, ^{238}U and ^{239}Pu," Nuclear Science Eng., 62,
        pp 636-651, 1977.

153.   Evans, A. E. and Krick, M. S., Trans. American
        Nuclear Soc. 23, pp 491-492, 1976.

154.   Evans, A. E. and Krick, M. S., "Equilibrium Delayed-
        Neutron Spectra from Fast Fission of ^{235}U, ^{238}U and
        ^{239}Pu," Nuclear Science Eng., 62, pp 652-659, 1977.

155.   Iwasaki, S., Yana, K., Sato, S., Sano, K., Hagiwara,
        M. and Sugijama, K., Proceedings of the Conference
        on Nuclear Cross-sections and Technology, Washington,
        pp 611-614, March, 1975.

156.   Chulick, E. T., Reeder, P. L., Bemis, C. E., Jr.,
        and Eichler, E., Nuclear Physics A 168, pp 250-258,
        1971.

157.   Saphier, D., Ilberg, D., Shalev, S. and Yiftah, S.,
        "Evaluated Delayed Neutron Spectra and Their Impor-
        tance in Reactor Calculations," Nuclear Science Eng.
        62, pp 660-694, 1977.

158. Rudstam, G., Proceedings of the Ninth International EMIS Conference on Electromagnetic Isotope Separators and Related Ion Accelerators, Israel, May, 1976.

159. Rudstam, G. and Lund, E., Physical Review, C 13, pp 321-330, 1976.

160. Lund, E. and Rudstam, G., Physical Review, C 13, pp 1544-1551, 1976.

161. Asghar, M., Gautheron, J. P., Bailleul, G., Bocquet, J. P., Greif, J., Schrader, H., Siegert, G., Ristori, C., Crançon, J. and Crawford, G. I., Nuclear Physics A 247, pp 359-376, 1975.

162. Asghar, M., Crançon, J., Gautheron, J. P. and Ristori, C., J. Inorganic Nuclear Chemistry 37, pp 2229-2233, 1975.

163. Ristori, C., CENG Grenoble, private communication, September 1976.

164. Kratz, J. V., Franz, H. and Herrmann, G., J. Inorganic Chemistry 35, pp 1407-1417, 1973.

165. Roeckl, E., Dittner, P. F., Klapisch, R., Thibault, C., Rigaud, C. and Prieels, R., Nuclear Physics A 222, pp 621-628, 1974.

166. Izak-Biran, T. and Amiel, S., Nuclear Science Eng. 57, pp 117-121, 1975.

167. Nir-El, Y. and Amiel, S., Proceedings of the Third International Conference on Nuclei Far From Stability, Cargese, CERN Report CERN 76-13, pp 322-326, May, 1976.

168. Shalev, S. and Rudstam, G., Physical Review Letters, 28, pp 687-690, 1972.

169. Shalev, S. and Rudstam, G., Nuclear Physics A, 230, pp 153-172, 1974.

170. Rudstam, G. and Shalev, S., Nuclear Physics A 235, pp 397-409, 1974.

171.   Shalev, S. and Rudstam, G., "Energy Spectra of
        Delayed Neutrons from Separated Fission Products,"
        Swedish Research Councils' Laboratory, Studsvik
        Report LF-64, 1975.

172.   Chrysochoides, N. G., Anoussis, J. N., Mitsonias, C.
        A. and Perricos, D. C., J. Nuclear Energy 25,
        pp 551-556, 1971.

173.   Franz, H., Kratz, J. V., Kratz, K. L., Rudolf, W.,
        Herrmann, G., Nuh, F. M., Prussin, S. G. and Shihab-
        Eldin, A. A., Physical Review Letters, 33, pp 859-
        862, 1974.

174.   Ray, P. K. and Kenney, E. S., Nuclear Instruments
        Methods 134, pp 559-564, 1976.

175.   Kellie, J. D., Kelvin Laboratory, University of Glas-
        gow, private communication, October, 1976.

176.   Crawford, G. I., Hall, S. J., Kellie, J. D., Asghar,
        M., Kratz, K. L., Proceedings of the International
        Conference on the Interactions of Neutrons with
        Nuclei, Lowell, Massachusetts, Paper PB2/H6, July,
        1976.

177.   Saphier, D., Ilberg, D., Shalev, S. and Yiftah, S.,
        Transactions American Nuclear Society 22, pp 671-672,
        1975.

178.   Rudstam, G., Proceedings of an Advisory Group Meeting
        on Transactinium Isotope Nuclear Data, Karlsruhe,
        November 1975, IAEA Report IAEA-186, Vol. 3, pp 305-
        308, 1976.

179.   Crawford, G. I., Hall, S. J., Kellie, J. D., Asghar,
        M., Bailleul, G., Bocquet, J. P., Chauvin, C.,
        Gautheron, J. P., Schrader, H., Siegert, G., Crançon,
        J. and Ristori, C., J. Physics A 7, L141-L143, 1974.

180.   Klapdor, H. V., Proceedings of the Third Internation-
        al Conference on Nuclei Far From Stability, Cargèse,
        CERN Report CERN76-13, pp 311-316, May, 1976.

181. Shihab-Eldin, A. A., Nuh, F. M., Halverson, W.,
      Prussin, S. G., Rudolph, W., Ohm, H., Kratz, K. L.,
      Ibid, pp 317-321.

182. Mihalczo, J. T., Nuclear Science Engineering 60,
      pp 262-275, 1976.

183. Lemmel, H. D., Proceedings of the Conference on
      Nuclear Cross-sections and Technology, Washington,
      pp 286-292, March, 1975.

184. de Saussure, G., Olsen, D. K. and Perez, R. B.,
      Nuclear Science Engineering 61, pp 496-506, 1976.

185. Czirr, J. B. and Sidhu, G. S., Nuclear Science Eng.
      57, pp 18-27, 1975.

186. Czirr, J. B. and Sidhu, G. S., Nuclear Science Eng.
      58, pp 371-376, 1975.

187. Czirr, J. B. and Sidhu, G. S., Nuclear Science Eng.
      60, pp 383-389, 1976.

188. Czirr, J. B. and Sidhu, G. S., Proceedings of the
      Conference on Nuclear Cross-sections and Technology,
      Washington, pp 546-548, March, 1975.

189. Reich, C. W., Op. Cit., Ref. 178, pp 265-304.

190. Barton, D. M., Diven, B. C., Hansen, G. E., Jarvis,
      G. A., Koontz, P. G. and Smith, R. K., Nuclear Science
      Engineering 60, pp 369-382, 1976.

191. Behrens, J. W., Carlson, G. W. and Bauer, R. W.,
      Proceedings of the Conference on Nuclear Cross-sec-
      tions and Technology, Washington, pp 591-596, March,
      1975.

192. Behrens, J. W. and Carlson, G. W., "Measurement of
      the Neutron-induced Fission Cross-section of ^{241}Pu
      Relative to ^{235}U from 0.001 to 30 MeV," Lawrence Liver-
      more Laboratory Report UCRL-51925, October, 1975.

193. Blons, J., Mazur, C. and Paya, D., Proceedings of the
      Conference on Nuclear Cross-sections and Technology,
      Washington, pp 642-645, March, 1975.

194.   Caner, M. and Yiftah, S., Nuclear Science Eng.
        59, pp 46-50, 1976.

195.   Coates, M. S., Gayther, D. B. and Pattenden, D. B.,
        Proceedings of the Conference on Nuclear Cross-
        sections and Technology, Washington, pp 568-571,
        March, 1975.

196.   Difilippo, F. C., Perez, R. B., de Saussure, G.,
        Olsen, D. K. and Ingle, R. W., Op. Cit., Paper
        PBl/J3, Ref. 176.

197.   Gayther, D. B., Op. Cit., Ref. 195, pp 564-567.

198.   Gilliam, D. M. and Knoll, G. F., Op. Cit., Ref. 195,
        pp 635-636.

199.   Lindner, M., Nagle, R. J. and Landrum, J. H., Nuclear
        Science Engineering 59, pp 381-394, 1976.

200.   Meadows, J. W., Nuclear Science Engineering 58,
        pp 255-257, 1975.

201.   Pearlstein, S. and Moxon, M. C., U. K. Nuclear Data
        Progress Report UKNDC(76)P80, Editor D. B. Gayther,
        AERE, Harwell, England, August, 1976.

202.   Spencer, R. R. and Kaeppeler, F., Op. Cit., Ref. 195,
        pp 620-622.

203.   Proceedings of an Advisory Group Meeting on Trans-
        actinium Isotope Nuclear Data, Karlsruhe, IAEA
        Report IAEA-186, 1976.

204.   Nunn, R. M., Op. Cit., ref. 203, Vol. 1, pp 7-16.

205.   WRENDA 76/77, World Request List for Nuclear Data,
        INDC(SEC)-55/URSF, August, 1976 (available from
        IAEA Nuclear Data Section).

206.   Barré, J. Y. and Bouchard, J., Proceedings of an
        Advisory Group Meeting on Transactinium Isotope
        Nuclear Data, Karlsruhe, November, 1975, IAEA Report
        IAEA-186, Vol. 1, pp 113-137, 1976.

207. Hjaerne, L., Chairman of the Working Group on Waste Management and Isotope Applications, Proceedings of an Advisory Group Meeting on Transactinium Isotope Nuclear Data, Karlsruhe, November, 1975, IAEA Report IAEA-186, Vol. 1, pp 113-137, 1976.

208. Kuesters, H. and Lavolic, M., Op. Cit., ref. 207, pp 139-166.

209. Bell, M. J., ORIGEN, The ORNL Isotope Generation and Depletion Code, Oak Ridge National Laboratory Report ORNL-4628, May, 1973.

210. Harte, G. A., CEGB Berkeley Nuclear Laboratories Gloucestershire, England, private communication, August, 1976.

211. Harte, G. A., CEGB Berkeley Nuclear Laboratories Report RD/B/N3564, 1976.

212. Benjamin, R. W., Proceedings of an Advisory Group Meeting on Transactinium Isotope Nuclear Data, Karlsruhe, November, 1975, IAEA Report IAEA-186, Vol. 2, pp 1-70, 1976.

213. James, G. D., Op. Cit ref. 212, pp 71-113.

214. Moore, M. S., Op. Cit, ref. 212, pp 161-185.

215. Igarasi, S., Op. Cit. ref. 212, Vol. 3, pp 1-163.

216. Yiftah, S., Gur, Y., and Caner, M., Op. Cit., ref. 215, pp 165-194.

217. Baranov, S. A., Zelenkov, A. G. and Kulakov, V. M., Op. Cit, ref. 215, pp 249-263.

218. Lynn, J. E., Op. Cit., ref. 215, pp 201-235.

219. Barré, J. Y., Op. Cit., ref. 206, pp 17-24.

220. Benjamin, R. W., Vandervelde, V. D., Gorrell, T. C. and McCrosson, F. J., Op. Cit. 195, pp 224-228.

221.   Benjamin, R. W., McCrosson, F. J., Gorrell, T. C.
       and Vandervelde, V. D., "A Consistent Set of Heavy
       Actinide Multigroup Cross-sections," Savannah River
       Laboratory Report DP-1394, December, 1975.

222.   Gayther, D. B., Thomas, B. W., Endecott, D. A. J.,
       Jolly, J. E., "Measurement of the Neutron Capture
       Cross-section of ^{241}Am using a Large Liquid Scin-
       tillator," U. K. Nuclear Data Progress Report,
       UKNDC(76)P80, D. B. Gayther, AERE, Harwell, Editor,
       pp 3-4, August, 1976.

223.   McKean, I. E. and Crouch, E. A. C., "Half-life of
       ^{241}Pu," U. K. Nuclear Data Progress Report UKNDC(76)-
       P80, D. B. Gayther, AERE, Harwell, Editor, pp 41-42,
       August, 1976.

224.   Mewissen, L., Poortmans, F., Rohr, G., Theobald, J.,
       Weigmann, H. and Vanpraet, G., Proceedings of the
       Conference on Nuclear Cross-sections and Technology,
       Washington, pp 729-732, March, 1975.

225.   Sweet, D. W., "Cross-sections of Actinides," U. K.
       Nuclear Data Progress Report UKNDC(76)P80, D. B.
       Gayther, AERE, Harwell, Editor, p 32, August, 1976.

226.   Wagemans, C. and Deruytter, A. J., Nuclear Science
       Engineering 60, pp 44-52, 1976.

227.   Weston, L. W. and Todd, J. H., Op. Cit., Ref. 224,
       pp 229-231.

228.   Wilkins, M., "The Half-life of ^{241}Pu," AERE, Harwell
       Report AERE-R7906, December, 1974.

229.   Glover, K. M., AERE, Harwell, private communication,
       November, 1976.

230.   Strohm, W. W. and Jordan, K. C., Transactions American
       Nuclear Society 18, p 185, 1974.

231.   Fischer, E. A., "Integral Measurements of Effective
       Delayed Neutron Fractions," Nuclear Science and Eng.
       62, p 105, 1977.

THE ANALYSIS OF REACTOR NOISE:
MEASURING STATISTICAL FLUCTUATIONS IN NUCLEAR SYSTEMS

N. Pacilio, A. Colombino, R. Mosiello,
F. Norelli, and V. M. Jorio

Ricerche ed Esperienze di Diagnostica
e Sicurezza Tramite Analisi di Rumore
Dipartimento Ricerca Tecnologica di Base ed Avanzata
Comitato Nazionale per l'Energia Nucleare
Casella Postale 2400, 00100 Rome, Italy

I. INTRODUCTION

A. Historical Review

The first observation of the statistical fluctuations
of the number of neutrons in a reactor was the now famous
Rossi-alpha experiment, named for Bruno Rossi who suggested
and set up the measuring apparatus during the early days of
Los Alamos. The original theoretical studies were formulated by Fermi, Feynman and de Hoffmann in 1944 (1) and again
by de Hoffman (1946) (2) during the same period and issued in
two Los Alamos reports: LADC-269 and LADC-256, respectively.
Like most of the scientific and literary production of the
time, the above-mentioned reports remained classified for
several years.

The first unclassified material appeared later, and was
a contribution by de Hoffman (1949) (3) to the second volume
of The Science and Engineering of Nuclear Power. The article
was substantially an elaboration of the material contained
in the LADC-256 report. One must wait seven more years for
the unclassified publication of the matter treated in the
LADC-269 report. It was issued in the Journal of Nuclear

67

Energy signed by Feynman, de Hoffmann and Serber (1956) (4).
John Orndoff (1957) (5) revived at Los Alamos the Rossi-Alpha
experiment, on the scheme of the original apparatus of 13
years earlier, and published his results in Nuclear Science
and Engineering. Since then, the door of reactor-noise analysis was finally opened officially.

B.   What is 'Noise'?

    Noise is indeed a versatile, multi-meaning word that
indicated something different in correspondence to the technical framework with which one is dealing. In acoustics,
noise means (1) "sound, especially of a loud, harsh or con-
fused kind;" (2) "a nonharmonious or discordant group of
sounds," according to the Random House Dictionary of the
English Language, 1961 (6). In the American Institue of
Physics Handbook (McGraw-Hill, 1963) (7) noise is more properly identified. In fact, one may read there that (1)
"noise is an erratic intermittent or statistically random
oscillation;" (2) "noise is the fluctuation in the output of
the detector when the incident radiation is steady."

    Further clarifications of noise and fluctuations are
given by Van der Ziel (1954) (8). He states that "the term
spontaneous fluctuations, although perhaps theoretically the
most appropriate, is not commonly used in practice; usually,
it is simply called noise. This latter term was derived as
a result of the acoustical effects accompanying spontaneous
electric fluctuations in receivers. The descriptive name
noise is used, actually, even when an acoustical effect is
not involved, as for instance, in television".

    From the angle of quantum mechanics, Osborn and Yip
(1963) (9) point out that "neutron noise is more properly
the field of studies of neutron multiplet densities other
than the singlet density in nuclear reactor systems." In
the branch of nuclear physics, as Cohn (1971) (10) sketches
with concise and brilliant phrasing, "reactor noise is neutron
fluctuations. The basic cause of the statistical fluctuations
in the neutron population is the randomness of all neutronic
reactions. There would be fluctuations even if all fissions
produced the same number of neutrons".

    The general problem of studying fluctuation phenomena
is introduced by Landau and Lifshitz (1959) (11) who affirm
that "it has already been emphasized several times that the

physical quantities that describe a macroscopic body in
equilibrium are practically always equal to their mean values
with very great accuracy. However, small as they are, deviations from these values do occur (the quantities fluctuate,
as one says), and the question arises of finding the probability distribution for such fluctuations".

C.   Reactor Noise

      "Reactor noise", jargon for "fluctuations in the output
signal from a neutron detector or other sensor inside or near-
by a nuclear reactor", is a well-established branch of studies
and applications in the field of nuclear reactor physics. The
final goal of the analysis of reactor noise is that of measuring time constants and dynamic characteristics, assessing
correct operation, diagnosing expected malfunctions and calling an early warning of unexpected abnormalities in the reactor system under observation, without any need for external
excitation or perturbation of the system itself. The latter
feature makes noise analysis a valuable technique, especially
in nuclear power plants, because it does not disturb the
energy production.

      Noise measurements for the securing of design data and
safety monitoring determine a situation that has not only a
scientific relevance but also a remarkable social impact.
The low cost of noise monitoring, in comparison with the high
cost of plant failures and/or shutdowns, plays a favorable
role in support of noise analysis.

D.   Reactor Noise Theory

      The evolution of the state of a nuclear reactor may be
assimilated to a stochastic process and may be mathematically
described by a set of stochastic equations. The major part
of problems of reactor physics can be solved in terms of the
behavior of the mean values of the state variables; e.g.,
number of neutrons, fuel temperature, coolant temperature
pressure and flow, void fraction. This attitude reduces the
study of the stochastic process to a mere deterministic analysis.

      The motivation for a special interest in the fluctuations
of the state variables is that these fluctuations contain information; i.e., more information than that contained in the

behavior of the mean values. A very appealing example of
the validity of the latter statement is given by Williams
(1974) (12) who deals with one of the simplest stochastic processes, the birth-death process. He shows that, for constant
birth and death probability per unit time and population individual, the time analysis of the mean value of the population contains information about the balance between birth
and death rates, and only the time analysis of the fluctuations (i.e., variance) of the population leads to separate
measurement of the two probabilities.

E.   Reactor Neutron Noise Theory

The evolution of the state of a zero-power reactor is
generally described by the time behavior of the number of
neutrons. If the problems are set and solved in terms of
the mean value of the number of neutrons, this area of studies is named *reactor kinetics.*

If the interest is extended to the fluctuations of the
number of neutrons, the amplified area of research may be
called *reactor stochastic kinetics.* Stacey (1969) (13) reported the advancements in the field of reactor stochastic
kinetic theory, mostly in the case of energy and space independent reactor models: he also outlined very appropriately
the ramification of the theory into different formalisms operated by the specialists facing the problem of neutron population fluctuations. His list may be supplemented and developed in order to show how the physical mathematics of
reactor neutron noise theory grew up in past and recent
years.

FORWARD STOCHASTIC METHOD

Harris (1958), Pluta (1962), Matthes (1962), Dalfes (1963),
Kemeny and Murgatroyd (1964), Pluta (1964), Dalfes (1966),
Matthes (1966), Kobayashi (1968), Clarke et al. (1968),
Kurihara and Sekiya (1969), Stacey (1969), Williams (1969),
Smith (1970), Pacilio (1972), Jorio and Pacilio (1973),
Isnard (1973), Lewins (1974), Arcipiani and Marseguerra
(1974), Norelli et al. (1974), Genoud (1975), Matthey (1975),
Colombino et al. (1975), Ruby and Wang (1975), Pacilio et al.
(1976) (13-38).

BACKWARD STOCHASTIC METHOD

Pál (1958), Bell (1965), Williams (1967), Zolotor (1968), Stacey (1969), Arcipiani and Marseguerra (1974), Furuhashi (1975), Marseguerra (1975) (*13,32,39-44*).

LANGEVIN METHOD

Dalfes (1963), Akcasu and Osborn (1966), Sheff and Albrecht (1966), Nieto (1969), Stacey (1969), Williams (1974) (*13,17, 41,45-47*).

BRANCHING PROCESS THEORY

Zolotukhin and Mogilner (1963), Babala (1966), Saito and Taji (1967) (*48-50*).

GREEN'S FUNCTION TYPE KERNEL METHOD

Harris (1965), Sheff and Albrecht (1966) (*46,51*).

QUANTUM LIOUVILLE EQUATION APPROACH

Osborn and Natelson (1965), Burger (1969) (*52,53*).

An ancestor of all the mathematical procedures is an early work by Courant and Wallace (1947) (*54*) who set detailed probability balance equations and formulated a differential equation substantially equivalent to the one obtainable via the forward stochastic method.

Each one of the above-mentioned approaches was useful in the past for describing some particular experimental procedure; often, the purpose of the treatment did not intend any specific message for the experimentalist. The reference list of reactor neutron noise studies is indeed full of elegant and clever theoretical works and contributions that could not reach the world of reactor physicists, characterized by entirely different interests and problems.

F.   Reactor Neutron Noise Analysis Techniques

One of the most impressive facts concerning reactor neutron noise analysis techniques is the multiplication and branching of the early (Rossi-alpha experiment, Feynman-variance experiment) pioneering procedures into several

special measurements that differ in methodology, data collection and reduction, and ultimate interpretation in terms
of reactor parameters.

The experimental methods of reactor noise analysis are
today represented by several (ca 20) different techniques
that may be classified in five major categories:

> Analysis of probability profiles of neutron counts
>
> Analysis of moments of neutron count distributions
>
> Analysis of correlation among neutron counts in
> the time domain
>
> Analysis of correlation among neutron counting
> level polarities in the time domain
>
> Analysis of reactor power level in the frequency
> domain

One of the disadvantageous consequences of this proliferation is the tendency to interpret each type of reactor
noise experiment according to either a sort of heuristic
theory made up on purpose or one of the mathematical theories
introduced in Section E: so the overall impression is that
each heuristic or mathematical treatment could lead to a
limited number of experimental techniques, and no one of
them could be adopted for working out an adequately complete
and organic theory of ALL the methods. The unsatisfactory
nature of this position is the motivation for the unified
development of the present review.

G.   Survey of Literature on Reactor Noise

The proceedings of five multinational meetings, two
extensive review publications and four books illustrate the
evolution of noise analysis theory and techniques. Basic
and detailed material on reactor noise may be found in the
proceedings of the meetings held in Florida (1963, 1966),
Petten, 1967, Tokyo/Kyoto, 1968, and Rome, 1974 (55-58).
They are labeled accordingly in the bibliography of this
presentation. Excellent coverage of the subject has been
prepared by Seifritz and Stegemann (59) and by Saito (60).
The first survey is complemented by an extended reference
list on noise analysis experimental techniques; the second

review is characterized by a strictly theoretical approach
and stretches out to power reactor noise problems. The four
books have been authored by Thie, 1963 (*61*), Pacilio, 1969
(*62*), Uhrig, 1970 (*63*), and Williams, 1974 (*25*). Motivations,
issues and purposes of the books are explained by the authors.

In his preface Thie states: "The use of reactor noise
as a means of obtaining information on nuclear reactors
appears to be gaining in popularity. The techniques for
collecting and using such data are not well known, however,
and published reports are widely scattered. It is hoped
that collecting pertinent information together will
not only save investigators loss of valuable time developing methods (which, in the end, may be found similar to
those already in use) but also, considerably facilitate
future work in the field. This book is primarily directed
to the scientist or engineer who wishes to perform and/or
interpret various types of reactor experiments and should
therefore appeal to (1) individuals engaged in reactor testing, who may find this information of value in expediting
tests; (2) research and development personnel in projects
involving problems in dynamics, for whom the analogies and
the examples discussed here may well apply to their work;
and (3) students of nuclear engineering who may find this
book a guide to a course of experiments in reactor kinetics."

In the presentation of his book, Uhrig writes: "This
book is an outgrowth of several years' experience in teaching a course dealing with the application of random noise
theory to nuclear reactor systems, as well as research and
industrial consulting work in the field. It is designed to
serve the dual purpose of supplementing a course in nuclear
reactor noise at the first-year graduate level as a reference book for practicing engineers and scientists interested
in applying random noise techniques to nuclear reactor systems. There has been a deliberate attempt to make the book
as self-contained as possible."

The philosophy implied by Williams in his book is explained with the following words: "The book itself is an
attempt to complement the works of Thie and Uhrig, which
are the only existing publications on random processes in
nuclear reactors of any depth written in the English language.
The aim of this book, therefore, can be summed up by the
word *pragmatic*. In other words, it describes the problems
that a nuclear engineer may meet which involve random

fluctuations, and sets out in detail how they may be interpreted in terms of various models of the reactor system."

In the introduction of his book, Pacilio affirms that
"In many laboratories around the world, groups of scientists
work on reactor-noise analysis on a fulltime schedule.
Their contribution is clearly visible in the continually
improving degree of sophistication of the theoretical treatment, in the highly automated versions of experimental techniques and data processing codes, and in a number of special-
purpose instruments now commercially available."

The need for an organic and unified theory of the matter is emphasized as a major issue in the same introduction:
"Today a bibliography on reactor-noise analysis is no longer
acceptable as a guide to the state of the art, since the
quantity of published literature would make a neophite
approach to the subject extremely confusing."

H.   In Search of a Unified Theory for Reactor Noise

After a long journey through the elaborate list of
published material on reactor noise, the same old question
comes back again: Is a common theoretical origin available
for all the reactor neutron noise analysis techniques?
This presentation answers the question; it is intentionally
directed to explore, expand and unify reactor neutron noise
studies with the purpose of setting an organic theory that
can generate ALL the measuring methods.

Motivations, purposes and destinations of the monograph
are more or less those mentioned in Section G, by the
authors of the four books; but one basic issue is enforced
here: the introduction of an organic and progressive theory
that builds models for interpreting experiments of noise
analysis. All models are energy-independent; some of them
are also space-independent. The structure of the article
is as linear and as self-contained as possible.

First, a theory is presented for describing reactor
neutron noise as a stochastic phenomenon with neutrons and
neutron counts from a multidetector apparatus as state variables. The reactor system is completely characterized in
the time evolution of mean values and fluctuations of its
state variables by a multivariate joint distribution. This

is shown to generate all the neutron noise analysis techniques developed in 30 years of experience with nuclear reactor experimental physics.

Second, new state variables (e.g., precursors, neutrons
and neutron counts from different spatial zones) are added
for constructing more complex models. Because of the analytical complication, an alternative mathematical procedure
must be adopted. Nevertheless, the reactor system is again
characterized in the time evolution of mean values and fluctuations of its state variables by variances, covariances,
auto- and cross-correlation functions of neutron counts.

Finally, a liaison is created between stochastic parameters and the conventional physical quantities that appear
in reactor kinetics.

The whole effort is dedicated to the experimenters;
the ad hoc treatment of the theory can give them useful and
practical indications concerning the way the experiments
should be performed and important formulae upon which to
work. The article is also oriented to scientists in
national laboratories, and teachers and students in academic
institutions who expressed the exigency and the opportunity
for a common ground in the introduction of reactor neutron
noise to newcomers.

II.   THEORY

A.   Markovian Stochastic Processes in Nuclear Systems

A multiplying medium is a system in which a population
of individuals lives and propagates; individuals are generated, generate other individuals, leak out and die. For a
nuclear multiplying system, the well-known demographic phenomena of birth, immigration, death, emigration and transform
are interpreted by the processes of fission, source emission,
capture, leakage and detection, respectively. All of them
affect a population of particles; i.e., neutrons, precursors
of delayed neutrons, detected neutrons.

In a stochastic theory, the state of the system is
characterized by the number of particles of each type present
at a given time instant. Transitions of the system from one
state to another occur because of processes; a process is de-

fined by the probability of occurrence per unit time and by
its effect on the population of particles. Particles are
exposed to each process according to probability laws. Thus,
the ensemble of nuclear phenomena may be assimilated to a
stochastic process. In a stochastic theory, the time evolution of the system is never uniquely determined, but the
prediction is based on relations among state probabilities.

The stochastic processes in a nuclear multiplying system can be considered a la Markov. In a Markovian stochastic process, the behavior of the system is assumed to be
memoryless; i.e., its future depends only on its present
state and not on the past.

The simplest procedure in dealing with a population
evolution in a nuclear system is to adopt quantities such
as fission probability, source-emission probability, capture probability, etc., and define them time-independently.
Given these elementary probabilities, one can write down
differential equations, first order in time, whose solutions
specify the population size probability distribution at any
instant of time. These equations are usually named *detailed
probability balance equations*.

B.   The Detailed Probability Balance Equations

As a meaningful example of how to face the reactor
neutron noise general formulation, detailed probability
balance equations are given for a point, monokinetic zero-
power reactor model, in which a single group of precursors
of delayed neutrons are considered. If the *forward stochastic method* is adopted, a balance equation is written for
the probability that there are N neutrons and C precursors
at the time t+dt, and that there have been R detections
from time 0 to time t+dt. Some authors express this concept
by introducing a pseudo particle named detectron (=detected
neutron) and by replacing the corresponding sentence with
"there are R detectrons at time t+dt". The concept of detectron and the pseudo particle itself will be maintained
for the rest of the article.

The state probability may be written as P(N,C,R,t+dt)
and expressed in terms of the various possible states at
time t and all the possible and mutually exclusive transitions.

For the chosen reactor model, the following causes of transition or nontransition must be taken into account:

(SE) Source Emission
(DP) Decay of Precursor
(A)   Loss by Capture or Leakage
(F)   Fission
(D)   Detection
      Nothing happened

If the probabilities are introduced for each of the six processes, so that

S is the probability per unit time that a source neutron enters the reactor;

λ is the probability per unit time that a delayed neutron is emitted by the precursor C;

μ_c is the probability per unit time that a neutron undergoes capture or leaks away from the reactor;

μ_f is the probability per unit time that a neutron induces a fission reaction;

η is the probability per unit time that a neutron is detected;

$1-\{S+\lambda C+N(\mu_c + \mu_f+\eta)\}dt$ is the probability that nothing occurred in the time interval $t \to t+dt$, for small dt;

the following balance equations can be written:

$$
\begin{aligned}
P(N,C,R,t+dt) = \; & P(N-1,C,R,t)Sdt + \\
& + P(N-1,C+1,R,t)\lambda[C+1]dt + \\
& + P(N+1,C,R,t)\mu_c[N+1]dt + \\
& + \sum_n\sum_m P(N-n+1,C-m,R,t)\mu_f[N-n+1]p(n,m)dt + \\
& + P(N+1,C,R-1,t)\eta[N+1]dt + \\
& + P(N,C,R,t)\{1-[S+\lambda C+(\mu_c+\mu_f+\eta)N]dt\}
\end{aligned}
$$

Here, $p(n,m)$ is the probability that a fission reaction produces n neutrons and m precursors of delayed neutrons. If

the limit is calculated for dt→0, the following set of ordinary differential equations is obtained:

$$\frac{dP(N,C,R,t)}{dt} = [P(N-1,C,R,t)-P(N,C,R,t)]S +$$
$$+ [P(N-1,C+1,R,t)[C+1]-P(N,C,R,t)C]\lambda +$$
$$+ [P(N+1,C,R,t)[N+1]-P(N,C,R,t)N]\mu_c +$$
$$+ [\sum_n\sum_m P(N-\eta+1,C-m,R,t)[N-n+1]p(n,m) -$$
$$- P(N,C,R,t)N]\mu_f +$$
$$+ [P(N+1,C,R-1,t)[N+1]-P(N,C,R,t)N]\eta.$$

C.   State Probabilities and State Probability Generating Functions

In this section, the concept of state probability generating function is introduced; considerable use of it will
be made in the rest of the presentation. Although the generating function method may constitute an additional difficulty in studying stochastic processes, the initial extra
effort is counterbalanced by substantial and convenient
simplification in the mathematical treatment. The main advantage is in the handling of a single function to represent a whole collection of individual items; i.e., the
infinite set of state probabilities, and in the task of
solving a single differential equation instead of an infinite system of detailed probability balance equations.

If the following function is introduced:

$$F(x,y,z,t) = \sum_{N=0}^{\infty} \sum_{C=0}^{\infty} \sum_{R=0}^{\infty} P(N,C,R,t)x^N y^C z^R$$

$F(x,y,z,t)$ is named the generating function of the probability distribution $P(N,C,R,t)$. In fact, the relationship
holds

$$P(N,C,R,t) = \left(\frac{1}{N!C!R!}\right) \left(\frac{\partial^{N+C+R} F}{\partial x^N \partial y^C \partial z^R}\right)_{x=y=z=0}$$

The variables x, y and z are auxiliaries associated to
each type particle (neutrons, precursors and detectrons, respectively) and are defined in the interval [0,1]. F(x,y,

z,t) is also the generating function of the factorial moments of the state variables N,C and R. In fact, if expanded in power series of (x-1), (y-1) and (z-1), the generating function becomes

$$F(x,y,z,t) = \sum_{i=0}^{\infty} \sum_{j=0}^{\infty} \sum_{k=0}^{\infty} \frac{M_{ijk}}{i!j!k!} (x-1)^i (y-1)^j (z-1)^k$$

The relationship

$$M_{ijk} = \left(\frac{\partial^{i+j+k} F}{\partial x^i \partial y^j \partial z^k} \right)_{x=y=z=1} = i!j!k! \; \overline{\binom{N}{i}\binom{C}{j}\binom{R}{k}}$$

gives the factorial moments in the case of one group of neutrons, precursors and detectrons.

Furthermore, if the following function is introduced

$$f(x,y) = \sum_{n=0}^{\infty} \sum_{m=0}^{\infty} p(n,m) x^n y^m$$

f(x,y) is named the generating function of the probability distribution p(n,m). The relationship holds

$$p(n,m) = \left(\frac{1}{n!m!} \right) \left(\frac{\partial^{n+m} f}{\partial x^n \partial y^m} \right)_{x=y=0}$$

The auxiliary variables x and y have been defined previously. f(x,y) is also the generating function of the factorial moments of the state variables n and m. If expanded in power series of (x-1) and (y-1), the generating function becomes

$$f(x,y) = \sum_{i=0}^{\infty} \sum_{j=0}^{\infty} \frac{m_{ij}}{i!j!} (x-1)^i (y-1)^j$$

The relationship

$$m_{ij} = \left(\frac{\partial^{i+j} f}{\partial x^i \partial y^j} \right)_{x=y=1} = i!j! \; \overline{\binom{n}{i}\binom{m}{j}}$$

gives the factorial moments in the case of one group of neutrons and precursors emitted in a fission reaction.

By the introduction of the two probability generating functions, the set of detailed probability balance equations transforms into a single, partial differential equation:

$$\frac{\partial F(x,y,z,t)}{\partial t} = S[x-1]F + \{\mu_c[1-x] + \mu_f[f(x,y)-x] +$$

$$+ \eta[z-x]\} \frac{\partial F}{\partial x} + \lambda[x-y]\frac{\partial F}{\partial y}$$

The generating function also may be interpreted as a partition function of the system; i.e., an indicator of the presence and multiplicity of the various types of particles for a certain state. An expressive example is given by the glossary of Table I.

TABLE I

STOCHASTIC GLOSSARY FOR DESCRIBING POSSIBLE
STATES OF THE NUCLEAR SYSTEM

No. Particles	Associated Expression of the Generating Function
No particle	1
1 neutron	x
1 precursor	y
1 detectron	z
1 neutron and 1 precursor	xy
1 neutron and 1 detectron	xz
1 precursor and 1 detectron	yz
N neutrons	x^N
C precursors	y^C
R detectrons	z^R

Solutions with more than one group of precursors can
be obtained for this and similar problems with only additional
algebraic complexity (see, for example, Williams, 1974 (12).
Laplace solutions are related to the conventional transfer
function arising from the presence of the first order equations or behavior in the mean.

D.   The Bartlett Formalism

Once the stochastic glossary has been established, it
may be used to describe the state transitions induced by the
different processes via the Bartlett (64) formalism, after the
assumption is made that reactor neutron noise is a Markov process. In his book, Bartlett states that "the appropriate
equation for many Markov chains can be written down at once
in terms of the possible transitions" without going through
the tedious and time-consuming routine involving the detailed
probability balance equations. In the case of reactor neutron noise, the Bartlett formalism transforms into Table II.

TABLE II

STATE TRANSITIONS IN THE BARTLETT FORMALISM
(for the meaning of symbols labeling
the various processes, see Section IIB)

Process	Initial State	Final State	Stochastic Operator
(SE)	1	x	1
(DP)	y	x	$\partial/\partial y$
(A)	x	1	$\partial/\partial x$
(F)	x	$f(x,y)$	$\partial/\partial x$
(D)	x	z	$\partial/\partial x$

The physical implications of such a formulation for the initial and final state are fully explained by the stochastic glossary of Table I. The role of the state transition stochastic operator is that of defining the particle that gets lost in the transition; i.e., the operator implies the partial derivative of the state probability generating function $F(x,y,z,t)$ with respect to the dummy variable associated to the particle that induces, undergoes or is exposed to the corresponding process. If the chance of transition is independent of the number of particles; see, for instance, process (SE); the stochastic operator becomes unitary.

After the Bartlett formalism, a differential equation for reactor neutron noise can be written by suitably assembling the elements of the following scheme:

(1)	(2)	(3)	(4)
(SE)	S	$x-1$	F
(DP)	λ	$x-y$	$\partial F/\partial y$
(A)	μ_c	$1-x$	$\partial F/\partial x$
(F)	μ_f	$f(x,y)-x$	$\partial F/\partial x$
(D)	η	$z-x$	$\partial F/\partial x$

where the numbers stand for:

(1) process

(2) probability of occurrence of the process per unit time and per particle (except for S which is only a probability per unit time)

(3) state transition in the terminology of the stochastic glossary; i.e., final state minus initial state

(4) dependence on the state probability generating function

The few elementary rules of compilation are:

1.   the lhs member of the time-differential equation
      consists of the partial derivative of $F(x,y,z,t)$
      with respect to the time variable t;

2.   the rhs member of the time-differential equation
      is obtained by adding vertically the results of
      the horizontal products of columns (2), (3) and
      (4) for each process in the scheme;

3.   explicit initial conditions must be coupled with
      the time-differential equation in order to pin-
      point the proper solution.

The final result is completely equivalent to the differential equation obtained via the detailed probability balance method; i.e.,

$$\frac{\partial F}{\partial t} = S[x-1]F + \{\mu_c[1-x]+\mu_f[f(x,y)-x]+\eta[z-x]\frac{\partial F}{\partial x} +$$

$$+ \lambda[x-y]\frac{\partial F}{\partial y} ;$$

$$F(x,y,z,0)=F_0.$$

E.   Models of Nuclear Systems: Sketch and Formats

In the rest of the presentation, a number of models of
nuclear systems are studied. As previously stated, a system is characterized by the presence and multiplicity of
particles of each type; therefore, the first feature of the
system is described by a box with the following structure:

PARTICLE	NUMBER OF PARTICLES	ASSOCIATED VARIABLE

The system is also characterized by all the possible
processes that may occur and cause state transitions. This
further feature of the system is described by a Bartlett
scheme, as represented by a box:

(1) PROCESS	(2) PROBABILITY	(3) TRANSITION	(4) DEPENDENCE ON PGF

The time evolution of the system is hence outlined by the partial differential equation obtained from the Bartlett scheme:

$$\frac{\partial F}{\partial t} =$$

coupled with the initial condition:

$$F(.,o)=F_o$$

There are two mathematical procedures possible at this point.

If the partial differential equation can be solved analytically, the solution is given in terms of

$$F(.,t)=$$

This may be done, subject to approximating $f(x)$ as developed in Sections IIIA and IIIB.

An alternative mathematical procedure may be adopted. It is based on the derivation of a set of time differential equations in the first and second order factorial moments of the state variables as developed in Sections IIIC and IIIF.

Both procedures lead to autocorrelation and crosscorrelation functions.

If u is the net production rate of some particle (so that the expected number of particles in the time interval $0 \rightarrow t$ is $\bar{U}=ut$) the expected number of pairs of particles in the same interval $0 \rightarrow t$ can be defined as:

$$\frac{1}{2} \overline{U(U-1)} = \int_o^t dt_2 \int_o^{t_2} \pi(t_1 t_2) u \, dt_1 \qquad 0 \leqslant t_1 \leqslant t_2 \leqslant t$$

where udt_1 is the probability that the first particle of
              the pair is counted in the time interval $t_1 \rightarrow t_1 + dt_1$;

              $\pi(t_1, t_2) dt_2$ is the conditional probability that the
              second particle of the pair is counted in the time
              interval $t_2 \rightarrow t_2 + dt_2$, given a first count in $t_1 \rightarrow t_1$
              $+ dt_1$.

Under the hypothesis of an ergodic process (see Section IIIF)
$\pi(t_1 t_2)$ does not depend on the choice of time t_1 but ONLY on
the time lapse $t_2 - t_1$.

The function $u\pi(t_1, t_2)$ is defined as the <u>autocorrelation</u>
function AC(U,U) of the state variable U and can be written
as $\phi_{uu}(t_2 - t_1)$. Its integral definition can also be converted into a differential relationship; i.e.,

$$AC(U,U) = \frac{1}{2} \overline{\dot{U}(\dot{U}-1)} = \phi_{uu}(t_2 - t_1) \text{ (observed)} = u\pi$$

Furthermore, if u and v are the net production rates of
two types (1 and 2) of particles (so that the expected number of particles of the two types in the time interval $0 \rightarrow t$
is $\bar{U} = ut$ and $\bar{V} = vt$, respectively) the expected number of mixed
pairs of the two particles in the same interval $0 \rightarrow t$ can be
defined as:

$$\frac{1}{2} \overline{UV} = \int_0^t dt_2 \int_0^{t_2} \pi_v(t_1, t_2) u dt_1 = \int_0^t dt_2 \int_0^{t_2} \pi_u(t_1, t_2) v dt_1$$

$$0 \leqslant t_1 \leqslant t_2 \leqslant t$$

where

              udt_1 is the probability that the first particle of the
              pair is of type 1 and is counted in the time interval
              $t_1 \rightarrow t_1 + dt_1$;

              $\pi_v(t_1, t_2) dt_2$ is the conditional probability that the
              second particle of the pair is of type 2 and is counted
              in the time interval $t_2 \rightarrow t_2 + dt_2$, given a first count of
              type 1 in the time interval $t_1 \rightarrow t_1 + dt_1$;

vdt_1 is the probability that the first particle of the pair is of type 2 and is counted in the time interval $t_1 \to t_1 + dt_1$;

$\pi_u(t_1, t_2) dt_2$ is the conditional probability that the second particle of the pair is of type 1 and is counted in the time interval t_2 $t_2 + dt_2$, given a first count of type 2 in the time interval t_1 $t_1 + dt_1$.

Under the hypothesis of an ergodic process (see Section IIIF) $\pi_u(t_1, t_2)$ and $\pi_v(t_1, t_2)$ do not depend on the choice of time t, but ONLY on the time lapse $t_2 - t_1$.

The function $u\pi_v(t_1, t_2) = v \pi_u(t_1, t_2)$ is defined as the crosscorrelation function $XC(U,V)$ of the state variables U and V and can be written as $\phi_{uv}(t_2 - t_1)$. Its integral definition also can be converted into a differential relationship; i.e.,

$$XC(U,V) = \frac{1}{2} \overset{..}{UV} = \phi_{uv}(t_2 - t_1) \text{ (observed)} = u\pi_v = v\pi_m.$$

In the present definition of auto- and crosscorrelation functions, we have not followed the practice of some authorities of (a) defining the variation about a zero mean value; or (b) subsequently normalizing the functions to give unity at $t_2 - t_1 = 0$. This latter step is sometimes regarded as the difference between the unnormalized variance function and the normalized correlation function. Our usage intends to be more general and admits a more direct relation of the theory to the experimental results being analyzed.*

F.   The Ergodic Hypothesis

Experimenters know very well that the first fundamental assumption for all reactor neutron noise analysis techniques is that the reactor is in steady-state operating conditions; i.e., the neutron population evolution is expected to be stationary.

---

* With the standard normalizations not used here, the autocorrelation function is the conditional probability π, itself.

This corresponds to inserting the neutron source into
a subcritical system and waiting for the neutron population
to evolve into equilibrium with the source; the equilibrium
is reached within several neutron chain lifetimes. Once the
stationary condition is set up, reactor noise measurements
can be triggered.

The physical and mathematical implications of the so-
called ergodic hypothesis are described suggestively by Ash
(65). He writes, "Assume a random time series of neutron
fluctuations partitioned into n strips, T sec. in length.
What a suitable length is will become apparent. If the
strips are examined, it will be found that they certainly
are not replicas of each other; however, if the series is a
stationary time series, then they are statistical replicas
of each other. The stationarity property implies that each
of the n strips has the same statistical properties, such as
the same mean (over the length of the sample) and variance.
That is, their statistics are time-invariant.

"Now consider the ensemble (collection) of the n time
strips, one under the other, to form a two-dimensional array
of data. Since each strip, because of its stationary property, contains identical statistical information, it is as if
the neutron data were taken from an ensemble of n identical
reactors and each strip were a record from each reactor.
First, consider the time average of some property of the
series taken over any one of the strips in time; e.g., the
time average of the square of the deviation from the mean
neutron level; i.e., the variance over the time. Second,
consider the two-dimensional array of amplitude data from
the n strips and calculate the variance, but take each datum
from a given column of the array. This calculation gives the
variance over the ensemble of n hypothetical reactors. If
the individual strips possess the stationary property, the
ensemble variance, or any other ensemble average (like the
ensemble mean) will be independent of the column of the
array that is chosen to obtain the data. Further, that
the ensemble average is equal to the time average enunciates
the ergodic property. A proof of the ergodic hypothesis has
been found for only a limited number of statistical processes. Mathematically speaking, processes that are of physical interest are only quasiergodic. This distinction can
be ignored for our needs."

The ergodic hypothesis also means the time invariance
of the state probability generating function for the population of neutrons (and precursors if they are taken into
account in the nuclear system under study). This corresponds, finally, to the circumstance that all the moments of
the population of neutrons (and precursors) exhibit vanish-
ing time derivatives in the set of differential equations.

In conclusion, three further points should be made.
First, if delayed neutron precursors are considered, the
settling time of reactor with source may, in principle, be
longer than a few lifetimes or generation/reproduction
times; in practice, however, the prompt behavior dominates
the approach to an equilibrium condition. Second, the argument has been given in terms of the population of neutrons
and not detectrons; these latter are not stationary in the
sense used so far, since evidently, the population of detectrons is monotonically increasing. However, the stationary
assumption corresponds to a constant expectation of the rate
of creation of detectrons and analogous measures of detectron behavior. Third, the stationary hypothesis is not valid
in the special case of a just critical reactor, in absence of
source; it may be demonstrated by counter-example that in
this particular case, the behavior of the system is not ergodic.

G.   Nuclear Systems with Neutrons

Suppose only neutrons are present in the nuclear system.

PARTICLE	NUMBER OF PARTICLES	ASSOCIATED VARIABLE
Neutron	N	x

BARTLETT SCHEME

(1)	(2)	(3)	(4)
Source Emission	S	$x-1$	F
Capture and Leakage	μ_c	$1-x$	$\partial F/\partial x$
Fission	μ_f	$f(x)-x$	$\partial F/\partial x$

$$\frac{\partial F(x,t)}{\partial t} = S[x-1]F + \{\mu_c[1-x] + \mu_f[f(x)-x]\}\frac{\partial F}{\partial x} \; ; \; F(x,0)=1$$

The initial condition corresponds to the presence of no neutrons at the time of insertion of the source into the system.

In the case of the special approximation of $f(x)$ quadratic in $(x-1)$; that is,

$$f(x) = 1 + f'(1)[x-1] + \frac{1}{2} f''(1)[x-1]^2$$

the solution of the time differential equation has been shown by Smith (26,27) to be

$$F(x,t) = [1+Y_1(1-x)(1-e^{-\alpha t})]^{-S/\alpha Y_1}$$

where

$$\alpha = \mu_c + \mu_f - \mu_f f'(1)$$

$$Y_1 = \mu_f f''(1)/2\alpha$$

Limits, implications and consequences of the truncation of the generating function of the distribution of neutrons emitted by fission have been widely discussed by Pacilio et al. (38).

The ergodic condition for the neutron distribution; i.e., the time invariance of the state probability generating function and the equilibrium of the neutron population with the source can be imposed by awaiting several neutron chain lifetimes ($t \gg 1/\alpha$); that is, by calculating the following limit:

$$\lim_{\alpha t \to \infty} F(x,t) = [1+Y_1(1-x)]^{-S/\alpha Y_1}$$

The same result can be obtained by setting the lhs member of the time differential equation equal to zero and by integrating the corresponding equation.

III. MODELS FOR INTERPRETING EXPERIMENTS

A. Nuclear Systems with Neutrons and Detectrons

Suppose that neutrons and detectrons from H sensors are present in the nuclear system.

PARTICLE	NUMBER OF PARTICLES	ASSOCIATED VARIABLE
Neutron	N	x
Detectron 1	R_1	z_1
Detectron 2	R_2	z_2
...	..	..
Detectron H	R_H	z_H

BARTLETT SCHEME

(1)	(2)	(3)	(4)
Source emission	S	$x-1$	F
Capture and Leakage	μ_c	$1-x$	$\partial F/\partial x$
Fission	μ_f	$f(x)-x$	$\partial F/\partial x$
Detection by sensor 1	η_1	z_1-x	$\partial F/\partial x$
Detection by sensor 2	η_2	z_2-x	$\partial F/\partial x$
...	..	...	...
Detection by sensor H	η_H	z_H-x	$\partial F/\partial x$

$$\frac{\partial F}{\partial t}(x,z_1,z_2,\ldots,z_H,t) = S[x-1]F + \{\mu_c[1-x]+\mu_f[f(x)-x]$$

$$+ \sum_{i=1}^{H} \eta_i [z_i-x] \frac{\partial F}{\partial x}$$

If the truncation of $f(x)$ introduced in Section IIG is recalled, the time differential equation also can be written as

$$\frac{\partial F}{\partial t} = S[x-1]F + \alpha Y_1 [x-x_1][x-x_2]\frac{\partial F}{\partial x}$$

where

$$\alpha = \mu_c + \mu_f + \eta - \mu_f f'(1) \qquad\qquad Y_i^* = 2Y_1 n_i/\alpha$$

$$\eta = \sum_{i=1}^{H} n_i \qquad\qquad x_1 = 1+[1+Z]/2Y_1$$

$$Y_1 = \mu_f f''(1)/2\alpha \qquad\qquad x_2 = 1+[1-Z]/2Y_1$$

$$Z = \sqrt{1+2 \sum_{i=1}^{H} Y_i^* [1-z_i]}$$

The initial condition implies the ergodic hypothesis; i.e.,

$$F(x,z_1,z_2,\ldots,z_H,0) = [1+Y_1[1-x]^{-S/\alpha Y_1}$$

The solution of the time differential equation has been shown by Pacilio et al. (38) to be

$$F(x,z_1 z_e,\ldots,z_H,t) =$$

$$= \{1+Y_1[1-x]+ [(Y_1/2Z)(1-x)(1-Z)+(1/4Z)^2](1-e^{-\alpha Z t})\}^{-S/\alpha Y_1}$$

$$*\exp (S/2Y_1)(1-Z)t$$

The probability distribution $P(N,R_1,R_2,\ldots,R_H,t)$ can be obtained from the generating function via the following formula

---

Note: It might be objected that this definition of α is different from the one adopted in Section IIG; but the presence of the term η is either immaterial ($\eta=\epsilon\mu_f$ with $\epsilon< 10^{-3}$) or suitably compensated by diminishing μ_c via a calibration, for instance, at criticality.

$$P(N,R_1,R_2,\ldots,R_H,t) =$$

$$\frac{1}{N!R_1!R_2!\ldots R_H!}\left[\frac{\partial^{N+R_1+R_2+\ldots+R_H} F}{\partial x^n \partial z_1^{R_1} \partial z_2^{R_2}\ldots \partial z_H^{R_H}}\right]_{y=z_1=z_2=\ldots=z_H=0}$$

The major problem in this operation lies in the task of deriving a composite function a large number of times with respect to its variables. Faa di Bruno (66) proposed a closed form expression for the nth order derivative of a differentiable r-fold composite function based on the partitions of the positive integer, n. Routti and Szeless (67) presented compact formulae for computing higher order derivatives of composite functions; in particular, probability generating functions of reactor neutron noise. Mosiello (68) improved the Routti and Szeless approach by introducing two recursion formalisms for the determination of the nth order derivative of a composite function strictly tied up to the idea that the operation of deriving a composite function is merely an analysis problem of combining.

The marginal distribution of neutrons is obtained for $z_1=z_2=\ldots=z_H=Z=1$. Its form is

$$F(x,1,1,\ldots,1,t) = [1 + Y_1[1-x]]^{-S/\alpha Y_1}$$

and confirms the ergodic hypothesis; i.e., the steady state distribution of neutrons. This distribution is well known in the literature as the Negative Binomial, Polya, Pascal or Poisson-Logarithmic distribution.

The marginal distribution of detectrons is obtainable for x=1. Its form is

$$F(1,Z,t)=$$
$$= 1+(1/4Z)(1-Z)^2(1-e^{-\alpha Z t})^{-S/\alpha Y_1} * \exp[(S/2Y_1)(1-Z)t] = G(Z,t).$$

When reduced to a single detector experiment; i.e., if Z=z, the distribution with generating function $G(z,t)$ can be easily shown to be identical to the PMZBB distribution.

This distribution has been introduced independently by Pal
(39), Mogilner and Zolotukhin (69), Bell (70,71) and Babala
(49) and is named after the initials of the authors.

Mogilner and Zolotukhin approximated the PMZBB distribution with a Negative Binomial profile, Pacilio and Jorio (72)
proposed another alternative consisting of the Poisson-Rad-
ical distribution, Zingoni et al. (73) introduced new profiles matching the one from the PMZBB distribution, based on
the family of Poisson-Algebraic probability group. It is
interesting to note that all the proposed approximations
belong to the class of generalized Poisson distributions
introduced by Gurland (74).

The analytical solution allows complete answers to be
given to a series of questions. First of all, it is evident
that the PMZBB distribution does not satisfy any balance or
differential equation. It is, in fact, the marginal distribution of a multivariate joint profile associated with a
stochastic process in which the state variables are the
number of neutrons and detectrons of various types.

The solution, in terms of a multivariate distribution of
neutrons and detectrons, instead of a univariate distribution
of the number of counts from a single detector such as the
PMZBB profile, helps a better understanding of the nature of
the reactor noise experiments because it states unequivocally
the condition of the neutron population during the real measurements.

Finally, the multivariate joint distribution offers a
sound theoretical background for the interpretation of cross-
correlation technique results based on the elaboration of
signals coming from a multidetector measuring apparatus.
Since cross-correlation techniques are getting more and more
popular today, the single-detector experiment described by
the PMZBB distribution does not meet adequately the needs
and the requirements expressed by noise analysts.

B. A Unified Theory of Reactor Neutron Noise Analysis
    Techniques

In this section, a unified theory of all the reactor
neutron noise analysis techniques contained in the five
major categories mentioned in Section IF will be given.
It consists of:

a. naming the technique

b. indicating the observable values of state
   probabilities, distribution moments and
   other related quantities in the formalism
   adopted in the present article

c. expressing the mathematical link between
   the physical quantities of point (b) and
   the corresponding value of the generating
   function of the marginal distribution of
   detectrons as derived from the analytical
   solution of the equation in Section IIIA

d. offering additional information and helpful
   remarks about particular relations between
   experimental conditions and values of state
   and auxiliary variables

The techniques are named according to the nomenclature
introduced by Pacilio (62).

The formulation for the probability of a state consisting of the presence of R (or R_1 and R_2) detectrons at time
t is

P(R,t)   or $P(R_1,R_2,t)$

The corresponding state probability generating functions
are, respectively,

G(z,t)   or $G(z_1,z_2,t)$

where the explicit expression of the two generating functions
is given in Section IIIA for the special case of one or two
detectron groups.

The role of the auxiliary variables z' and Z also must
be recalled in this circumstance. They are, respectively,

$$z' = \sqrt{1+2y_1^*[1-z]}$$

$$Z = \sqrt{1+2Y_1^*[1-z_1]+2Y_2^*[1-z_2]}$$

Most of the time the generating functions appear in
their derived expression, with respect to variables z, z_1

and z_2, calculated in the extreme points ($=0$ or $=1$) of the auxiliary variable definition interval. The formalism is of the following type; e.g.,

$$G'(1,t) = \left(\frac{\partial G(z,t)}{\partial z}\right)_{z=1} = \overline{R};$$

$$\frac{1}{2}G''(0,t) = \frac{1}{2}\left(\frac{\partial^2 G(z,t)}{\partial z^2}\right)_{z=0} = P(2,t)$$

$$G''_{z_1,z_2}(1,1,t) = \left(\frac{\partial^2 G(z_1,z_2,t)}{\partial z_1 \partial z_2}\right)_{z_1=z_2=1} = \overline{R_1 R_2}$$

C. Nuclear Systems with Neutrons and Detectrons: An Alternative Mathematical Procedure

If the nuclear system considered in Section IIIA is recalled once again, the alternative mathematical treatment in terms of moments for the partial differential equation can be proposed. This evidently is useful where exact analytical solutions are not available, but may be preferred, for low-order moments, when the experimental results are to be analyzed in terms of moments anyway. The equations in this form can be derived directly from the original detailed probability balance equations, although the Bartlett formalism again is more compact.

The factorial moments M_{ijk} have been defined at Section IIIC for a neutron-precursor-detectron system.

Consider now the Bartlett equation for the neutron-multidetectron model:

$$\frac{\partial F(x,z_1,z_2,\ldots,z_H,t)}{\partial t} = S[x-1]F + \{\mu_c[1-x]+\mu_f[f(x)-x] +$$

$$+ \sum_{i=1}^{\Sigma} n_i[z_i-x]\} \frac{\partial F}{\partial x}\,] \;; \qquad F(x,z_1,z_2,\ldots,z_H,0) = F_o$$

and take partial derivatives with respect to $x,z_1,z_2,\ldots,z_H$ in turn. On evaluation at $x=z_1=z_2=\ldots=z_H=1$, one obtains the equations for the first order factorial moments (the mean behavior).

ANALYSIS OF PROBABILITY PROFILES OF NEUTRON COUNTS

METHOD	OBSERVED PROBABILITY	FORMULA FOR DATA INTERPRETATION	NOTES
Zero-count probability	$P(0,t)$	$G(0,t)$	$\begin{cases} z=0 \\ z'=\sqrt{1+2Y_f^2} \end{cases}$
Zero-count/One-count probability	$\dfrac{\overline{R}P(0,t)}{P(1,t)}$	$\dfrac{G'(1,t)G(0,t)}{G'(0,t)}$	$\begin{cases} z=0 \\ z'=\sqrt{1+2Y_f^2} \\ z=1 \\ z'=1 \end{cases}$
Furuhashi & Izumi probability set	$\dfrac{2P(0,t)P(2,t)}{[P(1,t)]^2}$	$\dfrac{G(0,t)G''(0,t)}{[G'(0,t)]^2}$	$\begin{cases} z=0 \\ z'=\sqrt{1+2Y_f^2} \end{cases}$
Even- or odd-count probability	$\displaystyle\sum_0^\infty P(2R,t)-\sum_0^\infty P(2R+1,t)$	$G(-1,t)$	$\begin{cases} z=-1 \\ z'=\sqrt{1+4Y_f^2} \end{cases}$
Probability distribution	$\{P(R,t)\}$	$\dfrac{1}{R!}\left[\dfrac{\partial^R G(z,t)}{\partial z^R}\right]_{z=0}$	$\begin{cases} z=0 \\ z'=\sqrt{1+2Y_f^2} \end{cases}$

ANALYSIS OF MOMENTS OF NEUTRON COUNT DISTRIBUTION

METHOD	OBSERVED MOMENT	FORMULA FOR DATA INTERPRETATION	NOTES
Feynman-variance	$\dfrac{\overline{R^2}-(\overline{R})^2}{\overline{R}}$	$\dfrac{G''(1,t)+G'(1,t)-[G'(1,t)]^2}{G'(1,t)}$	$\begin{cases} z=1 \\ z'=1 \end{cases}$
Expected number of neutron pairs	$\dfrac{\overline{R(R-1)}}{(\overline{R})^2}$	$\dfrac{G''(1,t)}{[G'(1,t)]^2}$	$\begin{cases} z=1 \\ z'=1 \end{cases}$
Third-order moment	$\dfrac{\overline{R(R-1)(R-2)}-3\overline{R}\,\overline{R(R-1)}+2(\overline{R})^3}{\overline{R}}$	$\dfrac{G'''(1,t)-3G'(1,t)G''(1,t)+2[G'(1,t)]^3}{G'(1,t)}$	$\begin{cases} z=1 \\ z'=1 \end{cases}$
Expected number of correlated neutron triplets	$\dfrac{\overline{R(R-1)(R-2)}-3\overline{R}[\overline{R(R-1)}-(\overline{R}^2)]-\overline{R})^3}{(\overline{R})^3}$	$\dfrac{G'''(1,t)-3G'(1,t)G''(1,t)+2[G'(1,t)]^3}{[G'(1,t)]^3}$	$\begin{cases} z=1 \\ z'=1 \end{cases}$
Covariance	$\overline{R_1 R_2}-\overline{R_1}\,\overline{R_2}$	$G''_{z_1 z_2}(1,1,t)-G'_{z_1}(1,1,t)\,G'_{z_2}(1,1,t)$	$\begin{cases} z_1=z_2=1 \\ Z=1 \end{cases}$

ANALYSIS OF CORRELATION AMONG NEUTRON COUNTS IN THE TIME DOMAIN

METHOD	OBSERVED DATA	FORMULA FOR DATA INTERPRETATION	NOTES
Auto-correlation a.k.a. Rossi-alpha technique	$\phi_{RR}(\tau) = \overline{R(t)R(t+\tau)}$	$(1/2)\dfrac{d^2 G''(1,\tau)}{d\tau^2}$	$\begin{cases} z=1 \\ z'=1 \end{cases}$
Cross-correlation	$\phi_{R_1R_2}(\tau) = \overline{R_1(t)R_2(t+\tau)}$	$(1/2)\dfrac{d^2 G''_{z_1z_2}(1,1,\tau)}{d\tau^2}$	$\begin{cases} z_1=z_2=1 \\ Z=1 \end{cases}$
Time interval of each neutron count from a random origin a.k.a. OP-technique	$Y(t)$	$-\dfrac{dG(0,t)}{dt}$	$\begin{cases} z=0 \\ z'=\sqrt{1+2Y_1^{\#}} \end{cases}$
Time interval between contiguous neutron counts a.k.a. PP-technique	$p(t)$	$\dfrac{t}{G'(1,t)}\dfrac{d^2 G(0,t)}{dt^2}$	$\begin{cases} z=0 \\ z'=\sqrt{1+2Y_1^{\#}} \end{cases}$ $\begin{cases} z=1 \\ z'=1 \end{cases}$

ANALYSIS OF CORRELATION AMONG NEUTRON COUNTING LEVEL POLARITIES IN THE TIME DOMAIN

METHOD	OBSERVED DATA	FORMULA FOR DATA INTERPRETATION	NOTES
Polarity auto-correlation*	$sign[C(t)-\bar{C}]\,sign[C(t+\tau)-\bar{C}]$ (+)	$(2/\pi)arcsin\dfrac{\phi_{RR}(\tau)}{\phi_{RR}(0)}$	$\phi_{RR}(t)=\dfrac{1}{2}\dfrac{d^2G(1,t)}{dt^2}$ $\begin{cases}z=1\\z'=1\end{cases}$
Polarity cross-correlation*	$sign[C_1(t)-\bar{C}_1]\,sign[C_2(t+\tau)-\bar{C}_2]$	$(2/\pi)arcsin\dfrac{\phi_{R_1R_2}(\tau)}{\phi_{R_1R_2}(0)}$	$\phi_{R_1R_2}(t)=\dfrac{1}{2}\dfrac{d^2G''_{z_1z_2}(1,1,t)}{dt^2}$ $\begin{cases}z_1=z_2=1\\Z=1\end{cases}$
Covariance via polarity sampling†	$P_1+P_3-(P_2+P_4)$ (++)	$(2/\pi)arcsin[G''_{z_1z_2}(1,1,t)-G'_{z_1}(1,1,b)G'_{z_2}(1,1,t)]$	$\begin{cases}z_1=z_2=1\\Z=1\end{cases}$

* $C(t)$ is the instantaneous neutron counting level; $\phi_{RR}(t)/\phi_{RR}(o)$ is the strictly normalized correlation function.

† $X1=sign[R_1(t)-\bar{R_1}]$, $X2=sign[R_2(t)-\bar{R_2}]$
$P1=P1(X1,X2)$ is the probability of a joint event $(+,+)$
$P2=P2(X1,X2)$ is the probability of a joint event $(-,+)$
$P3=P3(X1,X2)$ is the probability of a joint event $(-,-)$
$P4=P4(X1,X2)$ is the probability of a joint event $(+,-)$

ANALYSIS OF THE REACTOR POWER LEVEL IN THE FREQUENCY DOMAIN

METHOD	OBSERVED DATA	FORMULA FOR DATA INTERPRETATION	NOTES
Auto-power spectral density	$\overline{\phi}_{RR}(\omega) = \int_{-\infty}^{\infty} e^{-i\omega t}\, \phi_{RR}(t)\, dt$	$\phi_{RR}(t) = \dfrac{1}{2}\, \dfrac{d^2 G''(1,t)}{dt^2}$	$\begin{cases} z = 1 \\ z' = 1 \end{cases}$
Cross-power spectral density	$\overline{\phi}_{R_1 R_2}(\omega) = \int_{-\infty}^{\infty} e^{-i\omega t}\, \phi_{R_1 R_2}(t)\, dt$	$\phi_{R_1 R_2}(t) = \dfrac{1}{2}\, \dfrac{d^2 G''_{z_1 z_2}(1,1,t)}{dt^2}$	$\begin{cases} z_1 = z_2 = 1 \\ Z = 1 \end{cases}$

$$\bar{N} = -\alpha\bar{N} + S \qquad\qquad \alpha = \mu_c + \mu_f + \eta - \mu_f F'(1) \qquad (C1)$$

$$\bar{R}_i = \eta_i \bar{N} \qquad\qquad \eta = \sum_{i=1}^{H} \eta_i \qquad i = 1,2,\dots,H \qquad (C2)$$

These equations are also known as <u>singlet kinetic equations.</u>

A second partial derivation with appropriate cross-terms with respect to $x, z_1, z_2, \dots, z_H$ will lead to equations for the second order factorial moments or the <u>doublet kinetic equations.</u> The following table gives the definitions of the first and second order factorial moments from the state probability generating function.

First Order Factorial Moments:

$$\left(\frac{\partial F}{\partial x} \right)_{x=z_1=z_2=\dots=z_H=1} = \bar{N}$$

$$\left(\frac{\partial F}{\partial z_i} \right)_{x=z_1=z_2=\dots z_H=1} = \bar{R}_i$$

Second Order Factorial Moments:

$$\left(\frac{\partial^2 F}{\partial x^2} \right)_{x=z_1=z_2=\dots=z_H=1} = \overline{N(N-1)}$$

$$\left(\frac{\partial^2 F}{\partial x \partial z_i} \right)_{x=z_1=z_2=\dots=z_H=1} = \overline{NR_i}$$

$$\left(\frac{\partial^2 F}{\partial z_i^2} \right)_{x=z_1=z_2=\dots=z_H=1} = \overline{R_i(R_i-1)}$$

$$\left(\frac{\partial^2 F}{\partial z_i \partial z_j} \right)_{x=z_1=z_2=\dots=z_H=1} = \overline{R_i(R_j-1)}$$

However, it is rather more convenient to express the results in terms of the factorial cumulants $T(N,R,t)$. The table that follows gives the relations between second-order factorial cumulants and factorial moments. At first order, there is no difference between them.

$$T(N,N,t) = \overline{N(N-1)} - (\bar{N})^2$$

$$T(R_i,R_i,t) = \overline{R_i(R_i-1)} - (\overline{R_i})^2$$

$$T(N,R_i,t) = \overline{NR_i} - \overline{N} \cdot \overline{R_i}$$

$$T(R_i,R_j,t) = \overline{R_iR_j} - \overline{R_i} \cdot \overline{R_j}$$

One can in principle proceed to higher orders, but for the present purposes, this effort is not necessary. It is interesting to note that $\frac{1}{2}T(N,N,t)$ is the expected number of correlated neutron pairs; i.e., the expected number $(\bar{N})$ of all the possible couples minus the expected number $(\bar{N})^2/2$ of uncorrelated pairs. The same holds for detectron pairs and other particles in general.

$\frac{1}{2}T(N,R,t)$ is then the expected number of correlated neutron, detectron pairs; i.e., the expected number $\frac{1}{2}\overline{NR}$ of all the possible pairs minus the expected number $\frac{1}{2}\overline{N} \cdot \overline{R}$ of uncorrelated couples.

Time derivatives of factorial cumulants are related to time derivatives of factorial moments via the following formulae:

$$\dot{T}(N,N,t) = \overline{N(N-1)}^{\cdot} - 2\,\overline{N} \cdot \dot{\overline{N}}$$

$$\dot{T}(N,R,t) = \dot{\overline{NR}} - \dot{\overline{N}}\,\overline{R} - \overline{N} \cdot \dot{\overline{R}}$$

$$\dot{T}(R,R,t) = \overline{R(R-1)}^{\cdot} - 2\,\overline{R} \cdot \dot{\overline{R}}$$

After some manipulation and substitution with the singlet kinetic equations, the doublet kinetic equations can be written in terms of factorial cumulants as:

$$\tfrac{1}{2}\,\dot{T}(N,N,t) = -\,\alpha T(N,N,t) + \alpha Y_1\bar{N} \qquad \alpha Y_1 = \tfrac{1}{2}\mu_f f''(1) \quad (C3)$$

$$\dot{T}(N,R_i,t) = -\alpha T(N,R_i,t) + \eta_i T(N,N,t) \quad T(N,R,0) = 0 \quad (C4)$$

$$\tfrac{1}{2}\,\dot{T}(R_i,R_i,t) = \eta_i T(N,R_i,t) \qquad\qquad\qquad T(R,R,0) = 0 \quad (C5)$$

$$\dot{T}(R_i,R_j,t) = \eta_j T(N,R_i,t) + \eta_i T(N,R_j,t); \quad i=1,2,\ldots,H \quad (C6)$$

If the ergodic hypothesis is assumed; i.e.,

$$\dot{\bar{N}} = \dot{T}(N,N) = 0$$

the analytical solutions for the mean values, the factorial cumulants and the auto- and cross-correlation functions of the number of detectrons are:

$$\bar{N} = S/\alpha \qquad\qquad\qquad\qquad \alpha Y_i^* = 2\eta_i Y_1 \qquad (C7)$$

$$\bar{R}_i = \eta_i\bar{N}t \qquad\qquad\qquad \mathcal{F}(x) = 1 - \frac{1-e^{-x}}{x} \quad (C8)$$

$$T(R_i,R_i,t) = \overline{R_i}\,Y_i^*\mathcal{F}(\alpha t) \qquad\qquad i = 1,2,\ldots,H \quad (C9)$$

$$T(R_i,R_i,t) = \bar{R}_i Y_j^*\mathcal{F}(\alpha t) = \bar{R}_j Y_i^*\mathcal{F}(\alpha t) \qquad\qquad\qquad (C10)$$

$$AC(R_i,R_i) = \eta_i^2\bar{N}[N^- + Y_1 e^{-\alpha t}] \qquad = \phi_{R_iR_i} \qquad (C11)$$

$$XC(R_i,R_j) = \eta_i\eta_j\bar{N}[N^- + Y_1 e^{-\alpha t}] \qquad = \phi_{R_iR_j} \qquad (C12)$$

Some mathematical and physical implications in the above equations may be outlined. They are:

> Equation C1 is the well-known prompt-neutron kinetic equation where α, named Rossi-alpha after Bruno Rossi is the self-reduction rate of the number of prompt-neutrons.

Equations C2, C5, C6 and C8 represent a standard
feature of noise analysis theory; no matter how
complicated the system is or how large the number
of particles, equations dealing with the mean value and the time derivative of the factorial moments (or factorial cumulants) of the number of
detectrons remain always the same.

If the ergodic hypothesis is assumed, the differential relations in Equations C1 and C3 become
algebraic and lead to time-independent solutions
for $\bar{N}$ and $T(N,N)$; e.g., Equation C7.

If the ergodic hypothesis is assumed, Equation C4
acquires the same structure of Equation C1; i.e.,
a self-reduction rate term characterized by the
Rossi-α plus a time-independent source term,
$\eta_i T(N,N)$.

Equations C9 and C10 reproduce the original
Feynman formulae for the variance and covariance
experiments, while Equations C11 and C12 lead to
the classical Rossi-alpha formulae for one-
detector and two-detector measuring apparati.

The discrete indexes i and j are referred to
sensors; quantities marked by asterisk are
related to detectrons.

$\mathcal{F}(x)$ is the Feynman function.

D.   Nuclear Systems with Precursors, Neutrons and Detectrons

Suppose that precursors, neutrons and detectrons from
H sensors are present in the nuclear system.

PARTICLE	NO. PARTICLES	ASSOCIATED VARIABLE
Neutron	N	x
Precursor	C	y
Detectron 1	R_1	z_1
Detectron 2	R_2	z_2
...		
Detectron H	R_H	z_H

BARTLETT SCHEME

(1)	(2)	(3)	(4)
Source emission	S	$x-1$	F
Capture and Leakage	μ_c	$1-x$	$\partial F/\partial x$
Fission	μ_f	$f(x,y)-x$	$\partial F/\partial x$
Precursor decay	λ	$x-y$	$\partial F/\partial y$
Detection by Sensor 1	η_1	z_1-x	$\partial F/\partial x$
Detection by Sensor 2	η_2	z_2-x	$\partial F/\partial x$
...	..	...	...
Detection by Sensor H	η_H	z_H-x	$\partial F/\partial x$

$$\frac{\partial F(x,y,z_1,z_2,\ldots,z_H,t)}{\partial t} = S[x-1]F+\{\mu_c[1-x]+\mu_f[f(x,y)-x] +$$

$$\sum_{i=1}^{H} \eta_i[z_i-x]\}\,\frac{\partial F}{\partial x} + \lambda\,[x-y]\frac{\partial F}{\partial y} \; ; \quad F(x,y,z_1 z_2,\ldots,z_H,0) = 0$$

The singlet kinetic equations are:

$$\dot{\bar{N}} = -\alpha\bar{N} + \lambda\bar{C} + S\; ; \qquad \alpha = \mu_c+\mu_f+\eta-\mu_f m_{10} \tag{D1}$$

$$\dot{\bar{C}} = \alpha_c\bar{N}-\lambda\bar{C} \qquad ; \qquad \alpha_c = \mu_f m_{01} \tag{D2}$$

$$\dot{\bar{R_i}} = \eta_i\bar{N} \qquad ; \qquad \alpha Y_1 = \frac{1}{2}\mu_f m_{20} \quad i,j,=1,2,\ldots,H \tag{D3}$$

The doublet kinetic equations are:

$$\frac{1}{2}\dot{T}(N,N,t) = -\alpha T(N,N,t)+\lambda T(N,C,t)+ \frac{1}{2}\mu_f m_{20}\bar{N} \tag{D4}$$

$$\frac{1}{2}\dot{T}(C,C,t) = -\lambda T(C,C,t)+\alpha_c T(N,C,t)+ \frac{1}{2}\mu_f m_{02}\bar{N} \tag{D5}$$

$$\dot{T}(N,C,t) = - [\alpha+\lambda]T(N,C,t)+\alpha_c T(N,N,t)+\lambda T(C,C,t)+\mu_f m_{11}\bar{N} \tag{D6}$$

$$\dot{T}(N,R_i,t) = -\alpha T(N,R_i,t) + \lambda T(C,R_i,t) + \eta_i T(N,N,t) \tag{D7}$$

$$\dot{T}(C,R_i,t) = -\lambda T(C,R_i,t) + \alpha_c T(N,R_i,t) + \eta_i T(N,C,t) \tag{D8}$$

$$\frac{1}{2}\dot{T}(R_i,R_i,t) = \eta_i T(N,R_i,t) \tag{D9}$$

$$\dot{T}(R_i,R_j,t) = \eta_j T(N,R_i,t) + \eta_i T(N,R_j,t) \tag{D10}$$

The initial conditions are:

$$T(N,R,0) = T(C,R,0) = T(R,R,0) = 0$$

If the ergodic hypothesis is assumed; i.e.,

$$\dot{\bar{N}} = \dot{\bar{C}} = \dot{T}(N,N) = \dot{T}(N,C) = \dot{T}(C,C) = 0$$

the analytical solutions for the mean value, the factorial cumulants and the auto-and cross-correlation functions of the number of detectrons are:

$$\bar{N} = S/(\alpha-\alpha_c) \tag{D11}$$

$$\bar{R}_i = \eta_i \bar{N} t \tag{D12}$$

$$T(R_i,R_i,t) = \bar{R}_i Y^*_{1i} \mathscr{F}(\alpha t) + \bar{R}_i Y^*_{2i} \mathscr{F}[(\lambda/\alpha)(\alpha-\alpha_c)t] \tag{D13}$$

$$T(R_i,R_j,t) = \bar{R}_i Y^*_{1j} \mathscr{F}(\alpha t) + \bar{R}_i Y^*_{2j} \mathscr{F}[\lambda/\alpha)(\alpha-\alpha_c)t] =$$

$$= \bar{R}_j Y^*_{1i} \mathscr{F}(\alpha t) + \bar{R}_j Y^*_{2i} \mathscr{F}[(\lambda/\alpha)(\alpha-\alpha_c)t] \tag{D14}$$

$$AC(R_i,R_i) = \eta^2_i \bar{N}[\bar{N}+Y_1 e^{-\alpha t}+(\lambda/\alpha)Y_2 e^{-(\lambda/\alpha)(\alpha-\alpha_c)t}] = \phi_{R_i R_i} \tag{D15}$$

$$XC(R_i,R_j) = \eta_i \eta_j \bar{N}[\bar{N}+Y_1 e^{-\alpha t}+(\lambda/\alpha)Y_2 e^{-(\lambda/\alpha)(\alpha-\alpha_c)t}] = \phi_{R_i R_j} \tag{D16}$$

where

$$\mathscr{F}(x) = 1 - \frac{1-e^{-x}}{x}$$

$$(\alpha - \alpha_c)Y_2 = \frac{1}{2}\mu_f[m_{20} + 2m_{11} + m_{02}]$$

$$\alpha Y^*_{1i} = 2\eta_i Y_1$$

$$(\alpha - \alpha_c)Y^*_{2i} = 2\eta_i Y_2$$

$$i,j = 1,2,\ldots,H$$

Some mathematical and physical implications in Equation D1 to Equation D16 may be outlined; they are:

> Equations D1 and D2 are the well-known prompt-neutron and delayed neutron precursor kinetic equations, where the Rossi-α is the self-reduction rate of the number of prompt neutrons, α_c is the value of the Rossi-α at criticality, and λ is the self-reduction rate of the number of precursors.

> Equations D3, D9, D10 and D12 are analogous to Equations C2, C5, C6 and C8, and exhibit the same characteristics.

> If the ergodic hypothesis is assumed, the differential relations in Equations D1, D2, D4, D5 and D6 become algebraic and lead to time-independent solutions for $\bar{N}$, $\bar{C}$, $T(N,N)$, $T(N,C)$ and $T(C,C)$; e.g., Equation D11.

> If the ergodic hypothesis is assumed, the system of Equations D7 and D8 acquire the same structure of the system composed by Equations D1 and D2, in the non-ergodic condition; i.e., the matrix

$$\begin{pmatrix} -\alpha & \lambda \\ \alpha_c & -\lambda \end{pmatrix}$$

> plus time-independent source terms $\eta_i T(N,N)$ and $\eta_i T(N,C)$, alternatively.

Equations D13 and D14 extend the original Feynman formulae for the variance and covariance experiments to include the presence of a single group of delayed-neutron precursors, while Equations D15 and D16 adapt the same extension to the classical Rossi-alpha formulae for a single- and double-detector experiment.

The discrete indices i and j are referred to sensors, subindices 1 and 2 correspond to prompt-neutrons and to the global population of neutrons, respectively, quantities marked by asterisk are related to detectrons.

The model treated in this section degenerates into the model treated in Section IIIC if the presence of precursors is neglected. This corresponds to the following three limits:

$$\alpha_c \to 0 \qquad\qquad \lim_{x \to 0} \mathscr{F}(x) = 0 \qquad\qquad \lambda \to 0$$

Because of the first, Equation D11 degenerates into Equation C7; because of the second, Equations D13 and D14 degenerate into Equations C9 and C10, respectively; because of the third, Equations D15 and D16 degenerate into Equations C11 and C12, respectively.

E. Coupled Reactor Systems: Neutrons and Detectrons in a Symmetrical Two-Zone Reactor

Suppose that neutrons and detectrons of a single species per zone are present in a symmetrical two-zone reactor.

PARTICLE	NO. PARTICLES	ASSOCIATED VARIABLE
Neutron in zone 1	N_1	x_1
Neutron in zone 2	N_2	x_2
Detectron in zone 1	R_1	z_1
Detectron in zone 2	R_2	z_2

BARTLETT SCHEME

(1)	(2)	(3)	(4)
Source emission in Zone 1	S	x_1-1	F
Capture and leakage in Zone 1	$\mu_c-\mu_t$	$1-x_1$	$\partial F/\partial x_1$
Fission in Zone 1	μ_f	$f(x_1)-x_1$	$\partial F/\partial x_1$
Transfer from Zone 1 to 2	μ_t	x_2-x_1	$\partial F/\partial x_1$
Detection in Zone 1	n_1	z_1-x_1	$\partial F/\partial x_1$
Source emission in Zone 2	S	x_2-1	F
Capture and leakage in Zone 2	$\mu_c-\mu_t$	$1-x_2$	$\partial F/\partial x_2$
Fission in Zone 2	μ_f	$f(x_2)-x_2$	$\partial F/\partial x_2$
Transfer from Zone 2 to 1	μ_t	x_1-x_2	$\partial F/\partial x_2$
Detection in Zone 2	n_2	z_2-x_2	$\partial F/\partial x_2$

$$\frac{\partial F}{\partial t}(x_1,z_1,x_2,z_2,t) =$$

$$= S[x_1-1]F+\{(\mu_c-\mu_t)(1-x_1)+\mu_f[f(x_1)-x_1]+\mu_t[x_2-x_1]+n_1[z_1-x_1]\}$$

$$\frac{\partial F}{\partial x_1} + S[x_2-1]F+\{(\mu_c-\mu_t)(1-x_2)+\mu_f[f(x_2)-x_2]+\mu_f[x_1-x_2] +$$

$$n_2[z_2-x_2]\}\frac{\partial F}{\partial x_2} \; ; \; F(x_1,z_1,x_2,z_2,0) = F_o$$

The singlet kinetic equations are:

$$\dot{\bar{N}}_1 = -\alpha\bar{N}_1+\mu_t\bar{N}_2+S \qquad\qquad (E1)$$

$$\dot{\bar{N}}_2 = \mu_t\bar{N}_1-\alpha\bar{N}_2+S \qquad\qquad (E2)$$

$$\dot{\bar{R}}_1 = \eta_1 \bar{N}_1 \tag{E3}$$

$$\dot{\bar{R}}_2 = \eta_2 \bar{N}_2 \tag{E4}$$

The doublet kinetic equations are:

$$\frac{1}{2}\dot{T}(N_1,N_1,t) = -\alpha\, T(N_1,N_1,t) + \mu_t T(N_1,N_2,t) + \frac{1}{2}\mu_f f''_{x_1 x_1}(1)\bar{N}_1 \tag{E5}$$

$$\frac{1}{2}\dot{T}(N_2,N_2,t) = -\alpha T(N_2,N_2,t) + \mu_t T(N_1,N_2,t) + \frac{1}{2}\mu_f f''_{x_2 x_2}(1)\bar{N}_2 \tag{E6}$$

$$\dot{T}(N_1,N_2,t) = -2\alpha T(N_1,N_2,t) + \mu_t T(N_1,N_1,t) + \mu_t T(N_2,N_2,t) \tag{E7}$$

$$\dot{T}(N_1,R_i,t) = -\alpha T(N_1,R_i,t) + \mu_t T(N_2,R_i,t) + \eta_i T(N_1,N_i,t) \tag{E8}$$

$$\dot{T}(N_2,R_i,t) = -\alpha T(N_2,R_i,t) + \mu_t T(N_1,R_i,t) + \eta_i T(N_2,N_i,t) \tag{E9}$$

$$\frac{1}{2}\dot{T}(R_i,R_i,t) = \eta_i T(N_i,R_i,t) \tag{E10}$$

$$\dot{T}(R_1,R_2,t) = \eta_1 T(N_1,R_2,t) + \eta_2 T(N_2,R_1,t) \tag{E11}$$

The initial conditions are:

$$T(N,R,0) = T(R,R,0) = 0$$

If the ergodic hypothesis is assumed; i.e.,

$$\dot{\bar{N}}_1 = \dot{\bar{N}}_2 = \dot{T}(N_1,N_1) = \dot{T}(N_1,N_2) = \dot{T}(N_2,N_2) = 0$$

the analytical solutions for the mean value, the factorial cumulants and the auto- and cross-correlation functions of the number of detectrons are:

$$\bar{N}_1 = \bar{N}_2 = \bar{N} = S/(\alpha-\mu_t) \tag{E12}$$

$$\bar{R}_i = \eta_i \bar{N} t; \quad (i = 1,2) \tag{E13}$$

$$T(R_1, R_1, t) = \bar{R}_1 Y^*_{a1} \mathscr{F}[(\alpha-\mu_t)t] + \bar{R}_1 Y^*_{b1} \mathscr{F}[(\alpha+\mu_t)t] \tag{E14}$$

$$T(R_2, R_2, t) = \bar{R}_2 Y^*_{a2} \mathscr{F}[(\alpha-\mu_t)] + \bar{R}_2 Y^*_{b2} \mathscr{F}[(\alpha+\mu_t)t] \tag{E15}$$

$$T(R_1, R_2, t) = \bar{R}_1 Y^*_{a2} \mathscr{F}[(\alpha-\mu_t)] - \bar{R}_1 Y^*_{b2} \mathscr{F}[(\alpha+\mu_t)t] = \tag{E16}$$

$$= \bar{R}_2 Y^*_{a1} \mathscr{F}[(\alpha-\mu_t)t] - \bar{R}_2 Y^*_{b1} \mathscr{F}[\alpha+\mu_t)]$$

$$AC(R_1, R_1) = \eta_1^2 \bar{N}[\bar{N} + \tfrac{1}{2} Y_a e^{-(\alpha-\mu_t)t} + \tfrac{1}{2} Y_b e^{-(\alpha_t+\mu_t)t}] = \phi_{R_1 R_1} \tag{E17}$$

$$AC(R_2, R_2) = \eta_2^2 \bar{N}[\bar{N} + \tfrac{1}{2} Y_a e^{-(\alpha-\mu_t)t} + \tfrac{1}{2} Y_b e^{-(\alpha_t+\mu_t)t}] = \phi_{R_2 R_2} \tag{E18}$$

$$XC(R_1, R_2) = \eta_1 \eta_2 \bar{N}[\bar{N} + \tfrac{1}{2} Y_a e^{-(\alpha-\mu_t)t} - \tfrac{1}{2} Y_b e^{-(\alpha+\mu_t)t}] = \tag{E19}$$
$$= \phi_{R_1 R_2}$$

where

$$(\alpha-\mu_t)Y_a = (\alpha+\mu_t)Y_b = \tfrac{1}{2} \mu_f f''_{xx}(1)$$
$$(\alpha-\mu_t)Y^*_{ai} = \eta_i Y_a$$
$$(\alpha+\mu_t)Y^*_{bi} = \eta_i Y_b$$

Some mathematical and physical implications in Equations E1 to E19 may be outlined; they are:

> Equations E1 and E2 are prompt-neutron kinetic equations for a symmetrical two-zone reactor whereas the Rossi-α is the self-reduction rate of the number of prompt-neutrons in each of the two identical zones, when decoupled (i.e., taken far apart) from the other.

> Equations E3 and E4, E10, E11 and E13 are analogous to Equations C2, C5, C6 and C8, and exhibit the same characteristics.

If the ergodic hypothesis is assumed, the differential relations in Equations E1, E2, E5, E6 and
E7 become algebraic and lead to time-independent
solutions for $\bar{N}_1$, $\bar{N}_2$, $T(N_1,N_1)$, $T(N_2,N_2)$ and $T(N_1,N_2)$; e.g., Equation E12.

If the ergodic hypothesis is assumed, the system
of Equations E8 and E9 acquires the same structure
of the system composed by Equations E1 and E2 in
its non-ergodic version; i.e., the matrix

$$\begin{pmatrix} -\alpha & \mu_t \\ \mu_t & -\alpha \end{pmatrix}$$

plus time-independent source terms $\eta_i T(N_1,N_i)$
and $\eta_i T(N_2,N_i)$ for $i=1,2$, alternatively.

Equations E14, E15 and E16 extend the original
Feynman formulae for the variance and covariance
experiments. They consist of a bimodal behavior
characterized by two time eigenvalues: $(\alpha-\mu_t)$
for the fundamental mode; named for simplicity
mode A; and $(\alpha + \mu_t)$ for the superior mode;
named mode B. In Equation E16, a negative
amplitude for mode B of the covariance must
be pinpointed; Equations E17, E18 and E19 extend
the classical Rossi-alpha formulae to a bimodal
(mode A and B) decay characterized by two time
eigenvalues; i.e., $(\alpha-\mu_t)$ and $(\alpha+\mu_t)$, respectively. In Equation E19, a negative amplitude
for mode B of the cross-correlation function
must be pinpointed.

Subindices 1 and 2 correspond to reactor zones
and sensors (since there is only one sensor per
zone; subindices A and B must be attributed to
their respective mode amplitudes; quantities
marked by asterisk are related to detectrons.

The model treated in this section degenerates
into the model treated in Section IIIC if a

decoupling operation is performed on the system;
i.e., the two symmetrical zones are separated
and set at no interaction distance. This
corresponds to the following limits:

$$\mu_t \to 0 \qquad Y_a, Y_b \to Y_1 \qquad Y^*_{ai}, Y^*_{bi} \to Y^*_{li}$$

Because of them, Equation E12 degenerates into
Equation C7, Equations E14 and E15 degenerate
into Equation C9, the joint factorial cumulant
of second-order $T(R_1, R_2)$ vanishes and confirms
the independence of the two zones, Equations E17
and E18 degenerate into Equation C11, the cross-
correlation $XC(R_1, R_2)$ becomes a mere constant.

F. **Coupled Nuclear Systems: Neutrons and Detectrons in an
Asymmetrical Two-Zone Reactor**

Suppose that neutrons and detectrons are present in an
asymmetrical, two-zone reactor.

PARTICLE	NO. PARTICLES	ASSOCIATED VARIABLE
Neutron in Zone 1	N_1	x_1
Neutron in Zone 2	N_2	x_2
Detectron in Zone 1	R_1	z_1
Detectron in Zone 2	R_2	z_2

$$\frac{\partial F}{\partial t}(x_1, z_1, x_2, z_2, t) =$$

$$= S_1[x_1 - 1]F + \{(\mu_{c1} - \mu_{t2})(1 - x_1) + \mu_{f1}[f(x_1) - x_1] + \eta_1[z_1 - x_1] +$$

$$\mu_{t2}[x_2 - x_1]\}\frac{\partial F}{\partial x_1} + S_2(x_2 - 1)F + \{(\mu_{c2} - \mu_{t1})(1 - x_2) + \mu_{f2}[f(x_2) - x_2] +$$

$$\eta_2[z_2 - x_2] + \mu_{t1}[x_1 - x_2]\}\frac{\partial F}{\partial x_2}; \qquad F(x_1, z_1, x_2, z_2, 0) = 0$$

BARTLETT SCHEME

(1)	(2)	(3)	(4)
Source emission in Zone 1	S_1	x_1-1	F
Capture and Leakage in Zone 1	$\mu_{c1}-\mu_{t2}$	$1-x_1$	$\partial F/\partial x_1$
Fission in Zone 1	μ_{f1}	$f(x_1)-x_1$	$\partial F/\partial x_1$
Transfer from Zone 1 to 2	μ_{t2}	x_2-x_1	$\partial F/\partial x_1$
Detection in Zone 1	η_1	z_1-x_1	$\partial F/\partial x_1$
Source emission in Zone 2	S_2	x_2-1	F
Capture and Leakage in Zone 2	$\mu_{c2}-\mu_{t1}$	$1-x_2$	$\partial F/\partial x_2$
Fission in Zone 2	μ_{f2}	$f(x_2)-x_2$	$\partial F/\partial x_2$
Transfer from Zone 2 to 1	μ_{t1}	s_1-x_2	$\partial F/\partial x_2$
Detection in Zone 2	η_2	z_2-x_2	$\partial F/\partial x_2$

The singlet kinetic equations are:

$$\dot{\bar{N}}_1 = -\alpha_1\bar{N}_1+\mu_{t1}\bar{N}_2+S_1 \qquad \text{where } \alpha_i = \mu_{ci}+\mu_{fi}+\eta_i-\mu_{fi}f'_{xi} \tag{1}$$

$$\tag{F1}$$

$$\dot{\bar{N}}_2 = \mu_{t2}\bar{N}_1 - \alpha_2\bar{N}_2+S_2 \qquad ; \qquad (i=1,2) \tag{F2}$$

$$\dot{\bar{R}}_1 = \eta_1\bar{N}_1 \tag{F3}$$

$$\dot{\bar{R}}_2 = \eta_2\bar{N}_2 \tag{F4}$$

The doublet kinetic equations are:

$$\frac{1}{2}\dot{T}(N_1,N_1,t) = -\alpha_1 T(N_1,N_1,t)+\mu_{t1}T(N_1,N_2,t)+\frac{1}{2}\mu_{f1}f''_{x_1x_1}(1)\bar{N}_1 \tag{F5}$$

$$\frac{1}{2} \dot{T}(N_2,N_2 t) = - \alpha_2 T(N_2,N_2,t) + \mu_{t2} T(N_1,N_2,t) +$$
$$\frac{1}{2} \mu_{f2} f''_{x_2 x_2}(1) \bar{N}_2 \quad (F6)$$

$$\dot{T}(N_1,N_2,t) = -[\alpha_1+\alpha_2] T(N_1,N_2,t) + \mu_{t2} T(N_1,N_1,t) +$$
$$\mu_{t1} T(N_2,N_2,t) \quad (F7)$$

$$\dot{T}(N_1,R_i,t) = -\alpha_1 T(N_1,R_i,t) + \mu_{t1} T(N_2,R_i,t) + \eta_i T(N_1,N_i,t)$$
$$(F8)$$

$$\dot{T}(N_2,R_i,t) = \mu_{t2} T(N_1,R_i,t) - \alpha_2 T(N_2,R_i,t) + \eta_i T(N_2,N_i,t)$$
$$(F9)$$

$$\frac{1}{2}\dot{T}(R_i,R_i,t) = \eta_i T(N_i,R_i,t) \quad (F10)$$

$$\dot{T}(R_1,R_2,t) = \eta_1 T(N_1,R_2,t) + \eta_2 T(N_2,R_1,t) \quad (i=1,2) \quad (F11)$$

The initial conditions are:

$$T(N,R,0) = T(R,R,0) = 0$$

If the ergodic hypothesis is assumed; i.e.,

$$\dot{\bar{N}}_1 = \dot{\bar{N}}_2 = \dot{T}(N_1,N_1) = \dot{T}(N_1,N_2) = \dot{T}(N_2,N_2) = 0$$

the analytical solutions for the mean values, the factorial cumulants and the auto-and cross-correlation functions of the number of detectrons are:

$$\bar{N}_1 = (\alpha_2 S_1 + \mu_{t1} S_2)/(\alpha_1 - \Gamma)(\alpha_2 + \Gamma) \quad (F12)$$

$$\bar{N}_2 = (\mu_{t2} S_1 + \alpha_1 S_2)/(\alpha_1 - \Gamma)(\alpha_2 + \Gamma) \quad (F13)$$

$$\bar{R}_i = \eta_i \bar{N}_i t \quad (i=1,2) \quad (F14)$$

$$T(R_1,R_1,t) = \bar{R}_1 A^* \mathscr{F}[(\alpha_1-\Gamma)t]+\bar{R}_1 B^* \mathscr{F}[(\alpha_2+\Gamma)t] \qquad (F15)$$

$$T(R_2,R_2,t) = \bar{R}_2 C^* \mathscr{F}[(\alpha_1-\Gamma)t]+\bar{R}_2 D^* \mathscr{F}[\alpha_2+\Gamma)t] \qquad (F16)$$

$$T(R_1,R_2,t) = \bar{R}_1 E_2^* \mathscr{F}[(\alpha_1-\Gamma)t]-\bar{R}_1 F_2^* \mathscr{F}[(\alpha_2+\Gamma)t]$$

$$= \bar{R}_2 E_1^* \mathscr{F}[(\alpha_1-\Gamma)t]-\bar{R}_2 F_1^* \mathscr{F}[(\alpha_2+\Gamma)t] \qquad (F17)$$

$$AC(R_1,R_1) = \eta_1^2 \bar{N}_1\{\bar{N}_1+ \tfrac{1}{2} Ae^{-(\alpha_1-\Gamma)t} + \tfrac{1}{2} Be^{-(\alpha_2+\Gamma)t}\}$$

$$= \phi_{R_1 R_1} \qquad (F18)$$

$$AC(R_2,R_2) = \eta_2^2 \bar{N}_2\{\bar{N}_2 + \tfrac{1}{2} Ce^{-(\alpha_1-\Gamma)t}+ \tfrac{1}{2} De^{-(\alpha_2+\Gamma)t}\}$$

$$= \phi_{R_2 R_2} \qquad (F19)$$

$$XC(R_1,R_2) = \eta_1\eta_2 \bar{N}_1\{\bar{N}_2+ \tfrac{1}{2} E_2 e^{-(\alpha_1-\Gamma)t}- \tfrac{1}{2} F_2 e^{-(\alpha_2+\Gamma)t}\}$$

$$= \eta_1\eta_2 \bar{N}_2\{\bar{N}_1 + \tfrac{1}{2} E_1 e^{-(\alpha_1-\Gamma)t}- \tfrac{1}{2} F_1 e^{-(\alpha_2+\Gamma)t}\}= \phi_{R_1 R_2} \qquad (F20)$$

where

$$\alpha_i= \mu_{ci}+ \mu_{fi}+ \eta_i- \mu_{fi}f'_{xi}(1) \qquad ; \qquad (i=1,2)$$

$$\alpha_1 < \alpha_2$$

$$(\alpha_1-\Gamma)(\alpha_2+\Gamma) = \alpha_1\alpha_2-\mu_{t2}\mu_{t1}$$

$$(\alpha_1-\Gamma)A = \frac{2}{[(\alpha_2+\Gamma)-(\alpha_1-\Gamma)](\alpha_1+\alpha_2)}$$

$$\{[\alpha_2^2-(\alpha_1-\Gamma)^2] \tfrac{1}{2} \mu_{f1}f''_{x_1 x_1}(1)+(\mu_{t2})^2 (\bar{N}_2/\bar{N}_1)\tfrac{1}{2} \mu_{f2}f''_{x_2 x_2}(1)\}$$

$$(\alpha_2+\Gamma)B = \frac{2}{[(\alpha_2+\Gamma)-(\alpha_1-\Gamma)](\alpha_1+\alpha_2)}$$

$$\{[(\alpha_2+\Gamma)^2-\alpha_2^2]\tfrac{1}{2}\mu_{f1}f''_{x_1x_1}(1)-(\mu_{t1})^2(\bar{N}_2/\bar{N}_1)\tfrac{1}{2}\mu_{f2}f''_{x_2x_2}(1)\}$$

$$(\alpha_1-\Gamma)A^* = \eta_1 A$$

$$(\alpha_2+\Gamma)B^* = \eta_1 B$$

$$(\alpha_1-\Gamma)C = \frac{2}{[(\alpha_2+\Gamma)-(\alpha_1-\Gamma)](\alpha_1+\alpha_2)}$$

$$\{(\mu_{t2})^2(\bar{N}_1/\bar{N}_2)\tfrac{1}{2}\mu_{f1}f''_{x_1x_1}(1)+[\alpha_1^2-(\alpha_1-\Gamma)^2]\tfrac{1}{2}\mu_{f2}f''_{x_2x_2}(1)\}$$

$$(\alpha_2+\Gamma)D = \frac{2}{[(\alpha_2+\Gamma)-(\alpha_1-\Gamma)](\alpha_1+\alpha_2)}$$

$$\{-(\mu_{t2})^2(\bar{N}_1/\bar{N}_2)\tfrac{1}{2}\mu_{f1}f''_{x_1x_1}(1)+[(\alpha_2+\Gamma)^2-\alpha_1^2]\tfrac{1}{2}\mu_{f2}f''_{x_2x_2}(1)\}$$

$$(\alpha_1-\Gamma)C^* = \eta_2 C$$

$$(\alpha_2+\Gamma)D^* = \eta_2 D$$

$$(\alpha_1-\Gamma)E_1 = (\alpha_2+\Gamma)F_1 = \frac{2}{[(\alpha_2+\Gamma)-(\alpha_1-\Gamma)](\alpha_1+\alpha_2)}$$

$$\{\mu_{t2}\alpha_2(\bar{N}_1/\bar{N}_2)\tfrac{1}{2}\mu_{f1}f''_{x_1x_1}(1)+\mu_{t1}\alpha_1\tfrac{1}{2}\mu_{f2}f''_{x_2x_2}(1)\}$$

$$(\alpha_1-\Gamma)E_2 = (\alpha_2+\Gamma)F_2 = \frac{2}{[(\alpha_2+\Gamma)-(\alpha_1-\Gamma)](\alpha_1+\alpha_2)}$$

$$\{\mu_{t2}\alpha_2\tfrac{1}{2}\mu_{f1}f''_{x_1x_1}(1)+\mu_{t1}\alpha_1(\bar{N}_2/\bar{N}_1)\tfrac{1}{2}\mu_{f2}f''_{x_2x_2}(1)\}$$

$$(\alpha_1 - \Gamma) E_i^* = \eta_i E_i$$

$$(\alpha_2 + \Gamma) F_i^* = \eta_i F_i$$

Some mathematical and physical implications in Equations F1 to F20 may be outlined; they are:

Equations F1 and F2 are prompt-neutron kinetic equations for an asymmetrical, two-zone reactor, whereas α_1 and α_2 are the self-reduction rates of the number of prompt-neutrons in their respective zone when decoupled from the other one.

Equations F3, F4, F10, F11 and F14 are analogous to Equations C2, C5, C6 and C8, and exhibit the same characteristics.

If the ergodic hypothesis is assumed, the differential relations in Equations F1, F2, F5, F6 and F7 become algebraic and lead to time-independent solutions for $\bar{N}_1$, $\bar{N}_2$, $T(N_1, N_1)$, $T(N_1, N_2)$ and $T(N_2, N_2)$; e.g., Equations F12 and F13.

If the ergodic hypothesis is assumed, the system of Equations F8 and F9 acquires the same structure of the system composed by Equations F1 and F2; i.e., the matrix

$$\begin{pmatrix} -\alpha_1 & \mu_{t1} \\ \mu_{t2} & -\alpha_2 \end{pmatrix}$$

plus time-independent source terms $\eta_i T(N_1, N_i)$ and $\eta_i T(N_2, N_i)$, alternately.

Equations F15, F16 and F17 extend the original Feynman formulae for the variance and covariance experiments. They consist of a bimodal behavior characterized by two time eigenvalues: $(\alpha_1 - \Gamma)$ for the fundamental mode (Mode A) and $(\alpha_2 + \Gamma)$ for the superior mode (Mode B). In Equation F17, a

negative amplitude for Mode B of the covariance
must be pinpointed. Equations F18, F19 and F20
extend the classical Rossi-alpha formulae to a
bimodal (Mode A and Mode B) decay characterized
by two time eigenvalues; i.e., $(\alpha_1-\Gamma)$ and $(\alpha_2+\Gamma)$,
respectively. In Equation F20, a negative ampli-
tude for Mode B of the cross-correlation function
must be pinpointed.

Subindices 1 and 2 correspond to reactor zones
and sensors (since there is only one sensor
per zone), constants A C E_i (i-1,2) refer to
Mode A, constants B D F_i (i=1,2) refer to Mode B,
quantities marked by asterisk are related to
detectrons.

The model treated in this section degenerates
into the model treated in Section IIIE if the
two zones are made symmetrical. This corresponds
to the following limits:

$$S_1, \; S_2 \rightarrow S$$
$$\alpha_1, \; \alpha_2 \rightarrow \alpha$$
$$\Gamma, \; \mu_{t2}, \mu_{t1} \rightarrow \mu_t$$
$$A, \quad C, \quad E_1, \; E_2 \rightarrow Y_a$$
$$B, \quad D, \quad F_1, \; F_2 \rightarrow Y_b$$
$$A*, \quad E_1^* \rightarrow Y_{a1}^*$$
$$C*, \quad E_2^* \rightarrow Y_{a2}^*$$
$$B*, \quad F_1^* \rightarrow Y_{b1}^*$$
$$D*, \quad F_2^* \rightarrow Y_{b2}^*$$

Because of them, the following degeneration scheme
is obtained:

$$F12 \text{ and } F13 \rightarrow E12$$
$$F15 \qquad \rightarrow F14$$

and in general,

$$Fk \qquad \rightarrow \qquad Ek-1$$

with k=15, 16,...,20.

IV. EPILOGUE

A. Stochastic Parameters and Kinetic Parameters

Finally, a one-to-one correspondence must be developed between stochastic parameters and kinetic parameters in use in nuclear reactor physics. The <u>primary</u> stochastic parameters are:

S Probability per unit time that a neutron emitted by the source enters the nuclear system

μ_c Probability per unit time that a neutron undergoes capture or leaks away from the system

μ_f Probability per unit time that a neutron induces a fission reaction

λ Probability per unit time that a precurser decay produces a neutron

$\mu_{t1}, \mu_{t2}, \mu_t$ Probability per unit time that a neutron transfers from one zone to the other

η Probability per unit time that a neutron is detected

f(x,y) Generating function of the probability distribution of neutron- and precursor-emission in a fission event

The <u>secondary</u> stochastic parameters are:

$\alpha = \mu_c + \mu_f + \eta - \mu_f f_x'(1,1)$ Balance between total probability of capture and leakage per unit time and production probability per unit time, limited to prompt-neutron only

$\frac{1}{2}\mu_f m_{20}$ Expected number of neutron pairs created by fissions per unit time

$\frac{1}{2}\mu_f m_{11}$ Expected number of neutron-precursor pairs created by fissions per unit time

$\frac{1}{2}\mu_f m_{02}$ Expected number of precursor pairs created by fissions per unit time

The kinetic parameters may be separated into four categories covering:

    a.   fission

    b.   lifetime and generation time of neutrons

    c.   multiplication and reactivity of the system

    d.   detection of neutrons

Kinetic parameters of category (a) depend only upon the generating function $f(x,y)$ and its derivatives:

m_{10}	Expected number of prompt-neutrons emitted by fission	$\bar{\nu}(1-\beta) = \bar{\nu}_p$
m_{01}	Expected number of precursors emitted by fission	$\bar{\nu}\beta$
$m_{10}+m_{01}$	Expected total number of prompt and delayed neutrons emitted by fission	$\bar{\nu}$
$m_{01}/(m_{10}+m_{01})$	Expected fraction of delayed neutron (and precursors)	β
m_{20}/m_{10}	Diven parameter	$\overline{\nu_p(\nu_p-1)}/(\bar{\nu}_p)^2$

Kinetic parameters of category (b) are:

$\mu_c+\mu_f+\eta$	Total probability of capture and leakage per unit time	μ

$1/\mu$	Neutron lifetime; i.e., expected time interval between birth and death	ℓ
$\mu_f m_{10}$	Probability per unit time that a neutron generates $\bar{\nu}_p$ prompt neutrons by fission	μ'_s
$1/\mu'_s$	Prompt generation time; i.e., expected time interval between the birth of a neutron and its procreation of another prompt neutron by fission	Λ'
$\mu_f m_{01}$	Probability per unit time that a neutron generates $\bar{\nu}\beta$ prompt neutrons by fission	α_c
$\mu_f(m_{10}+m_{01})$	Probability per unit time that a neutron generates $\bar{\nu}$ neutrons by fission	μ_g
$1/\mu_g$	Generation time; i.e., expected time interval between the birth of a neutron and its procreation of another neutron by fission	Λ

Kinetic parameters of category (c) are:

μ_g/μ	Multiplication factor; i.e., production-to-capture and leakage ratio	κ
μ'_g/μ	Prompt multiplication factor	$\kappa_p=\kappa(1-\beta)$
$(\kappa-1)\kappa$	Reactivity	ρ

ρ/β	Reactivity in β units (dollars)	$

α	Rossi-alpha	$(1-\kappa_p)/\ell$
		$(\beta-\rho)/\Lambda$
		$\beta(1-\$)/\Lambda$
		$\alpha_c(1-\$)$

Kinetic parameters of category (d) are:

$\mu_f\bar{N}$	Fission rate	$\dot{F}$

η/μ_f	Detection efficiency; i.e., ratio between detection rate and fission rate	ε

$\eta\bar{N}$	Detection rate	$\dot{F}\varepsilon$

$\eta\bar{N}t$	Expected number of detectrons at time t;i.e. expected number of neutrons detected in the time interval $0 \to t$	$\dot{F}\varepsilon t=\bar{R}$

The following equations interrelate multiplication factor, reactivity, neutron lifetime and generation time:

$$\Lambda = \ell/\kappa = \ell(1-\rho) \qquad \Lambda' = \ell'/\kappa_p$$

B.   Conclusions

Besides local and sectional comments and considerations, which help in understanding the implications contained in the differential equations and their solutions, some matters reflect the general philosophy of the presentation and are common to all the problems.

The basic elements may be summarized in the following points:

a. If the pseudoparticle detectron is introduced for a realistic interpretation of experiments, the complexity of the problem is the same in the case of a single detectron species as well as in the case of many detectron types: this is due to the fact that the detection process is a one-way transform from neutron to detectron with no chance of reversal. Time differential equations involved are limited to first order.

b. When a single group of precursors is introduced, the problem amplifies in complexity because the transition from neutron to precursor (via fission) and from precursor to neutron (via precursor decay) is a two-way transform. This circumstance elevates to the second the order of the time-differential equations involved.

c. The same circumstance described in point (b) occurs when neutrons from two different zones are taken into account and transits from one zone to the other are both possible; two-way transforms are inherent again.

A general check also can be applied for controlling and assessing the validity of the time-differential equations in the factorial cumulants; they are especially useful for the less familiar second-order relations.

A good example of the check is offered by Equations D4 and D6: The first one is

$$\frac{1}{2}\dot{T}(N,N) \;=\; -\alpha T(N,N) \;+\lambda T(N,C) + \frac{1}{2}\mu_f m_{20}\bar{N}$$

The lhs member of the equation represents the variation in time of the expected number of correlated neutron pairs. This number is diminishing in time due to the dieaway of single neutrons according to their self-reduction rate α, so that the contribution of the first member of the rhs is negative. Correlated neutron pairs (N,N) also can be created by pairs (N,C) in which a precursor decays according to its

self-reduction rate λ; this explains the structure of the
second member of the rhs and its positive connotation. Correlated neutron pairs can be finally created directly by a
fission event, being $\frac{1}{2}m_{20}$ the expected number of neutron
pairs emitted in a fission process. This explains the
structure of the third member of the rhs and its positive
contribution.

One might observe that no other transitions, as for
instance, $(C,C) \rightarrow (N,N)$, are admitted in our scheme because
there is no other way to generate correlated pairs (N,N)
from either a single or a pair configuration via a single
process.

The second example is even more composite; it is

$$\dot{T}(N,C) = -[\alpha+\lambda]T(N,C)+\alpha_c T(N,N)+\lambda T(C,C)+\mu_f m_{11}\bar{N}$$

where the lhs member of the equation represents the variation
in time of the expected number of correlated neutron-precursor
pairs. This number is diminishing in time due to the dieaway of both neutrons and precursors according to their self-
reduction rates α and λ; so the contribution of the first
member of the rhs is negative. Correlated neutron-precursor
pairs (N,C) also can be created by pairs (N,N) in which a
neutron induces fission, α_c being the rate of production of
single precursors by neutrons via fission, by pairs (C,C)
in which a precursor decays and generates a single neutron
and directly by a fission event, m_{11} being the expected
number of neutron-precursor pairs created in a fission process.

The logic of these two fine examples can be generalized
in a sort of rule-of-thumb capable of compiling components
of time-differential equations via an heuristic approach or
simply putting the user in condition of operating stringent
tests on the correctness of the formulae with which to work.

The general variation in time of the expected number
of correlated (U,U) and (U,V) pairs is:

$$\frac{1}{2}\dot{T}(U,U) = -[srr(U)]T(U,U)+\Sigma\mu(.,U)T(U,.)+\Sigma\mu(U,.)T(.,U)+$$

$$+ Source(U,U)$$

$$\dot{T}(U,V) = -[srr(U)+srr(V)]T(U,V)+\Sigma\mu(.,V)T(U,.)+\Sigma\mu(U,.)$$

$$T(.,V)+Source(U,V)$$

The meaning of nonexplicit symbols is:

U:	number of particles of type 1
V:	number of particles of type 2
srr:	self-reduction rate
Σ:	summation over all the admitted transitions
$\mu(U,.)$:	probability per unit time that a process creates a pair with type-1 particles in bra position
$\mu(.,V)$:	probability per unit time that a process creates a pair with type-2 particles in ket position
$T(U,.,t)$:	expected number of correlated pairs with type 1 particles in bra position
$T(.,V,t)$:	expected number of correlated pairs with type 2 particles in ket position

In conclusion, one might have noticed that the fulcrum upon which the entire treatment turns around is the time-differential equation in the state probability generating function. The integration of this equation gives in explicit terms the state probability profile and can answer all questions about the system under analysis.

But the integration is not always possible; see, for instance, problems faced in Sections C4, C5 and C6. In these crucial cases, the time-differential equation can be suitably derived and transformed into a set of time-differential equations in the factorial moments or factorial cumulants of the state variables. The latter equations can be solved analytically, at least for the few simple elementary models introduced in this presentation.

One might object that this is not a very broad basis, but it is surely sound and free of arbitrary mathematical trickery; furthermore, its intuitive and expressive composite structure can be easily developed and extended to deal with more advanced and sophisticated models.

ACKNOWLEDGMENTS

The authors thank Editor Jeffery Lewins for providing, through detailed correspondence, appropriate stimulating and constructive contributions dedicated to ameliorate the structure of the entire monograph.

The use of modified nomenclature, with respect to that adopted in previous articles and writings by the authors, has been suggested by Editor Jeffery Lewins in order to avoid confusion and misunderstanding with the established nomenclature of nuclear reactor kinetics.

REFERENCES

1. Fermi, E., Feynman, R. P., de Hoffmann, F., "Theory of the Criticality of the Water Boiler and the Determination of the Number of Delayed Neutrons," USAEC Report MDDC-383 (LADC-269), Los Alamos Scientific Laboratory, 1944.

2. de Hoffmann, F., "Intensity Fluctuations of a Neutron Chain Reaction," USAEC Report MDDC-382 (LADC-256), Los Alamos Scientific Laboratory, 1946.

3. de Hoffmann, F., "Statistical Aspects of Pile Theory," The Science and Engineering of Nuclear Power, Vol. $\underline{2}$, 1949.

4. Feynman, R. P., de Hoffmann, F., Serber, R., "Dispersion of the Neutron Emission in U-235 Fission," J. of Nuclear Energy $\underline{3}$, 64, 1956.

5. Orndoff, J., "Prompt-Neutron Periods of Metal Critical Assemblies," Nuclear Science Eng. $\underline{2}$, Page 450, 1957.

6. Random House, Dictionary of the English Language, 1961.

7. American Institute of Physics Handbook, McGraw-Hill, 1963.

8. Van Der Ziel, Noise, Prentice-Hall, Inc., 1954.

9. Osborn, R. K., Yip, S., "Physical Theory of Neutron
    Noise in Reactors and Reactorlike Systems," Noise
    Analysis in Nuclear Systems, USAEC TID-7679, 1963.

10. Cohn, C. E., Book Review, Nuclear Science Eng. 45,
    Page 104, 1971.

11. Landau, L. D., Lifshitz, E. M., Statistical Physics,
    Pergamon Press, 1959.

12. Williams, M. M. R., Random Processes in Nuclear Reac-
    tors, Pergamon Press, 1974.

13. Stacey, W. M., Jr., "Space-Time Nuclear Reactor Kin-
    etics," Nuclear Science and Technology, Vol. 5,
    Academic Press, 1969.

14. Harris, D. R., "Stochastic Fluctuations in a Power
    Reactor," USAEC Report WAPD-TM-190, 1958.

15. Pluta, P. R., "An Analysis of Nuclear Reactor Fluctu-
    ations by Methods of Stochastic Processes," Ph.D.
    Thesis, University of Michigan; and "An Analysis of
    Fluctuations in a Boiling Reactor by Methods of
    Stochastic Processes," APED-4071, General Electric
    Co. Report, 1962.

16. Matthes, W., "Statistical Fluctuations and Their
    Correlation in Reactor Neutron Distributions,"
    Nukleonik 4, Page 213, 1962.

17. Dalfes, A., "The Fokker-Plank and Langevin Equations
    of a Nuclear Reactor," Nukleonik 5, Page 348, 1963.

18. Kemeny, L. G., Murgatroyd, W., "Stochastic Models for
    Fission Reactors," Noise Analysis in Nuclear Systems,
    USAEC-TID-7679, 1964.

19. Pluta, P. R., "Probabilistic Analysis of Reactor
    Kinetics," Reactor Kinetics and Control, USAEC TID-7662,
    1964.

20. Dalfes, A., "The Random Processes of a Nuclear Reactor
    and Their Detection," Nukleonik 8, Page 94, 1966.

21.   Matthes, W., "Theory of Fluctuations in Neutron Fields,"
      Nukleonik 8, Page 87, 1966.

22.   Kobayashi, T., "A Study of the Zero-Probability Method,"
      J. Nuclear Science Technology (Tokyo) 5, Page 145, 1968.

23.   Clarke, W. G., Harris, D. R., Natelson, M., Waltar, J.
      F., "Variances and Covariances of Neutron and Precur-
      sor Populations in Time-varying Reactors," Nuclear
      Science Eng. 31, Page 440, 1968.

24.   Kurihara, K., Sekiya, T., "A Stochastic Operator
      Method of Calculating the Time Interval Distributions
      in Neutron Detection," J. Nuclear Science Technology,
      (Tokyo) 6, Page 28, 1969.

25.   Williams, M. M. R., "Probability Distributions in
      Simple Neutronic Systems Via the Fokker-Plank Equation,
      J. Nuclear Energy 23, Page 633, 1969.

26.   Smith, P. R., "The Probability Distribution of Prompt-
      Neutrons in an Infinite Homogeneous Multiplying Medium,"
      J. Nuclear Energy 24, Page 159, 1970.

27.   Smith, P. R., "Prompt-Neutron Probability Distributions
      in an Infinite Homogeneous Medium Due to a Uniformly
      Distributed Source," J. Nuclear Energy 24, Page 164,
      1970.

28.   Pacilio, N., "Reactor Neutron-Noise Analysis: State
      of the Art," CNEN Report RT/FI(72)11, 1972.

29.   Jorio, V. M., Pacilio, N., "Cinetica Stocastica: (1)
      Neutronica, CNEN Report RT/FI (73)4, 1973.

30.   Isnard, R., "Processus Aléatoires dans un Réacteur
      Nucléaire Compte Tenu des Effets de la Temperature
      du Compustibile: le partie. Etablisment du Model;"
      and "Processus Aléatoires dans un Réacteur Nucléaire
      Compte Tenu des Effects de la Temperature du Combus-
      tible 2e Partie. Determination Experimentel des
      Parametres du Réacteur," Rev. Physics Applications 8,
      7 and 19, 1973.

31. Lewins, J., On the Interpretation of Markov Processes
     in a Nuclear Reactor, Atomkernenergie 23, Page 173,
     1974.

32. Arcipiani, B., Marseguerra, M., "The Rigorous Expres-
     sion for the Variance of Neutron Counts from the For-
     ward and Backward Kilmogorov Equations," CNEN Report
     RT/FI(74)40, 1974.

33. Norelli, F., Jorio, V. M., Pacilio, N., "Stochastic
     Kinetics: Analytical Solutions of Detailed Proba-
     bility Balance Equations," CNEN Report RT/FI(74)44,
     1974.

34. Genoud, J. P., "A Stochastic Study of Coupled Reactor
     Systems," Annals Nuclear Energy 2, Page 85, 1975.

35. Matthey, M., "A Stochastic Study of Noise in Boiling
     Reactors," Annals Nuclear Energy 2, Page 271, 1975.

36. Colombino, A., Mosiello, R., Norelli, F., Jorio, V.
     M., Pacilio, N., "Cinetica Stocastica: (2) Rivela-
     tronica", CNEN Report RT/FI(75)5, 1975.

37. Ruby, L., Wang, H. K., "Some Additions to Reactor
     Noise Theory Which Include Delayed Neutrons," Nuclear
     Science Eng. 57, Page 86, 1975.

38. Pacilio, N., Jorio, V. M., Norelli, F., Mosiello, R.,
     Colombino, A., Zingoni, E., "Toward a Unified Theory
     of Reactor Neutron Noise Analysis Techniques," Annals
     Nuclear Energy 5, Page 239, 1976.

39. Pal, L. I., "On the Theory of Stochastic Processes in
     Nuclear Reactor," Nuovo Cimento, Supplement 7, Page
     25, 1958.

40. Bell, G. I., "On the Stochastic Theory of Neutron
     Transport," Nuclear Science Engineering 21, 290, 1965.

41. Williams, M. M. R., "Reactor Noise in Heterogeneous
     Systems: I, Plate-type Elements," Nuclear Science
     Eng. 30, Page 188, 1967.

42. Zolotar, B. F., "Monte Carlo Analysis of Nuclear
     Reactor Fluctuations Models," Nuclear Science Eng.
     31, 282, 1968.

43. Furuhashi, A., "Neutron Counting Statistics in the
     Family-First-Pulse-Triggered Gate (1)," J. Nuclear
     Science Technology (Tokyo) 12, Page 145, 1975.

44. Marseguerra, M., "Joint Probability Distributions
     of Neutron Counts in Reactor-Noise Analysis," Nuclear
     Science Eng. 56, Page 16, 1975.

45. Akcasu, A. Z., Osborn, R. K., "Application of Lange-
     vin's Technique to Space-and Energy-dependent Noise
     Analysis," Nuclear Science Eng. 26, Page 13, 1966.

46. Sheff, J. R., Albrecht, R. W., "The Space Dependence
     of Reactor Noise I Theory," Nuclear Science Eng. 24,
     Page 246, 1966.

47. Nieto, J. M., "Application of Langevin's Method to
     Some Reactor-related Fluctuation Problems," Doctoral
     Thesis, University of Michigan, 1969.

48. Zolotukhin, V. G., Mogilner, A. I., Two Reports:
     "Distribution of the Number of Counts by a Neutron
     Detector in a Reactor," Atomic Energy (USSR) 10,
     Page 379, 1961; and Op. Cit., 15, Page 11, 1963.

49. Babala, D., "Neutron Counting Statistics in Nuclear
     Reactors," Kjeller Report KR-114, 1966.

50. Saito, K., Taji, Y., "Theory of Branching Processes
     of Neutrons in a Multiplying Medium," Nuclear Science
     Eng. 30, Page 54, 1967.

51. Harris, D. R., "Neutron Fluctuations in a Reactor of
     Finite Size," Nuclear Science Eng. 21, Page 369, 1965.

52. Osborn, R. K., Natelson, M., "Kinetic Equations for
     Neutron Distributions," J. Nuclear Energy A/B 19,
     Page 610, 1965.

53.   Burger, J. R., "Probability Distributions of Neutrons
       and Precursors in a Stationary Reactor Without Tem-
       perature Feedback," USAEC Report WAPD-TM-855, 1969.

54.   Courant, E. D., Wallace, P. R., "Fluctuations of the
       Number of Neutrons in a Pile," Physical Review 72,
       Page 1038, 1947.

55.   Uhrig, R. E. (Editor), Two Reports: "Noise Analysis
       in Nuclear Systems," USAEC Report TID-7679, 1964
       (Florida, 1963); and "Neutron Noise, Waves and Pulse
       Propagation," USAEC Report CONF-660206, 1967, (Florida
       1966).

56.   Petten, "Statistical Methods in Experimental Reactor
       Kinetics and Related Techniques," Proceedings of a
       Symposium held at Petten, December 12-19, 1967, RCN-
       Report 98, 1968.

57.   Tokyo/Kyoto, Transactions of a Japan-United States
       Seminar on Nuclear Reactor Noise Analysis, held at
       Tokyo and Kyoto, September 2-7, 1968.

58.   Pacilio, N., Jorio, V. M., Colombino, A. (Editors),
       Two Reports: "From Critical Assemblies to Power
       Reactors," Proceedings of the European American
       Committee on Reactor Physics Specialist Meeting on
       Reactor Noise, Rome, October 21-25, 1974, Annual
       Nuclear Energy 2, No. 2-5, 1975; and "Reactor Noise:
       From Critical Assemblies to Power Plants," Proceedings
       of NEACRP Specialist Meeting on Reactor Noise (SMORN-
       1), CSN Casaccia, Rome, Italy, NEACRD-U-71, Serie
       Simposi, CNEN, March, 1975.

59.   Seifritz, W., Stegemann, D., "Reactor Noise Analysis,"
       Atomic Energy Review 9 (1), Page 129, 1971.

60.   Saito, K., "On the Theory of Power Reactor Noise,"
       Annals Nuclear Science Eng. 1, PP 31, 107, 209, 1974.

61.   Thie, J. A., Reactor Noise, Rowman & Littlefield, 1963.

62.   Pacilio, N., "Reactor Noise Analysis in the Time
       Domain," USAEC Critical Review Series TID-24512, 1969.

63.   Uhrig, R. E., <u>Random Noise Techniques in Nuclear</u>
       <u>Reactor Systems</u>, Ronald Press, 1970.

64.   Bartlett, M. S., <u>An Introduction to Stochastic Pro-</u>
       <u>cesses with Special Reference to Methods and Applica-</u>
       <u>tions</u>, Cambridge University Press, 1955 (3rd Ed., 1977).

65.   Ash, M., <u>Nuclear Reactor Kinetics</u>, McGraw-Hill, 1965.

66.   Faa' Di Bruno, M., "Note sur une Nouvelle Formule de
       Calcul Differentiel," Quarterly J. Pure Applied Math.
       <u>1</u>, Page 359, 1855.

67.   Routti, J. T., Szeless, A., "Partitions of Numbers
       and Application to Higher Order Derivatives of Com-
       posite Functions," Lawrence Radiation Laboratory,
       Berkeley, <u>UCRL-20215</u>, 1970.

68.   Mosiello, R., Two Reports: "Due Algoritmi per il
       Calcolo della Derivata n-esima di una Funzione
       Composta," CNEN Report <u>RT/FI(75)12</u>, and "DERN: Un
       Programma per il Calcolo della Derivata n-esima di
       una Funzione Composta," CNEN Report <u>RT/FI(75)13</u>; 1975.

69.   Mogilner, A. I., Zolotukhin, V. G., "The Statistical
       R-Method of Measuring the Kinetic Parameters of a
       Reactor," Atomic Energy (USSR) <u>10</u>, Page 377, 1961.

70.   Bell, G. I., "Probability Distribution of Neutrons
       and Precursors in a Multiplying Assembly," Annual
       Physics <u>21</u>, Page 243, 1963.

71.   Bell, G. I., Anderson, W. A., Galbraith, D., "Proba-
       bility Distribution of Neutrons and Precursors in a
       Multiplying Medium II," Nuclear Science Eng. <u>16</u>,
       Page 118, 1963.

72.   Pacilio, N., Jorio, V. M., "Discrete Probability
       Distributions in Nuclear Reactor Noise," CNEN Report
       <u>RT/FI(74) 45</u>, 1974.

73.   Zingoni, E., Pacilio N., Jorio, V. M., "The Poisson-
       Algebraic Distributions: Another Family of General-
       ized Poisson Profiles in Nuclear-Reactor Noise Analysis,"
       CNEN Report RT/FI(74) 26, 1974.

74.   Gurland, J., "A Method of Estimation for Some General-
       ized Poisson Distributions," Proceedings of the Inter-
       national Symposium on 'Discrete Distributions',
       Montreal, 1963.

ON-LINE COMPUTERS IN NUCLEAR POWER PLANTS
A REVIEW

M. W. JERVIS

Control and Instrumentation Technology Engineer
Central Electricity Generating Board
Barnwood, Gloucester GL4 7RS, England

ABSTRACT

Digital computers play an important part in the design
of nuclear power plants, and on-line computer systems now
occupy an important place in plant operation. The basic
reasons for introducing computers into power plants are considered, and the control and display facilities that they
are required to provide are discussed. Special reference
is made to security requirements and to the system configurations employed to provide high integrity, including the
case in which computers are used in reactor protection systems. Some examples of systems in service are described,
and their performance is discussed with special reference
to difficulties and inadequacies that have been experienced.
Implications of current trends in technology are discussed
and conclusions reached on future trends.

I. INTRODUCTION

The on-line computer forms a part of the total control
and instrumentation (C & I) system. This C & I system is
provided to enable the power plant to be operated so that
the required electrical power output characteristics are
obtained while ensuring that metallurgical, environmental
and other constraints are not violated.

135

This role can be interpreted as providing the following basic facilities illustrated in Figure 1:

1. Manual and automated sequence control to enable the operator to raise power to the point where automatic control is engaged.

2. Interlocks to prevent the establishing of certain combinations of operating conditions.

3. Operation of the plant under automatic control with an operating margin from constraints.

4. Indication to the operator of the correct operation of the automatic control by monitoring the controlled plant parameter.

5. Indication to the operator, by initiating alarms, if the operating conditions approach the constraint boundaries.

6. Automatic power cutback or trip, if particular constraints are violated.

7. Post trip review of behavior of the plant before and after the trip.

8. Management information required for efficient operation, and to meet licensing conditions.

The unique capability of digital computers to collect, process, record and display large amounts of data have caused them to be exploited as part of the control and instrumentation system for almost all nuclear and fossil-fired power plants. In this context, "power plant" is regarded as an installation which is intended primarily for the production of electrical energy.

In the early days of the application of computers, it was considered necessary to justify their cost relative to other methods, but there is now a realization that there is no alternative to computers to perform many essential tasks. In any case, the cost of computing equipment has been falling over the last 10 years, and so in many cases, a clear cost advantage in favor of computers can be demonstrated.

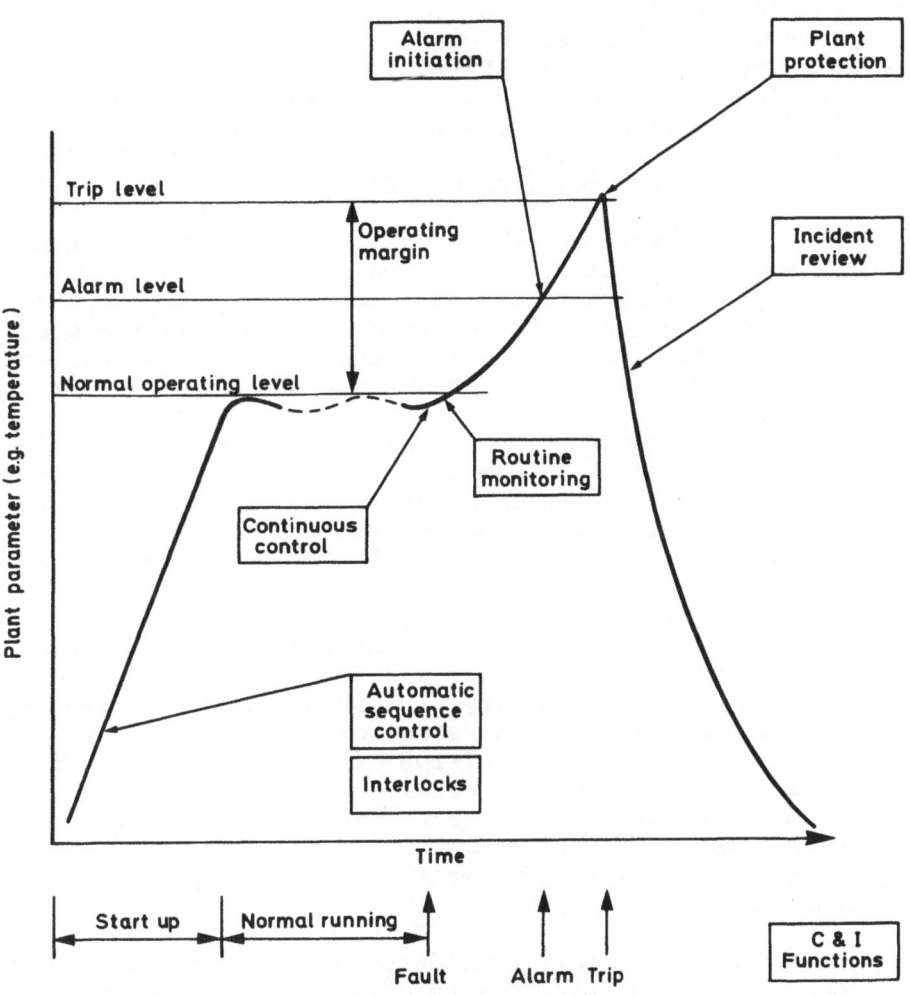

Figure 1

Examples of Control and Instrumentation Functions

The present review does not attempt to give an exhaustive survey of computers in nuclear installations; it gives a critical review of the background leading to the policies adopted, the problems that have arisen and the techniques used to provide the performance required, with emphasis on specific installations and their operating experience and with special reference to UK Central Electricity Generating Board (CEGB) nuclear power plants. It updates a previous review (1) which included fossil-fired plants and others dealing with nuclear power plants (2-6,12-15).

Any instrumentation system on a nuclear power plant has to deal with data from the very large numbers of transducers necessary to provide information on the 2D and 3D power distributions in reactors. For example, the Advanced Gas Cooled Reactor (AGR) has thermocouples in its 300 fuel channels and the Pressurized Water Reactor (PWR) has a large number of incore flux detectors, the output of these and many other transducers must be processed, displayed to the operator for short-term operation and logged for retention as long-term records or immediate post incident review and analysis. The magnitude of the amount of information to be handled is evident from the figures appearing in Table I. Comprehensive tables are given in Reference 5.

It soon became apparent that the sheer volume of quantitative information and alarms, and the complexity of the task was beyond the capability of conventional chart recorders and indicators, and that some form of automatic data logging was required. In addition, the transducers could be examined for "off normal" conditions, so offering the potentiality of exploiting a "management by exception" approach. Normal data were only presented "on demand" to the operator, but "off normal" situations were drawn to his attention.

In addition to these considerations, an important factor was the adoption of a policy of centralized control in power stations. For example, in the Canadian Pickering plant (16), controls needing alteration within 15 minutes of annunciation of an abnormal condition are located in the central control room.

Locally manned control boards on various parts of the plant, for example, turbine gauge boards, are eliminated

TABLE I

APPROXIMATE SIZES OF SOME COMPUTER SYSTEMS

Computer system described in Reference number	119,120 130,131	4	42	139 142	8 146
Type of reactor	AGR	PWR	PWR	CANDU	LMFBR
Country	UK	Germany	France	Canada	UK
Number of computers/reactor	3/2*	1	1	2	1
Size Core Store; words (w) or bytes (b)	64kw	48kw	64kb	32kw	32kw
Size Auxiliary Store; words (w) or bytes (b)	500kw (duplicated)	2x500kw (drums)	40Mb	256kw (fixed) + 585kw (moving arm)	600kw

Number of
  analog inputs
  digital inputs
	2400 3072	576 5280	600 3000	1418 512	2000 400

Number of cathode ray tube display units
  Monochrome
  Color
	6 –	4 –	– 3	10 –	5 –

See also Table II					

* 1 plus shared standby as in Figure 20.
  Figures refer to one reactor.

and the controls and indicators transferred to a central
control room. This has significant advantages in reducing
the number of operators required, and gives improved control
in a closely coupled combination of reactor, boiler, turbine
and electrical supply network; however, the concentration of
control and instrumentation in the central control room raises
acute ergonomic problems caused by the density of instrumentation. This congestion can be relieved by the use of a
computer system to perform some of the tasks performed by
conventional equipment. Generation of power from nuclear
reactors involves complex calculations as an essential part
of the operation. The availability of digital computers
allows such calculations to be made off-line or on-line as
appropriate.

In addition to these data processing display and recording applications, computers also can be used for closed
loop control, for execution of control algorithms, and for
the procedural control of the plant, commonly called sequence
control. Computers also can be used for interlock and protective functions.

The computer functions illustrated in Figure 2 must be
implemented with an integrity consistent with the application concerned, and these applications can be regarded
as falling into four categories.

The functions and application classes of computers in
nuclear power plants have been defined in a recently-produced
draft, International Electrotechnical Commission Standard (17)
and this offers a convenient way of considering the application of computers in nuclear power plants. These are listed
in Section II, A and B.

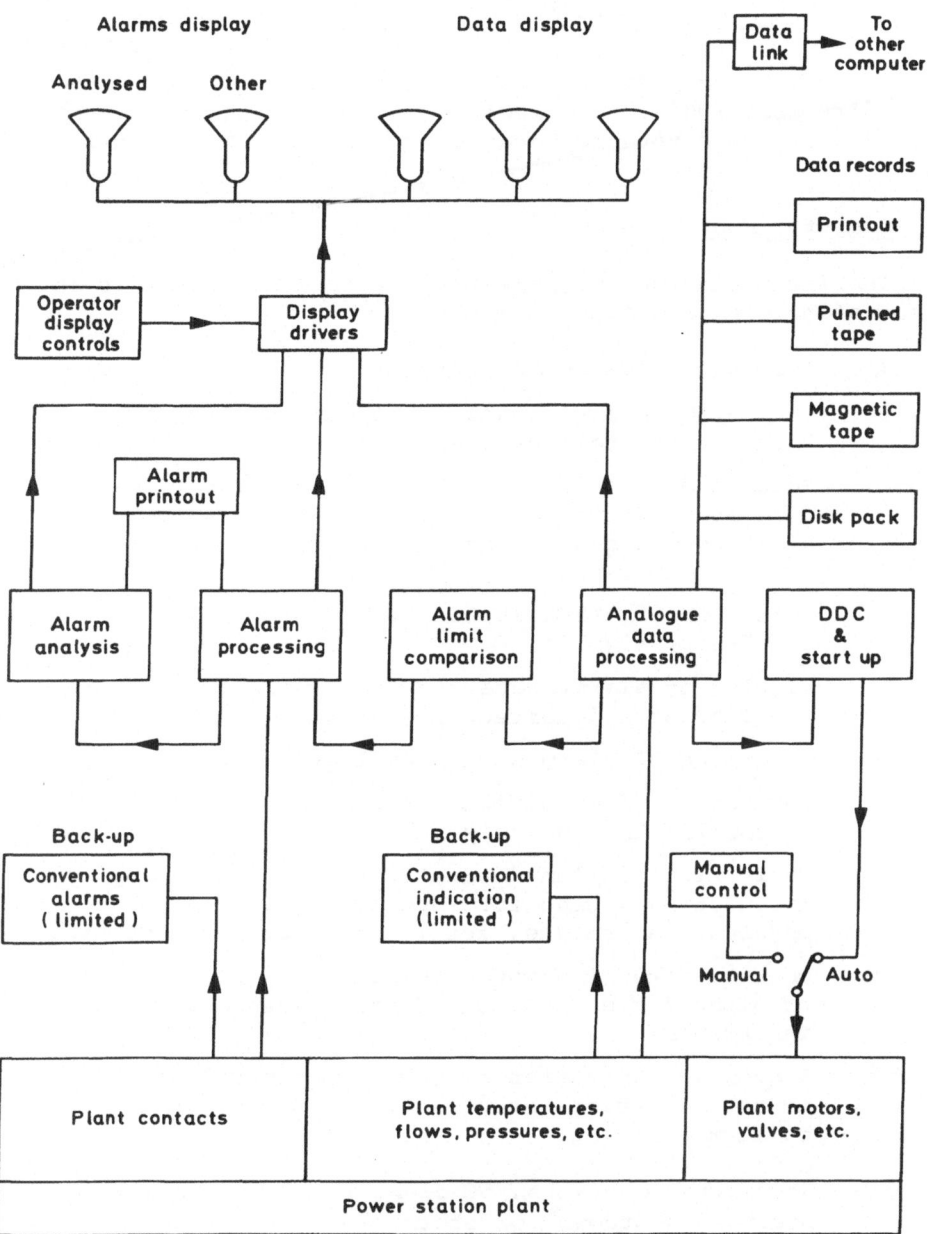

Figure 2

Block Diagram of Computer System Functions

II.   COMPUTER FUNCTIONS AND APPLICATION CLASSES

The following list is taken from the draft IEC Standard
(17) with some additional items. The terminology of the draft
Standard has been retained and the quotations are taken from
it.

A.   Computer Functions

The tasks that can be performed by on-line computers in
association with reactors include:

1.   Logging of operation states

2.   Display of plant signal states and values, to
     allow or aid correct operation

3.   Special display or recording methods for indication
     of histories, trends, reactor conditions, complex
     plant conditions or configurations

4.   Detection of alarm states from analog signals
     and binary signals

5.   Display of alarms detected and of alarms existing,
     to allow or aid correct operation

6.   Recording of alarm states as logs

7.   Derivation of significant alarms by alarm
     processing and analysis

8.   Derivation of plant operational information by
     calculations, used for instrumentation and for
     physics assessments, for records, or for licensing

9.   On-line determination of margins to trip and status
     of plant for evaluation of maximum acceptable load
     variations

10. Sequential operation of plant to control instru-
     mentation systems such as burst can detection,
     or flux scanning

11. Sequential operation of plant in association with
     startup, shutdown, or otherwise

12. Control of plant operation by implementation of
     control algorithms

13. Interlock functions where plant conditions are monitored and adverse operator action prevented directly

14. Protective functions where plant conditions are monitored and plant trips or other reactor protection system actions are taken

15. Functions where plant conditions are monitored and plant trips or other action taken for equipment safety or availability

16. Automatic testing of control or protection system functions

To these the author would add:

17. Acquisition of data for output to other computers for off-line processing.

18. Use of computers as an aid to commissioning.

In Section III, these functions are discussed using this list as a basis.

B. Application Classes (17)

"The tasks assigned to digital computers are classified into four application areas with varying requirements relating to reliability, redundancy and integrity.

Class 1
"Dedicated systems performing essential functions of the reactor protection system."

This class requires very special design approaches, with a redundant arrangement in a high-integrity configuration, particularly if the system is used for nuclear reactor protection.

Class 2
"Information or control agencies such that significant reactor operation is not possible without continuous availability of the essential functions of the computer system. No conventional standby equipment is provided for functions of this class."

In general, redundant computers will be necessary to meet the criteria of this class.

Class 3
"Information or control aids to operating staff which enhance the plant operation. A failure of the system may result in a degradation of overall plant performance, or may lead to a loss or reduction of reactor operational flexibility within a period of days."

In this class, it is generally acceptable that a single fault may cause loss of only one part of the total functions and that total loss is specifically confined to a particular area. Backup may be provided if required, to extend the period of reactor operation with part of the computer out of action, and a nonredundant computer configuration may be used.

Class 4
"Data logging systems used for plant operational recording and as an aid to conventional instrument and alarms. Normally, the signals involved can be checked by other means."

For Class 4 applications, it is tolerable for complete loss of facilities to occure because of a single failure. The implication is that a nonredundant configuration is acceptable; however, it also implies that there is adequate backup equipment to provide a service if the computer fails.

C. Implications of Functions and Application Classes

The functions listed in Section A are chosen to achieve the service required of the computer system and are best discussed by examples; some of these are discussed in Section III. The facilities affect the quantity of hardware and type of peripherals, and require software to drive them to give the facilities required.

The four application classifications given in Section IIB imply certain degrees of "security" and "reliability"; this is achieved by design features, specifically the provision

of redundancy and appropriate system configuration. It is
also convenient to discuss these classes in terms of specific
computer applications, with details of their systems and
configurations. These are discussed in Section V.

The computer functions are described in more detail in
Section III, and the application classes in Section IV.
These functions are considered in relation to Figure 1; the
basic system is shown in Figure 2.

III. FUNCTIONS

A. Permanent Records

General Requirements. All information available regarding
the plant during its complete life can be passed through the
computer system. It is theoretically possible to record all
these data and make them available for analysis, so making a
significant contribution to station availability and safety
by allowing operating margins to be continually reviewed.
This aspect is particularly important in cases wherein the
fatigue history of the plant must be assessed. Obviously,
it is not economic to record all information, typically in
the region of 2.5×10^8 readings per day per reactor unit,
but the computer gives a flexible means of selecting the
data to be recorded and permits high density, economic storage for subsequent analysis, either at the power plant site
or at a central computer facility. Furthermore, the data
can be chosen to cover very long periods with infrequent
sampling, or give fine detail over relatively short periods,
which may be of particular interest.

The requirements for permanent records can be classified
as follows:

      a.   regular logs for immediate and long-term
           operational use

      b.   post-incident records of selected information

      c.   records for long-term analysis

Regular Logs.   Typical logs are as follows:

a.   statutory records required by the terms of
     the Nuclear Site License

b.   long-term records required by the Station
     Operation Staff

c.   management information on a daily basis

d.   infrequent but detailed information; for
     example, for monitoring after maintenance
     operations or investigations

The computer system produces permanent records as printout using printers, so satisfying similar requirements to
manually updated log sheets and chart recorders; however, in
contrast to chart recorders with which information is
continuously and indiscriminately recorded, only records of
interest need be made by the computer. By careful selection
of the measurements to be recorded and their frequency of
printout, considerable economy can be made in the volume of
paper to be stored, and ease of retrieval compared with the
situation with chart recorders. Incidentally, the latter
have had a poor reliability record (18) and required excessive maintenance attention, although modern types are considerably better.

The computer system also permits logs to be compiled
from a selection of any of the analog inputs and print
them out at relatively frequent intervals. This facility
gives a record of items of plants of particular interest
over a special period; e.g., after maintenance. If used
with discretion, this facility can considerably reduce the
throughput of the routine logging system.

The data can be recorded in paper or magnetic tape form
which facilitates retrieval for analysis in a more efficient
manner than is possible with chart recorders; however, it
should be noted that the immediate indication of trends
given by chart recorders is not available from digital
printout although there are current developments that provide hard copy as graphs, and displays are available on
cathode ray tubes (crt's) as described in Section IIIB.

The amount of data to be recorded is reduced by producing frequent and regular logs of only the important and

critical measurements, the less important ones being logged less frequently and sometimes only on demand. Regular logging gives the background recording over the life of the station, the logging interval fixing the time resolution of the information that is related to the expected rate of change.

The records are provided either as printed log sheets or in other forms if further processing of the readings is envisaged. Typical records required comprise some 12 logs containing between 20 and 150 items printed out over periods ranging from 1 hour to 24 hours.

The printout of logs can be initiated manually on demand or at preset intervals to give the staff a review of the operating conditions of the plant. The format of the log is somewhat constrained by the printer characteristics; e.g., 80 characters per line; but this can be chosen to give a satisfactory layout. The speed at which logs are required dictates the type printer required.

If many different types of logs are required for different purposes by different departments, it is necessary to have a multiplicity of printers to avoid the complication of separating logs from a large printout, and to share the load to prevent excessive wear.

Incident History Reviews. In the event of an unusual plant situation, it is necessary to assist the operation by supplementing the coarse structure of the long-term records just described with shorter-term but finer structured information and shorter sampling intervals. These requirements can be divided into two parts:

    a.   information immediately made available to assist the operator in making the decision concerning continued running of the plant. If a shutdown is necessary or happens automatically, information is required concerning cause of the incident so an assessment can be made; that is, whether it can be restarted, or if further investigation is necessary. Data collected prior to the incident and extending after it is necessary because it may be that margins have been reduced during the period.

b.   information, which can have a longer access
time, required for subsequent analysis. This
requirement is discussed in the next section.

Traditionally, the short-term requirement (a)
is met by chart recorders, but these have
severe limitations imposed by slow chart speed
and timing accuracy. It is now common to employ computers for the purpose of providing
graphical displays, available on demand, on
the cathode ray tubes to give an indication of
the trend of before, during, and after an incident. The coverage is a function of the sampling time between points on the trace and the
number of points displayed. For most purposes,
30 minutes is chosen for the total time displayed; i.e., 180 points per parameter at 10 s
per point. Manually demanded printout of
alpha-numeric formats are also available for
reference.

An additional selection of values is written
on the auxiliary store, kept for a fixed
period of say 30 minutes, and then written
over. After a trip of a main plant item, e.g.,
a turbine, this stored information is read off
the store and passed to a paper tape punch to
produce a tape record. This partially satis-
fies requirement (b) for longer-term records.

On some stations, the 30 minute record is complemented by a three minute record, giving
finer detail of the transient occurring during
and after a trip. Both records are arranged
to continue a little beyond the time of the
trip so that a complete record is produced.
Economy of recording and tape are improved by
condensing the data before storage, e.g., by
recording only if a change is detected.

The conditions for initiation of these history
reviews must be chosen carefully, but include
obvious cases such as reactor trip, turbine
trip, and other major incidents that require
investigation.

If a certain auxiliary storage space is set
aside for this review facility, a decision
must be made concerning the number of variables
recorded, their sampling interval, and the time.
Some compromise is inevitable.

If details of electrical system faults that are
very fast need to be analyzed, it may be best
to deal with these separately. With electrical
faults, a time resolution of 10 ms or better
may be required, but this is usually unnecessary
for reactor or turbine plant.

Records for Long-term Analysis. For certain plant items
it is necessary to maintain long-term records of its performance and history in terms of temperature cycling , cumulative
nuclear irradiation , etc., which are later correlated with
other measurements and metallurgical examination. An example
of this is the nuclear reactor fuel endurance records which
are compared with the results of post-irradiation examination
of the fuel.

The amount of data to be recorded is such that paper tape
is inconvenient and so magnetic tape has been used in the
UK (CEGB) where there are special requirements for the continuous, long-term collection of fuel element temperature
cycling data. The use of magnetic tape also allows the continuous recording of data from the turbo-generator and other
important plants. The recording on the tape lasts approximately 48 hours, after which it is replaced by another and
the full tape sent to the CEGB Computing Center for processing, to reduce its volume for storage before transfer of the
condensed data to disc and return of the tape for reuse.
Particularly interesting tapes containing records of major
importance will be kept intact. This technique enables very
detailed records to be maintained in an economic form, the
records being available for analysis at the Computing Center
when necessary. If required, the tape can be played back at
the power station site using an off-line computer and hard
copy produced by the station printers.

This arrangement requires the availability of a large,
off-line data processing computer for reducing the volume
of data, and this condensation represents considerable computer time and cost. Depending on the rate charged, condensation can be more expensive than keeping the tapes.

One alternative is for the on-line machine to produce
the data in an already condensed form so that subsequent
processing is minimized; this method has also been adopted
in the UK. One method of condensation is to record only
significant changes (40).

As an alternative to the use of magnetic tape, the data
can be transmitted from the on-line machine to a computing
center by data link; this scheme has been adopted in the
latest UK AGR stations. This link also handles data from
other computers which augment the main one and assist commissioning by performing off-line analysis of the results
of commissioning tests of the type discussed in Section III J.

B. Display of Plant Data

General

Data collected and processed by the computer can , in
principle, be presented to the operator in the form of printed logs, but this method has some limitations including low
speed of printout , noise, and difficulty in presenting varying data. The characteristics of the digital computer facilitates the use of crts which overcome many of these limitations
and provide a fast and flexible method of displaying the
alpha-numeric (α-n) and graphics information. The widespread
use of crts is illustrated by Table II.

At the time that the earlier UK nuclear power plants were
being designed, only alpha-numeric displays were available,
but for the later stations, crts became available that were
capable of continuous line drawing so that pictorial displays
and graphs could be presented in monochrome (67-69).

The display system must provide indication of instantaneous values and trends of plant measurements satisfying the
same requirements as conventional instruments and recorders.
The digital computer can drive crt displays which permit
display in digital or analog form, the latter including
graphs and also pictorial displays such as mimic diagrams.

The quantity of information to be displayed, typically at
least 2,000 measurements per reactor, is so great that displays in alpha-numeric form, which take up minimum space ,
must be used for most of the information. The amount of
data presented on a crt that can be effectively comprehended

TABLE II

REFERENCES TO SOME CONTROL ROOM DESIGNS
EMPLOYING CRT DISPLAYS

Name of Plant or Equipment	Type of Reactor	Country	References
Oldbury	Magnox GCR	UK	97, 98
Wylfa	Magnox GCR	UK	99, 129
Hinkley Point 'B'	AGR	UK	119, 120, 130, 131
Dungeness 'B'	AGR	UK	10, 132, 133
Hartlepool	AGR	UK	121
Prototype Fast Reactor	LMFBR	UK	8, 146
Nuclenet	BWR-6	USA	27, 147
Advanced Control Room	PWR	USA	31
Plant Monitoring and Information System	PWR	USA	20
Browns Ferry	BWR	USA	32
Sequoyah	PWR	USA	32, 149
Watts Bar	PWR	USA	32
Bellefonte	PWR	USA	32
FFTF	LMFBR	USA	40
Gentilly	CANDU	Canada	29, 116, 143
Pickering	CANDU	Canada	16, 29, 30, 140, 141
Bruce	CANDU	Canada	29, 66, 142
Obrigheim		Germany	4
Biblis	PWR	Germany	4
TCI	PWR	France	42
HBWR	BWR	Norway	3, 22, 75, 76
Shimane	BWR	Japan	25

Abbreviations:

GCR   - Gas Cooled Reactor   PWR - Pressurized Water
        (CO_2 in UK)   Reactor

AGR   - Advanced Gas Cooled   BWR - Boiling Water Reactor
        Reactor

LMFBR - Liquid Metal Fast   HTGR - High Temperature
        Breeder Reactor   Gas Reactor

by the operator is limited, and there is no point in exceeding about 20 items per page or "format"; i.e., about 100 formats will accommodate 2,000 measurements; some will be alpha-numeric and others will be allocated to graphical or pictorial displays.

The data cannot be updated more rapidly than the scanning interval of the plant measurement; this must be chosen to be appropriate for the rates of change anticipated. This consideration must also be applied to any other limitations of the overall computer hardware and software. Typically, the update period is 5 s.

Permanent records of the displays can be obtained on demand by the operator. Time and date are indicated on both display and record. The change in seconds digit indicates that the display is being updated.

Alpha-numeric displays are used for the display of alarm messages, derived from the digital and analog inputs as shown in Figure 2, and for accessing any data held within the computer; e.g., alarm levels or inputs inhibited or deleted from scans.

Alpha-Numeric Displays

a.   Tabular. Typically, it is possible to position
an alpha-numeric character anywhere on a grid
comprising 24-30 lines of from 60 to 80 length
characters. There are facilities for drawing
attention to particular parts of the display
by flashing, and on modern equipment by reverse
video; i.e., black on white instead of white on
black.

Ergonomic aspects of the size and shape of
characters has been considered by many
authors (39,68-71). Typically, a character
height of 3-4 mm for a 1 m viewing distance
is adequate.

The layout of the data has an important effect
on the clarity of the display (19); work at CEGB
(72) emphasizes the need to structure the data
into blocks, avoiding lists of uniform appearance. An example is shown in Figure 3.

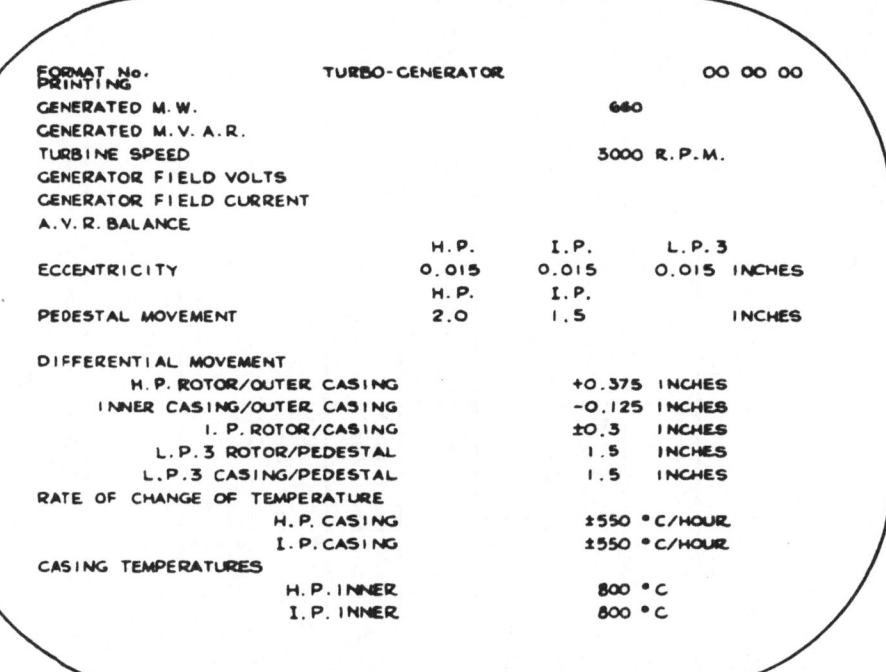

Figure 3

Tabular Alpha-Numeric Only Display

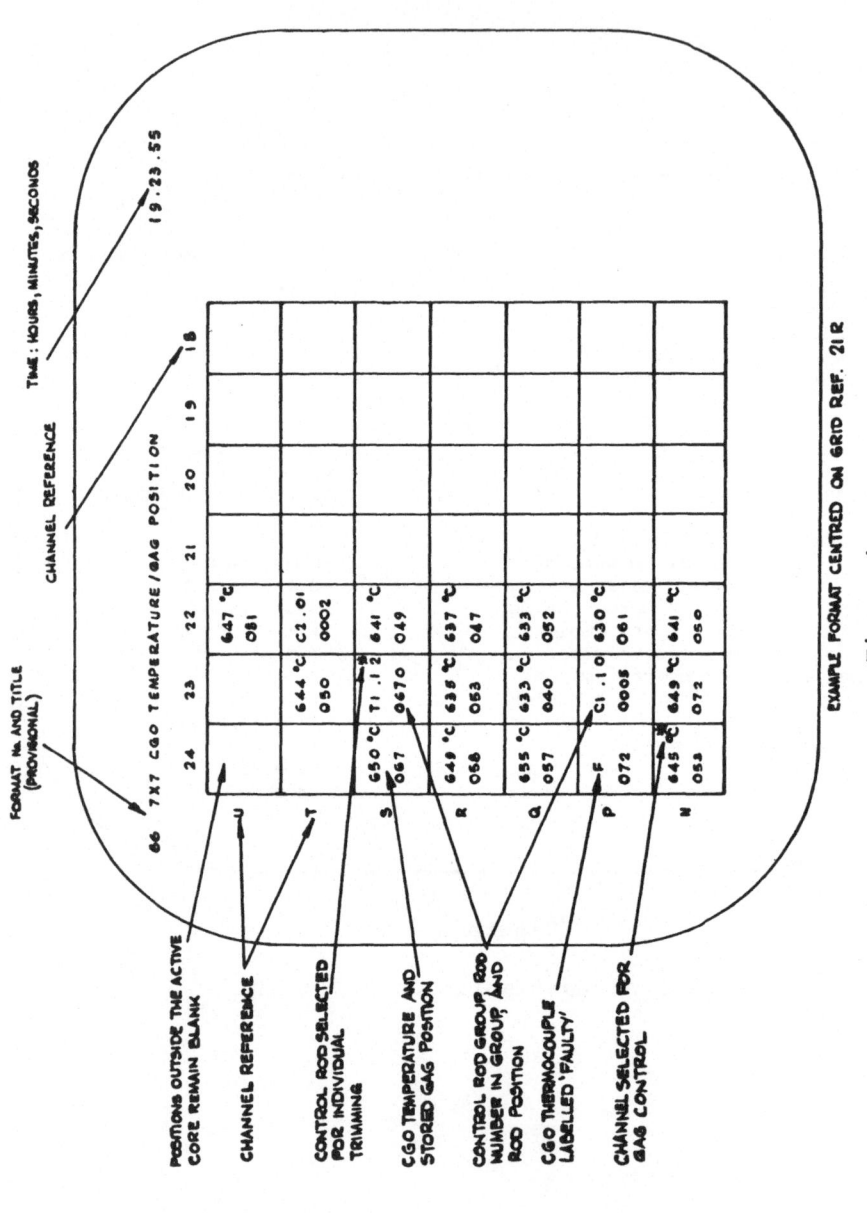

Alpha-Numeric Spatial Display

Figure 4

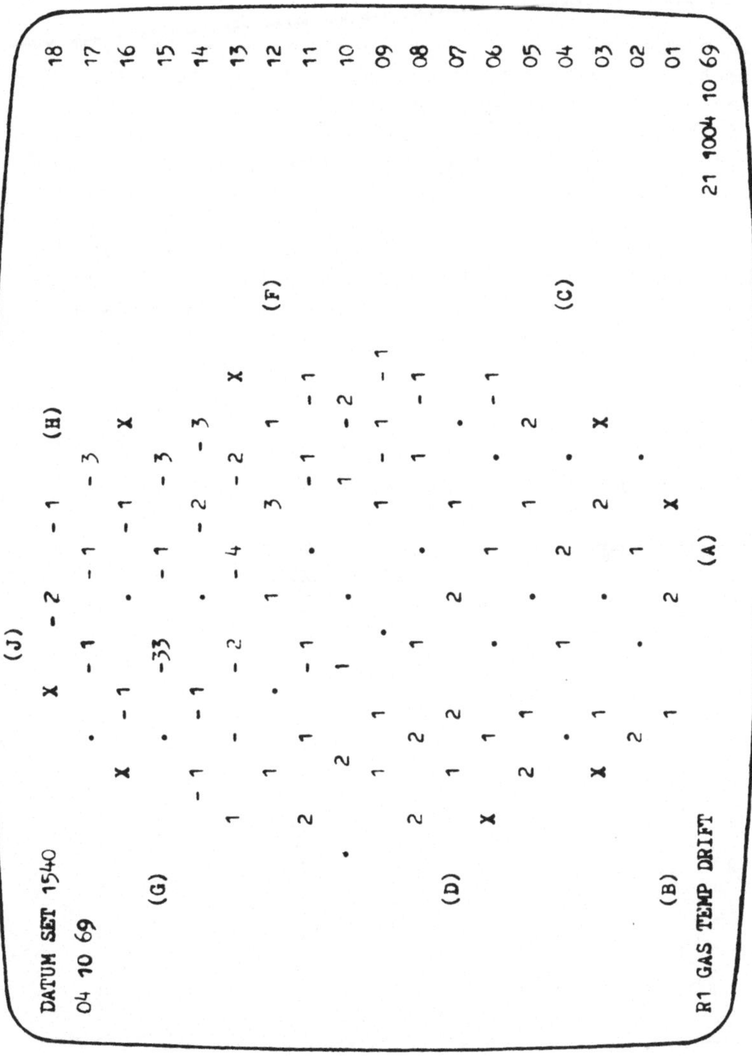

Figure 5

Spatial Display Giving Deviations from Set Datum

(A-J: reactor core sectors; X: no thermocouple provided; numbers are temperature differences in °C from manually set datum)

    b.   Spatial. Some examples are given in Figures 4 and
5, in references 10, 22, and 25, and in Table I.

    Trend Displays

    Display of a measurement plotted against time is used as
an alternative to conventional chart recorders (73,162). The
latter give poor resolution because of their limited chart
speeds and the chart synchronization is often found to be inadequate. In a post-incident inquiry, chart recordings tend
to be of only limited value.

    The design of graphical displays has been investigated by
CEGB (73). In general, it is desirable to show the trend as
a normal graph in which time is represented by the x-axis of
display and value by the y-axis, as in Figure 6.

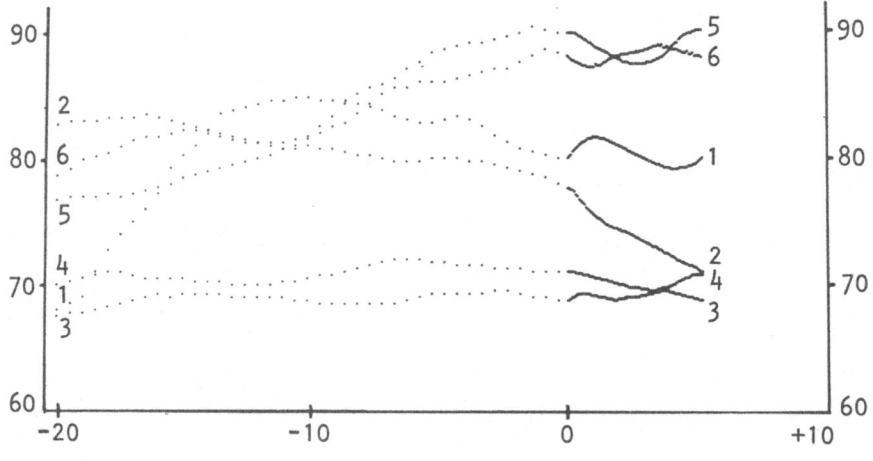

Figure 6

Trend Display of Parameter Against Time

It is ergonomically preferable that the past trend displayed remain steady, and that latest values are added to the right of the graph (17). The period of accumulation of past values should be chosen from considerations of expected rates of change and resolution required. Typically, the length of the record displayed is 30 minutes. Where a trend can be displayed, the computer record of past values should be available as a permanent record.

A trend can be indicated by displaying the current values relative to previous values on a histogram (162) or x-y plot of two plant parameters against each other as illustrated in Figure 7. Operating conditions are satisfactory if the points lie within a specified area, but plant damage occurs if they pass outside it (20).

Other Graphic Displays

If the hardware and software have the capability of drawing pictures, they can be used for plant mimics; analog representation; spatial distribution representation. These can, with advantage, incorporate α-n characters for identification of parts or give accurate quantitative information to supplement the graphics information. The latter may not be of high resolution, but gives easy comparison.

It will be seen that a fairly good resolution, say 256x256, is necessary to give an attractive representation and that the amount of detail is somewhat limited. Nevertheless, useful mimics can be designed and are in widespread use in electrical distribution applications. The use of electrical system mimics in power stations is not particularly attractive, since in general, a separate switching panel is usually provided in any case. This also can be arranged to give a mimic representation, although this panel can be replaced by a method of switching via the crt screen; for example, by a light pen or marker as will be discussed in the next section.

A mimic updated to indicate the progress of automatic sequence control of the plant at Hinkley Point 'B' is illustrated in Figure 8.

It may seem incongruous to display data in analog form on a computer-based crt system, but this can be most effective (74,75), for example, in the histogram type display (1,76) shown in Figures 9 and 10. It has been used to show control

rod penetrations in gas-cooled reactors and for temperature
profiles. It is particularly effective when combined with
α-n statements of the actual value to give high accuracy.

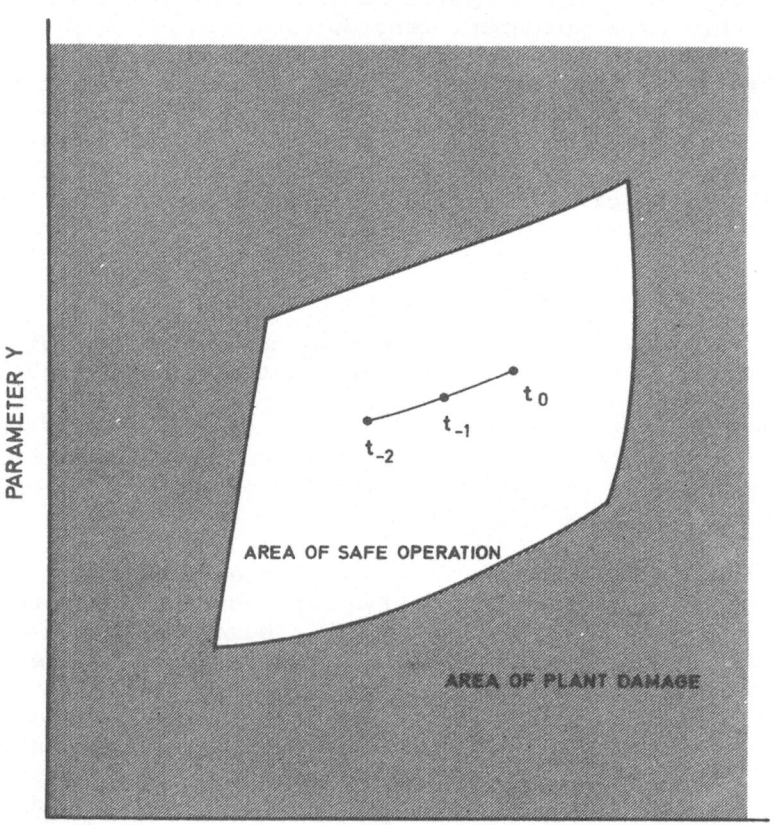

Figure 7

Plot of One Parameter Against Another, with Trend
Indication Against Previous Values, and Permitted
and Forbidden Operating Areas
(t_0, t_{-1}, t_{-2} - times for samples displayed)

The display of spatial distribution of measured values by crt can be particularly useful. It is very difficult to achieve this by other means, and a zoom capability can be introduced to give expansion of selected areas of, say, a horizontal cross-section through a reactor core to show either overall temperature distribution or fine structure, giving individual temperatures (10,133). This technique is used in the UK AGR and Magnox reactors (1), Light Water Reactors (LWRs) and Liquid Metal Fast Breeder Reactors (LMFBRs) (40).

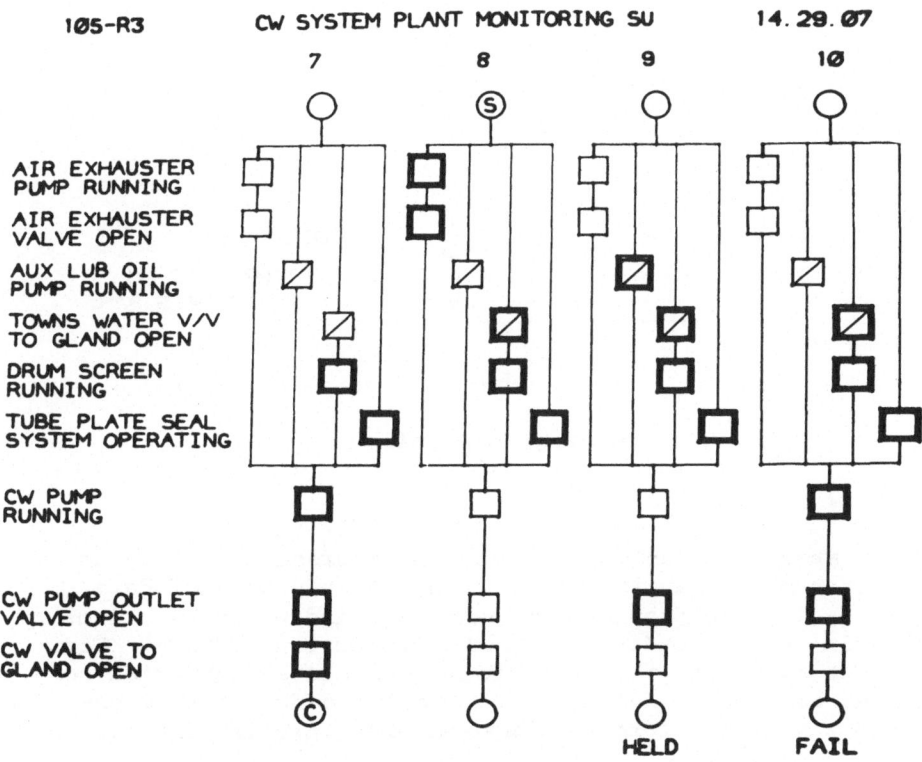

Figure 8

Sequence Control Monitoring Mimic

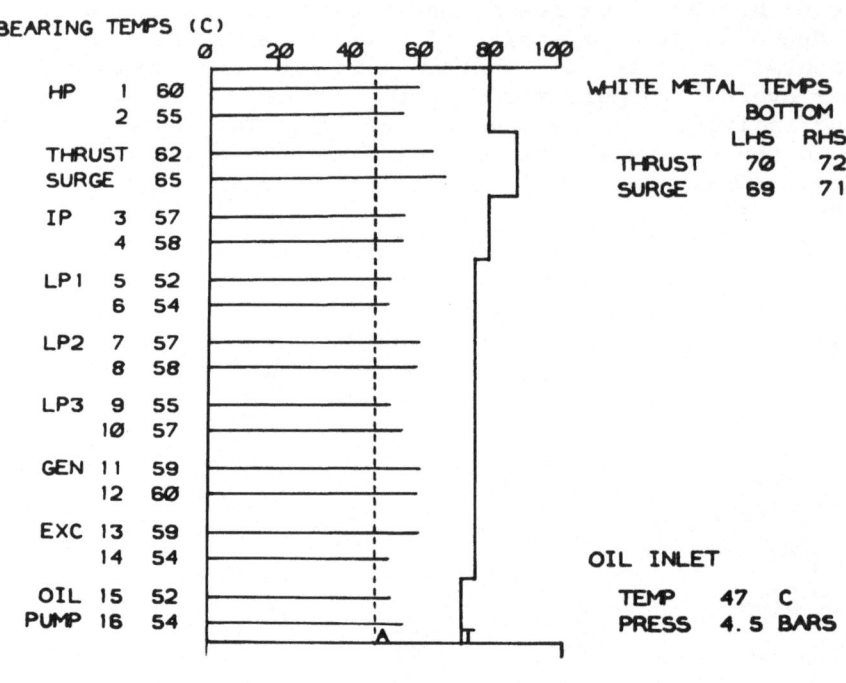

Figure 9

Histogram Display

The state of plant operation can be presented on a summary status display in which 12 important pieces of data
representing the plant performance are indicated by bright
spots in polar coordinates on the crt screen (23,89). For
normal conditions, the spots form a circle and abnormal conditions are easily detected. A similar effect can be obtained with lines forming a spoked wheel display (162).

Operator Selector Facilities

In a typical arrangement, the operator has several crts
available that he will wish to allocate temporarily, to particular displays selected from the total number available.

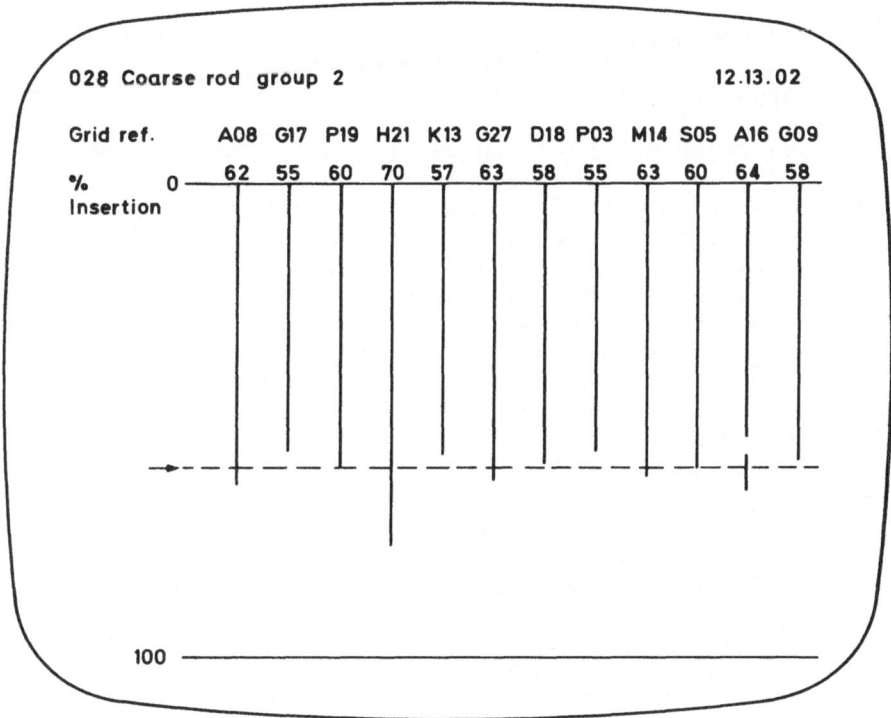

Figure 10

Control Rod Penetration Histogram Display

This selection will depend upon the operational conditions
of the plant and the operator's needs at the time. The
basic requirements are that:

a.   the combination of data on a given display
      should be matched to specific activities or
      situations; e.g., on-load fuel loading in
      an AGR

b.   the ergonomic design of the display is of a high
     standard.

c.   the mechanism of selection is easy to manipulate
     and the results of the selection process are
     apparent with minimum delay.

Various types of selection systems have been reviewed
(69,71) and some of these have been adopted in power plants.
The most popular is the keyboard or push-button matrix
(40,74,75), but the light pen and tracker ball also have
been used (74). If the keyboard is used, care must be
taken to avoid long and complicated sequences of button
operation to achieve the desired result (41). Some systems
have significant delay, of several seconds between operation
and change, to the display; the operators find this irrita-
ting. This can be improved by illuminating the button (164),
or lighting another lamp within about 1 s, to acknowledge
the command by the computer.

Selection of data displayed can be made automatically
by the software, this selection being made on the basis of
an interpretation of the alarms present (27,89). In this
way, the operator is assisted by having data presented to
him that is relevant to the development of fault situations.
The efficacy of such a scheme depends upon detailed knowledge
of plant faults, and raises similar problems to those that
arise with alarm analysis.

Use of Color CRTs

The availability of color television tubes and the
necessary hardware and software makes them attractive for
some displays (21-27,41,42,163).

Color gives an additional means of coding information
and for example, is useful in distinguishing graphs that
may cross; also, it can be used to identify fluids in a
mimic (163).

The exploitation of color requires caution, however,
and the effects can be unfortunate and tiring to the operator. The basic principles have been reviewed (28,90-92),
and briefly:

a. Color is a powerful attention getter and is very
    useful to identify special conditions; e.g.,
    alarm situations.

b. Color helps more in situations with random
    distributions than in structured situations.

c. Care must be taken in the juxtaposition of
    colors and areas of background of certain
    color combinations.

d. Color should be used only as redundant coding
    elements.

Color tubes are in use in the UK in fossil-fired stations
but the nuclear stations have monochrome. Little difficulty
is obtained in achieving sufficient coding in monochrome displays as by flashing, bright-up, reverse video and varying
symbols.

In this connection, it must be remembered that human memory can readily accept only approximately five levels of
coding, so color can only replace other methods rather than
be used in combination; however, color may provide more
efficient coding.

One difficulty that has been encountered is the standardization of colors on the crt with other color codes on the
plant and the maintenance of a rigorous rationale of color
coding.

C. Alarms

Detection of Alarm States from Analog and Binary Signals

Alarm states are detected by two methods:

a. by analog measurements passing outside alarm limits

b. binary on/off devices

In some cases, individual analog signals are processed
by electronic circuits having built-in alarm levels that
operate contacts which generate binary signals. These can
then be handled in the same way as other binary signals from,

say, protective circuits, limit switches, or fluid level
and pressure switches mounted directly on the plant. They
are of the (b) type shown in Figure 2.

With early technology, the provision of alarms on all
analog inputs has not been possible , due to the high cost
of individual circuits. The requirement can be met by
scanning the analog inputs and comparing their level with
a stored alarm level within the computer and so generating
alarms as shown in Figure 2.

A limitation in speed of response arises because of the
scanning interval, but this can be chosen to be appropriate
to the anticipated rate of change and margin to alarm level
expected.

The binary inputs are scanned by a semiconductor switch
and the analog inputs by a relay multiplexer. The multiplexers are the same ones that are used for acquiring data
for display, logging and control purposes.

Conventional Alarm Annunciators

Before the introduction of computer-based systems, alarm
conditions were drawn to the attention of the operator by
messages displayed in the back-lighted windows of fascia light
boxes, known as annunciators, the logic being implemented by
electromagnetic relays of semiconductor circuitry (93,94).
The number of alarms is large (1), typically greater than
2,000, so that for readability at a reasonable viewing distance, they occupy a considerable area and present a substantial problem to the control room designer.

With such large numbers, conventional lamp annunciator
fasciae are only marginally adequate and some means must
be adopted to reduce the number of lamp fasciae. Some reduction can be obtained by grouping, although individual
fasciae still must be provided at local points; this is
inconsistent with a centralized control policy. Another
approach is to use "coordinate selection"; e.g., by
numbering circulators 1-4, and bearings 1-8, 32 circulator
bearings can be covered by simultaneous operation of two
of 12 fasciae (95). This method has severe limitations if
several alarms occur.

A further disadvantage of lamp annunciators is that
they do not produce a permanent record, so they must be
supplemented by some form of logger to satisfy the requirements of alarm logging.

Out-of-service plants raise alarm signals that do not
represent fault conditions, and a means of suppressing these
is necessary to prevent their masking real alarms. However,
it is generally recognized to be difficult to decide when
this is appropriate (96); though the hardware is simple and
takes the form of an AND gate for alarm and suppression signals, which can be fitted locally at the plant.

Computer-Based Alarm System

The difficulty of reading large numbers of lamp fasciae
can be overcome by using crt displays. These have been exploited for this purpose in many power stations in North
America (1) and Europe (1,4). Economy of the display hardware
results from sharing it with other facilities; e.g., data,
sequence monitoring and graphical displays. The operator
then has a clear statement of alarms as a list on his desk
instead of a large number of small lamp fasciae distributed
round the control room panels.

For small numbers of alarms arriving infrequently, the
operator can accept the message and keep track of the situation without difficulty. Under these conditions, the system, while having facilities similar to the lamp annunciators, provides superior legibility and provides a permanent
record by a printer. A number of "pages" lists corresponding to a full tube area of alarms, is provided to accommodate the large number that may accumulate (97-100).

For higher rates of alarm arrival, the operator will
fall behind in accepting the messages so that a queue builds
up. If this becomes longer than the list which can be presented on the tube (typically, a maximum of 24 lines, but
some as low as four), the queue must be accommodated in a
buffer (100) or the use of a fast printer as discussed in
Section VII B.

This situation is different from the lamp fascia situation,
since with a crt list, the operator is constrained to examine
the messages in turn, while in the case of the fascia, he can
give priority to any message.

One alternative is to always present the latest message
at the bottom of the list or on a fresh page, so ensuring
that the latest situation is apparent. The former has been
adopted at the Stade Power Station, the buffer method being
provided for the desk operator.

While the lamp annunciator fasciae raise difficulties
in reading, their disposition can give assistance by pattern
recognition; i.e., certain alarm patterns on the panels are
recognized by the operator as being due to certain plant
conditions. This can to some extent be simulated on the
alarm display by the addition of suitable abbreviations or
codes, or by using "plant area" displays that show a limited
part of the plant at one time (1).

The crt is fundamentally capable of combining the advantages of ease of reading with plant group association, e.g.,
by mimic or other representation. There is scope for further investigation in this field, but the best method of
crt display of alarms does not yet appear to have been exploited. Recent developments in lamp annunciator fascia
design cause them to remain attractive, particularly when
integrated into a control desk divided into functional control areas containing associated indications and controls.

Design Principles for Computer-Based Alarm System

As with data display, the display to the operators must
be rapid and simple, and the operating sequences should follow as closely as is reasonable the established sequence of
conventional alarm annunciator equipment (93,94).

If paging is adopted, the operator should be able to
turn from one page to another by operation of a single control (17). The sudden detection of a large number of alarms
should not overload the system and adversely affect the system performance or cause loss of alarm information. The
system should operate normally with any number of existing
alarms, up to a predetermined maximum.

It is desirable that the display of an alarm take the
form of a reference code (such as the input address) and
a clear and unambiguous title. Where abbreviations are used,
they should be of established use within the overall plant
nomenclature. The use of abbreviations should be minimized,

but where a plant system normally is referred to by an abbreviation, that abbreviation should then be used always, and should also be used for the printout described in Section III C.

Standby Lamp Fascia System

As discussed in Sections II C and II D, a minimum amount of instrumentation must be provided to enable the plant to be safely shut down without the computer's being available; this minimum includes a number of standby alarm annunciators. Some of these are grouped to reduce the numbers, but in one station (100) the coverage is 800, although in CEGB stations it is nearer 100.

It is common practice for the standby system to be switched off while the crt system is operational, and only activated when the crt is not available (97). This appears to deny the operator channels of information. An integrated lamp annunciator/crt scheme might be better than two separate ones.

Alarm Analysis

In a typical nuclear power plant, the number of alarms handled by the computer system is large, usually in the region of 2,000. It may be desirable then that the most significant alarms caused by a fault should be detected and presented to the operator in a special way. The digital computer is very suitable for performing such functions. It has been used in several ways which include:

a. predefined classes of urgent and nonurgent alarms

b. dynamic checks when one alarm is detected, to judge its importance compared with other existing alarms, using predefined criteria

c. logic operations to group or deduce alarm conditions, using predefined logic processes for alarms

d. logic operations to condition the display of one alarm, dependent upon the state of other input signals and alarms

Reviews of these methods have been made (43,76). The one in use by CEGB is basically the tree method (d), in which analysis can present the operator with the root or "prime" cause of a group of related alarms. These prime causes can be displayed on a separate tube from the remaining consequential alarms that do not contribute to the initial alarm condition.

Descriptions are available on the Oldbury (97,98) and Wylfa (99) systems, and alarm analysis is provided on CEGB AGR stations. At Hinkley Point "B", a high-level "plant autocode" was used (101).

It is considered that alarm analysis can make an important contribution to safety, since the system permits a large reduction in the data assimilated and acted upon by the operator. The operator is presented with a display containing only the prime cause, i.e., significant alarms, although the others are available if required. At a time of stress, the operator is given a clear display uncluttered with nonessential information.

Unfortunately, the logical interrelations among 2,000 alarms can be very complex. Full alarm analysis necessitates a large investment in software. Associated hardware can outweigh the value to the operators. Furthermore, the analysis task takes considerable computer time and can cause unacceptable delay in the presentation of the result to the operator. In CEGB, there is a trend away from full alarm analysis and toward emphasis on good displays for the operator, to assist him in making his own mental analysis. However, simpler schemes of alarm suppression are being installed in fossil-fired stations. These may lead to satisfactory results without the complexity of full alarm analysis. For example, an out-of-service plant can be caused to supply a signal that suppresses associated alarms that would be misleading if left standing.

As a general principle, to allow for unforeseen plant circumstances, the operator should be able to display the primary plant conditions detected by the computer independently of the analysis process followed. This applies especially when automatic inhibition of alarms is used.

The development of alarm analysis techniques has been

continued at Munich (43) and Halden (44) where "plant disturbance analysis" implemented with a color crt uses tabular and mimic type displays, the latter showing the propa-
gation of the fault on a cause-consequence diagram. The
system is interactive with considerable operator involvement so it suffers from some of the disadvantages already
discussed, although automatic display selection by software
may ease this problem (89).

These developments do not significantly alter the conclusion that full alarm analysis, although attractive in
principle, is difficult to justify in view of the effort
involved.

Logging of Alarms

A permanent record of alarms in chronological order with
timing is a valuable extension to the post-incident review
of analog values discussed earlier (1,4,29,141).

If alarms are handled by a computer system and plain
English alarm titles are provided for the alarm display
system described in Section III E, it is a simple matter
for them to be printed out.

This log of alarms is virtually essential for effective
post-incident analysis and has proved its value on many
occasions in both nuclear and fossil-fired power plants.

The operators have stated a strong preference for plain
English rather than coded printout, but if any code is used
on the display, it should also appear on the printout.

Typically, a time resolution and accuracy of one second
can be provided.

D. On-Line Calculations

General

The computer can be used merely to acquire data, producing output that is subsequently processed off-line on a
separate computer system; however, the time lags in obtaining the results may not be acceptable. In such cases, the
processing can be performed on-line. This has the advantage

that the cost of separate processing is eliminated; this
may be significant, although the on-line processing will
represent an investment in hardware and software.

The on-line program can be run on a time-shared basis
with the other programs or on separate computers permanently
connected to the one acquiring the data.

It should be noted that the system will, in any case,
perform some on-line calculations, for example, thermocouple linearizing and square root extraction for flow
measurement.

Gas-Cooled Reactors

Because conditions in the UK gas-cooled reactors can be
calculated in the design stage, little on-line processing
of data regarding core conditions is necessary, apart from
making available axial neutron flux distributions measured
by the aeroball flux scanners. The distribution of channel
gas outlet temperatures is handled by the analog input
processing system using the normal, off-normal alarm system and crt system, but using special displays as discussed
in Section III D. In some Magnox stations, the fission product activity detected by the burst can detection system is
processed and small changes are measured relative to running
average, so improving the sensitivity of detection (103).

Light Water Reactors

In the BWR and PWR plants, comprehensive programs are
available for running off-line to establish major fuel
management control rod movement strategy (7,77). However,
the operator must be able to verify that the actual operating conditions are within prescribed limits, and a more
rapid access to processed information is then required.
This can involve (77) a realistic 3D simulation of nuclear,
thermal and hydraulic conditions, the calculated power distributions being updated from flux detector readings. A list
of such programs is given in the IEC Draft Standard (17).

Liquid Metal Fast Breeder Reactors

In one LMFBR plant (40) thermal power calculations are
made every 10 seconds by performing a mass heat balance.

The computer also estimates subcriticality during refueling, and warns the operator of any difference between calculated and measured reactivity.

Imposing a pseudo-random binary disturbance to a reactor system can be used to continuously monitor reactivity balance (45).

Radiation Monitoring

The output of many nuclear plant radiation detectors is in digital form and the pulses are counted by ratemeters that give an analog output. This output then can be logged and displayed by the computer system in a way similar to other plant measurements as discussed in Section III.

As in the case of burst fuel can detection on some UK Magnox reactors (103) and LMFBRs (40), there is some advantage in retaining the detector output in digital form and handling it digitally in the computer. In a large plant, the number of detectors is large, and a small computer, dedicated to radiation monitoring, can be justified (102).

The main advantage lies in the capability of on-line programs such as least-square analysis of trends, and pulse height analysis for simultaneous detection of multiple nuclides and displaying and logging them in a way that is of greatest value to the operator and the Regulatory Authority (32). crt displays can be used to give optimum information to the operator, for example, on a topographical basis to assist in emergencies (32,74).

Furthermore, radiation detectors are mounted at remote parts of the plant, so cabling costs are substantial. The use of computers with multiplexing near the detectors (104) can provide savings in cable costs and reduce the number of penetrations through containment, as illustrated in Figures 14 and 15.

In view of the importance of radiation monitoring in an emergency, system reliability must be acceptably high and this may necessitate a redundant arrangement.

Specification of the Calculations (IEC Draft Standard)

As with all computer software, the documentation must

include clear statements of the purpose of the calculations,
the relevant formulae, plant operating conditions, parameters,
signals and constants used. Furthermore, the results must
be reproducible with particular note being taken of the
plausibility of the data that may have been produced by defective transducers, and the plant state. Timing requirements for running the program should be defined (17).

E.   Sequence Control and Automated Startup

Automation of the startup of main and auxiliary plant
control (commonly referred to as sequence control) is provided to reduce the load on the operator during startup.
For example, 400 manual switchings without automatic control can be reduced to about 40 sequence groups. Wired
program electromagnetic relays or solid state electronic
logic is suitable for this duty, relays being used extensively throughout the world. Digital computers offer an
alternative and provide flexibility by having a program
that can be readily modified to take account of late plant
design changes or operating procedures, and such schemes
have been installed for automatic reactor and turbine
startup.

So far, this technique has not been used in the UK for
auxiliary plants. Most schemes are of the hard-wired logic
type, although computers are used to monitor the progress
of the sequence (1). A computer-based system, however, has
been included in the control scheme of the PFR fueling machine (8) described later.

The relatively low control rod withdrawal rates permitted on gas-cooled reactors and constraints on the rates of
change of temperature for the reactor structure cause the
time between shutdown and full power to be several hours.
To reduce operator loading involved in close manual control
during this period, automatic reactor startup has been provided in the UK AGR stations, the control being a combination
of sequential and variable modulating control.

In a typical system, the startup involves starting up to
low power, raising power to about 5% thermal, raising steam,
establishing modulating control systems, pipework warmup with
main turbine runup and loading, boiler feed pump turbine runup and loading to full power (10,121,133).

In the UK and elsewhere, it is recognized that automation of the runup of the turbine from barring speed to synchronizing speed, then synchronizing and loading up to the desired load, enables the turbine to be run up to a more consistent profile than can be done manually, and that over a number of starts, the cumulative plant damage is minimized. In CEGB, it is policy to fit automated run and loading to all turbines above 500 MWe rating. The automation system has, in the past, taken the form of hard-wired analog-type equipment, but it also can be implemented by a computer; and, as in the case of auxiliary plant sequence control, this has the advantage of flexibility. This has proved valuable when turbines have been rebladed, so altering their characteristics when turbines in a station are different from each other and require individual programs; or when runup characteristics are changed.

In the UK AGRs, the control is executed at a relatively low cost by using a computer that is time shared with other tasks, since little extra hardware is involved and many of the measurements are taken to the computer system for other reasons.

The computers used for applications of Section E are usually of the Class 3 kind, so manual control is provided to cover the occasions when the computer is not operational.

In general, computer programs are arranged in functional groups related to the sequential control tasks that permit independent plant commissioning and independent modifications to be made if required.

The programs usually incorporate prestart checks to ensure that the startup is not initiated until all the appropriate plant equipment is operational, and there are checks as the sequence proceeds with indications of malfunction.

Interlocks and plant protection is usually provided by additional hard-wired circuitry separate from the sequence equipment.

The control of reactor fueling machines involves a large number of digital signals used for control interlock and supervision. Digital computers can execute the complex logical operations necessary and can be used to supervise

the fuel handling (*8,46,141*) and provide a permanent record
of the fuel-handling process.   In the French St. Laurent GCR
(*105*), a computer was involved in a fuel meltout incident,
although the direct cause was operator error rather than a
computer fault.

F.   Continuous Control

General

Analog controllers have been used extensively in the
various automatic control loops in nuclear power plants.
The basic principle is illustrated in Figure 11, at least
one controller being required per control loop.   The necessary algorithm is executed by analog-type hardware in the
controller.

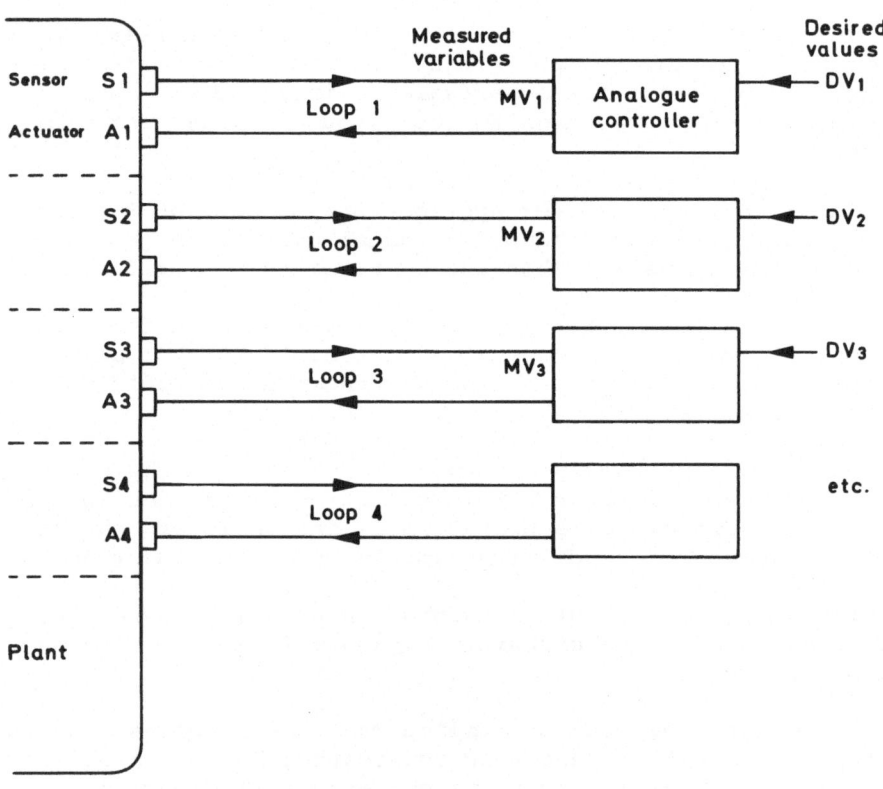

Figure 11

Basic Analog Control System

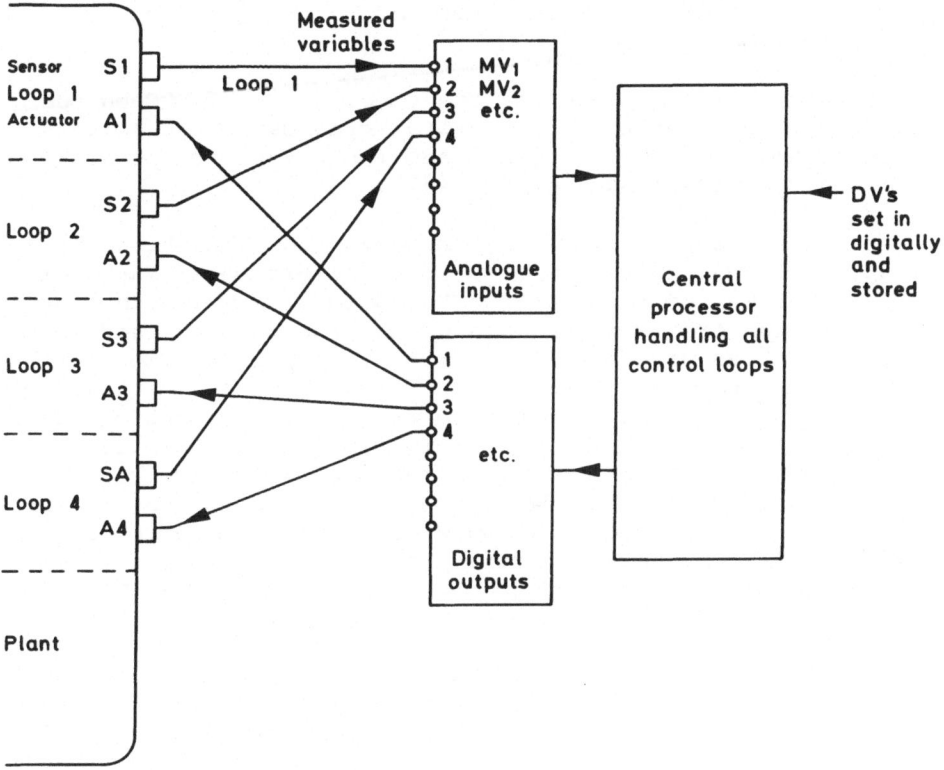

Figure 12

Basic Direct Digital Control Scheme

As an alternative, direct digital control (DDC) can be used in which a computer is time-shared among many loops as shown in Figure 12. The algorithm is executed digitally by the hardware and software of the digital computer. The DDC scheme has the advantage of:

a. implementation of more sophisticated control strategies than with analog schemes and execution of more complex algorithms (*3,4,47-49,106,107*)

b.   ease of modification and extension

c.   ease of reconfiguration and retuning of control
      loops

Claims are also made for the superiority of DDC on the
basis of lower failure rate, better diagnostic and self-
test methods, better stability and isolation from plants,
leading to high noise immunity (49).

In addition, particularly in cases when the number of
loops is large, there may be a cost difference in favor of
DDC. This was so for the UK AGR stations in which the computer system provided for data processing is also used to
perform DDC as discussed in Section VI. The incremental
cost of the necessary analog and digital inputs, digital outputs, core store and programming was lower than analog equipment necessary to perform an equivalent task. However, it
should be noted that because of the dramatic reduction in
the cost of computer hardware, this situation no longer applies. The present-day designer would think more in terms
of separate computers for data processing and DDC, and would
probably split up the DDC functions as discussed in Section
XIIIB.

With reference to the advantages of using more complex
algorithms, many authors have discussed the application of
modern control theory implemented by DDC (106).

In a less-sophisticated context, DDC permits control term
adaptation to be applied, and has been used where a peak
seeking, rather than averaging, of plant inputs was preferred.
This application also illustrates the advantages of easy ex-
tendability, discussed in Section VIA. The original scheme
included a five-zone arrangement for the control of spatial
instability in the reactor core. It was later necessary to
change to a 37-zone control system, and the DDC system enabled the modification to be made with relatively little disturbance to the project.

Reliability Considerations

As will be seen from Figure 12, a failure of the computer
involved in many control loops will cause control on all the
loops to be lost. This is a disadvantage, compared with

analog controllers where a single fault causes control on
only one loop to be lost. The reliability of the computer
is therefore a critical aspect, and the relation between the
failure of the control loops, the penalties for these failures
and the degree of backup must be carefully considered. The
considerations that apply are those relating to the application classes defined in Section IV.

An additional factor is that of safety. Good reactor
control has a beneficial effect on safety because the effect
of plant faults are reduced if designed operating levels are
held; however, faults in control systems can cause undesirable
transients, and these may be so severe that protection is invoked with resultant shutdown and economic penalties.

Aspects of DDC Systems

In general, the measured value of Figure 12 is in analog form and is scanned by a multiplexer. The sampling time
must be chosen to be less than the minimum dictated by the
dynamic performance of the loop. Very roughly, the sampling
interval is chosen to be shorter than the shortest time constant in the system. Desired values of the set points are
set in and stored digitally.

The software can provide control terms that include
digital filtering of the input (144), proportional gain,
integral and derivative action, dead bands and linearizing.
The algorithm can also include mathematical operations, and
the terms can be changed continuously by other inputs to
give adaptive control.

After computing the output, control output signals can
be fed to the plant in analog or digital form. In the case
of the channel gas outlet temperature control of the UK AGR,
the control is effected by variation of the mark space ratio
of power to the control rod actuator motors, which are of
the induction type. Some specific examples of applications
of DDC are given in Section VI.

G.   Interlocks

An interlock is an arrangement by which certain defined
actions are prevented for certain combinations of plant
states. Such interlocking is usually applied in the cir-

cuitry associated with operating solenoids, circuit breakers, etc., and is separate from any sequence control such as that described in Section IIIE, or plant protection described in Section IIIH, the latter being invoked as a last resort if the interlock does not cover the effects of the fault concerned.

Interlocks may be complex logically, but do not usually require any processing; and relay or semiconductor circuitry is quite adequate. However, there are some cases in which processing is required to obtain an interlock signal. In these cases, a computer-derived signal involving many plant signal sources and some processing is an advantage. Such signals are used in some plants, and often occur as part of automatic start or sequence programs. Since the use of interlocks implies hazard in respect to plant damage or safety to personnel, the interlock must have a high reliability and must be commensurate with the hazards involved. This necessitates an approach comparable with that adopted for protection as far as effecting the interlock; however, the spurious action of the interlock may be more tolerable than in the protection case, because it causes a holdup of a process rather than a plant trip, and the economic penalty is less severe.

H.   Reactor Protection

General

Plant protection involves detection of the dangerous conditions, performing logical decisions and initiating the trip action (108). For almost all plants, the protection has been effected by detection, using plant-mounted devices with digital outputs; e.g., pressure switches or analog methods; for example, temperature trip amplifiers. These methods have given excellent service, and in UK reactors, the fail-safe or fail-danger fault rate is extremely small and well within the acceptable limits to achieve a probability of < 1 in 10^7, over 5,000 hours, of failure to meet a demand to trip.

In some reactor systems, however, the number of trip sensors is very large, and the protection necessitates complex processing of the signals. In these cases, the conventional methods become difficult to apply; the use of digital computers as an alternative to conventional methods is worthy of serious consideration (109).

The digital approach has been implemented on some PWRs reactor trips being initiated when trip levels are exceeded for neutron flux, departure from nucleate boiling ratio and absolute value of axial power shaping (50). The last two are derived from several physical variables and require calculation. In the case of the LMFBR, computer-based systems become attractive both from the point of view of the need to execute algorithms (51) and the large number of reactor subassemblies that must be monitored. Computer-based systems have been used in France on PHENIX (52) and are proposed for SUPERPHENIX (53).

In principle, any computer-based system should meet the same design and performance criteria as those accepted for conventional protection systems. Although there are some differences between countries, the criteria are well understood and are formally defined (108-112); it is outside the scope of the present review to discuss them in detail.

In the conventional system, redundancy is combined with coincidence to trip on an "m out of n" basis, with n redundant channels being provided and a trip initiated if m inputs pass the trip level. A similar approach can be adopted for a typical computer-based scheme applied to an LMFBR, for example, on a 2 out of 4 basis.

Each reactor subassembly has four sensors, and each of these is scanned by a separate computer system, each computer dealing with one sensor in all subassemblies as shown in Figure 13. The four computers are combined at a 2 out of 4 gate to initiate a trip if any two computer "votes" agree on trip action. This arrangement may be replicated if more than one line of protection, e.g., temperature and flow, is required.

Such schemes have formed the basis for design studies in the UK (51,54,55,56,109); a similar scheme is used in the German BESSY scheme (57), although this uses 2 out of 3 voting strategy. Some schemes permit extensive internal system checks and also on-line modification of voting strategy (51), so enabling defective sensors to be eliminated from the protection strategy. This implies that defective sensors can be identified as such. Several schemes have been proposed to test sensors using electrical methods or by perturbing the plant and checking resulting effects by correlation (58).

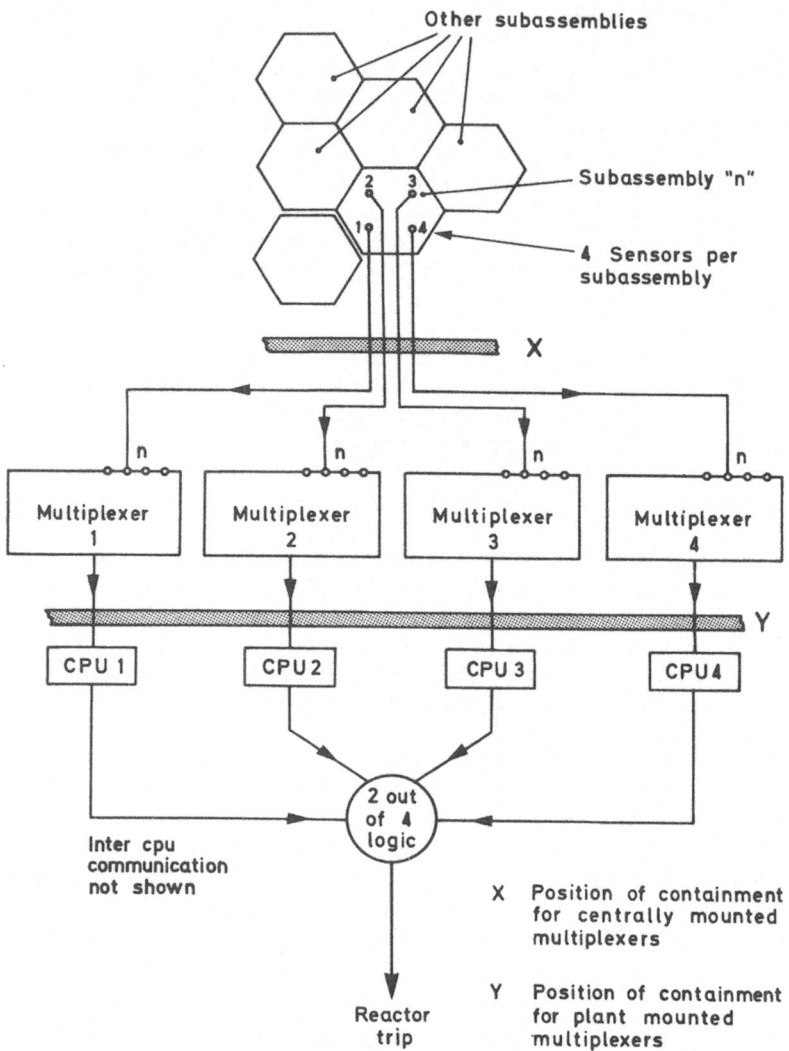

Figure 13

Computer-based Protection System
With "2 out of 4" Logic

In the case of trip transducers with an output in analog form, e.g., thermocouples in the fuel subassembly channel
outlets of an LMFBR, two arrangements can be used. The multiplexer can be of the analog type connected to the sensor, or
analog trip amplifiers with digital outputs can be provided
for each sensor. These digital outputs then can be scanned by
a digital multiplexer. The choice of method depends on the
relative failure rates and failure modes. The use of computers implies analog or digital multiplexers, and these can
be mounted near the plant, the connection to the computer
being made by relatively few wires. If the multiplexers are
mounted inside the reactor containment, the number of bio-
logical shield or containment penetrations can be drastically
reduced (40,56,57). Two schemes are illustrated in Figures
14 and 15. Feasibility of this method depends on the failure
rate of the multiplexers and their modes of failure.

Clearly, the acceptability of the use of computers depends upon the performance of the hardware and software.
Both these aspects present problems because the failure modes
are more difficult to analyze and predict than with analog
methods. The problem can be approached along the lines of
improved design and rigorous testing. These apply to both
hardware and software. The basic design criteria have been
considered (59-61,79-81,109), and the testing can form part
of an overall hardware and software test program such as
that proposed for BESSY (57).

In the case of hardware, the application of reliability
theory and prediction discussed in Section VIII gives some
indication of the likelihood of the hardwares meeting the
specified criteria. Theoretical work can be supported by
practical trials using automatic test systems (61,62,82).

A fundamental feature of "m out of n" systems is a
stringent requirement for independence of the n channels
so there is a very small probability of common mode failures
that would affect all channels. In conventional systems,
this is achieved by segregating hardware, and by rigorously
controlled administration of the testing of any modifications (108).

This procedure can be followed for the hardware part
of a computer-based protection system (109), although it can
be argued that the behavior of such a system under all fail-

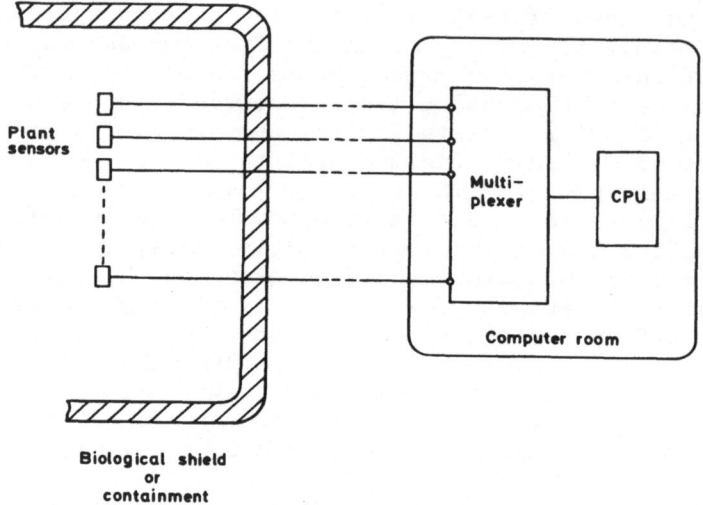

Figure 14

Computer System with Multiplexing in Computer Room

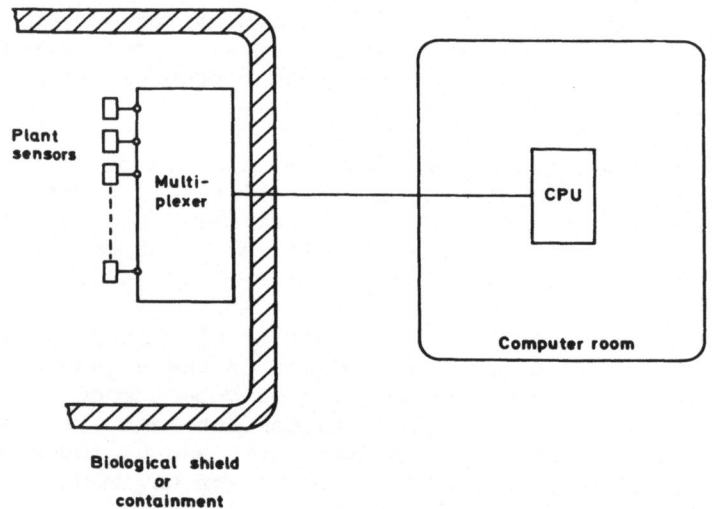

Figure 15
Computer System with Remote Multiplexing Inside Reactor
Containment

ure conditions is more difficult to establish than for conventional protection systems. The software raises special difficulties, (60, 79-82, 109) and it has been pointed out that for example, if the same programmer produced the software for all 'n' guard line computers, he might introduce common mode, perhaps fail-danger failures that would show themselves under some unexamined combination of plant fault and protection system failure. Various techniques are available to ameliorate this situation; for example, by using different software for each guard line, but a standardized approach has not yet been established. Thorough testing of the type proposed (82) should help to reduce the probability of common mode failures remaining undetected, but the size of the test rig and the time required to exercise it in all combinations may well present great difficulties.

Online Testing

Confidence in the reactor protection systems meeting the specified probability of satisfying demands on them for protection is obtained by proof testing. This testing, though tedious, is simple when confined to the electrical circuitry, but is difficult if the sensors are included (113, 114). Some indication of correct operation can be obtained, either by making small perturbations to small plant operating conditions and observing the results, or monitoring existing inherent fluctuations in the plant condition. By using online computers for correlation between several measurement channels (58), for example, neutron flux and temperature, such indication can be made. These techniques show promise of providing a proof-testing method.

In some cases, an electrical check can be applied to the sensors. Such a system has been proposed for the LMFBR subassembly output temperature thermocouple (51).

J.   Use of Computers During Commissioning

The closed-loop control system can be more effectively set to work, tuned, and optimized, if the dynamic performance of the plant is established in detail. The permanent online computer systems are not usually suitable for dynamic testing, being designed for normal operation during the life of the plant, with relatively low scanning speeds and peripherals suitable for normal full power operation.

Dynamic testing implies knowledge of movement of actuators with stroke times of tens of seconds; and logging of plant variables and actuator positions requires fast scanning speeds and special output equipment for recording the results for further detailed analyses, possibly using other computers.

Typically, the permanent computer is supplemented by a data logger that is retained for months or years after plant commissioning. It has facilities that include:

> slow analog scanners; scanning interval, 1 s
>
> fast analog scanners; scanning interval, about 0.1 s
>
> magnetic tape recorder, line printer, crt, plotter, data link to other computers

The equipment can be used for open-loop testing, system identification using pseudo-random binary sequence testing, and static and dynamic tests when the plant is on closed-loop control.

The automatic control system can be tested using open-loop methods to check the transfer function, but the tests can be more effective if supplemented by checks with the equipment connected with the loop closed through a plant simulator as shown in Figure 16. This simulator can take the form of analog amplifiers and components, or a digital computer fitted with suitable input and output peripherals (63, 115,116). For example, motorized potentiometers can be used to represent plant actuators, but actual actuators can be included if available, making use of signals from their position transmitters.

The simulation does not have to be detailed or accurate, a simple one being adequate for revealing faults in the control equipment. Faults then can be identified and rectified before the plant is available so that when the loops are closed on the actual plant, there is confidence that all faults in the control equipment have been eliminated (116).

The simulation technique is particularly valuable for DDC systems in which it is important to be able to check individual control loops that are time-shared in a computer system, and it can be used to establish a basis for contractual acceptance of hardware and software. Hardware can be readily checked by other methods, but complete system test-

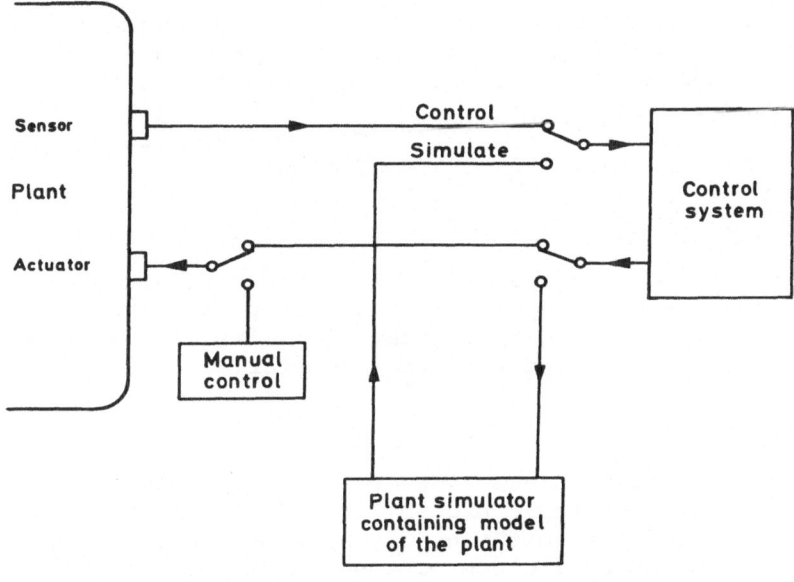

Figure 16

Use of Plant Simulator for Closed-Loop Testing

ing on a subsystem basis under realistic working conditions
is greatly facilitated if simulation is used.

It also enables the system hardware and software to be
thoroughly checked under a variety of working conditions including changeover to standby computer, power supply failure,
system overload, and the effects of failure of parts of the
system.

The loop can be closed via the control desk indications
and controls so that the operator has a realistic representation of how the plant will behave when under automatic control. This can be used to familiarize the operators with the
controls and their effects in terms of indications before the
reactor is in service.

K.   Live Test Facility

Instead of keeping spare parts of computer systems in a

store, they can be assembled into a working computer system.
This then can be used to test suspect modules and other parts
of the system during fault investigations and to test them
after repair.

IV. APPLICATIONS CLASSES

A. General

The reliability and security required of a computer system depends upon its application and in particular, the economic penalty resulting from nonavailability of some or all of
the facilities expected from the computer.

B. Class 1

At one extreme Class 1, where reactor protection is
involved, the acceptable probability of fail-danger situations
or spurious reactor shutdowns can be quantified. Some judgment can be made of the sum of money that justifiably should
be spent on redundancy or other techniques, to avoid spurious
shutdowns.

There is some reliability target that safety equipment
must meet, e.g., a 10^{-7} probability of not meeting a demand
for protection. There is no question of backup equipment,
since the computer is the protection system and has been used
because there is no reasonable alternative. However, it
should be noted that the computer does not have to handle
all the safety measurements and it may represent only part of
the total protection system and be complemented by analog
means. An example is the Fast Breeder Reactor, with fine
structure temperature protection on subassemblies by a computer, but gross core power protection by neutron flux using
analog amplifiers.

C. Class 2

This class covers cases wherein the computer is an essential part of the operation of the reactor. An example would
be a system with extensive DDC covering a large number of control loops that are essential to reactor operation. Another
case would be a reactor in which there is a positive power
coefficient and manual control of even a small number of
loops is not possible.

A failure of the computer system that causes control to be lost may then cause a situation in which the reactor must be shut down. This is very similar to the Class 1 situation discussed in Section IVB, and manual backup is not a significant factor. Availability of the reactor is directly related to availability of the computer, and some economic justification can be made.

In general, extensive DDC implies a situation demanding a low failure rate and a short time to repair. This usually necessitates redundancy of vital parts of the system. Such an arrangement has been adopted for the Canadian reactors and some UK AGR systems described in Section VI.

D. Class 3

Systems in this class provide a service which though important for effective control of the plant, can be interrupted for relatively short periods. The implication is that either:

a. it is acceptable for the service to be interrupted because it is not essential; e.g., some logs may be lost, or the post incident history review is not available at all times; or

b. the interruption causes effects that can be dealty with by other means; e.g., backup indication to computer-driven displays, or manual control backing up the DDC.

In the case of (a), an assessment is to be made of the importance of all facilities provided by the computer system, then to decide the length of time for which interruptions can be tolerated and the acceptable rate of occurrence of such interruptions. This can then be compared with the anticipated computer system performance. Typically, for a fairly complex system with no redundancy, the failure rate will be about once per 1,000 hours, and the length of the outage about one hour, i.e., an availability of 99.9%.

If this is not acceptable, the system must be provided with redundancy, or system (b) must be adopted with sufficient backup control and indications to run the reactor without the computer. Furthermore, these must be provided with

separate plant transducers and power supplies, and have a
breakdown frequency and outage time such that simultaneous
failure of computer and backup are reduced to an acceptably
low probability. For example, if the backup has an availability of 99.9%, with a computer availability of 99.9%, the
probability of simultaneous failure is 10^{-6}, i.e., a simultaneous failure lasting about 10 minutes in the 20 year life
of the plant.

E.   Class 4

This class covers equipment that provides facilities
that can be lost without serious consequences to reactor operation. This situation exists when the facilities, though
desirable, are not essential to operation or adequate backup
is provided. A relatively simple computer system with no redundancy is adequate and most logging and data processing
systems are of this type.

V.   COMPUTER HARDWARE AND SOFTWARE

A.   Aspects of System Design

The requirements listed in Section A are basically similar to other process control computer installations, the
main differences being the relatively large number of inputs
involved and the longer life expectancy of the power station;
typically, 20-30 years. This is much longer than say, chemical plants, so the problem of equipment obsolescence is more
important.

The basic requirement is data acquisition, processing,
then output to plant control, display, and printout equipment.
The numbers involved have been reviewed for some recent GEGB
stations (1) and for other European nuclear plants (4).

It is outside the scope of this review to discuss digital
computers in detail, and there is extensive literature on the
subject. However, it is appropriate to indicate the basic
subsystems and to discuss aspects relevant to the subject of
this review. Typical numbers and sizes of the subsystems are
given in Table I.

B.   Subsystems

The basic subsystems are:

Central Processor Unit (CPU) which contains a main fast
magnetic core store memory, organizes the system operation
and performs the logical operations. The CPU is connected
by a multiwire highway, to the peripheral units ('peripherals'). In a nuclear power station provided with the facilities shown in Figure 2, these peripherals typically comprise one or more of the following, typically connected as
shown in Figure 17.

Auxiliary Store. This provides the bulk storage that
is not economical to carry in the main store. It has the
disadvantage of relatively slow access time and usually is
in the form of a rotating magnetic drum or disk, one recent
development being a low-cost floppy disk. In modern systems,
the auxiliary store may be a mass core or semiconductor
store with no moving parts.

Analog Input Multiplexer System. This scans analog
information from the plant, amplifies and digitizes it so
it can be fed to the CPU.

Digital Input Multiplexer. This scans digital (i.e.,
on-off) information from plant contacts (e.g., alarm signals)
and feeds these to the CPU. The inputs include manual demands by operator switches.

Digital Output Unit. This converts signals from the
processor to output signals to control the plant by on-
off actions.

Cathode Ray Tube (crt) or Visual Display Unit (VDU)
System. This accepts digital information from the CPU and
displays it as alphanumeric, graphical, analog or pictorial
information (for the purpose of this review, crts and VDUs
can be considered indistinguishable).

Printers. These accept digital information from the CPU
and produce hard (typed) copy to record analog measurements
and digital information including alarm messages.

Paper Tape Punches. As with printers, but producing
punched paper tape for analysis off line and not in real time.

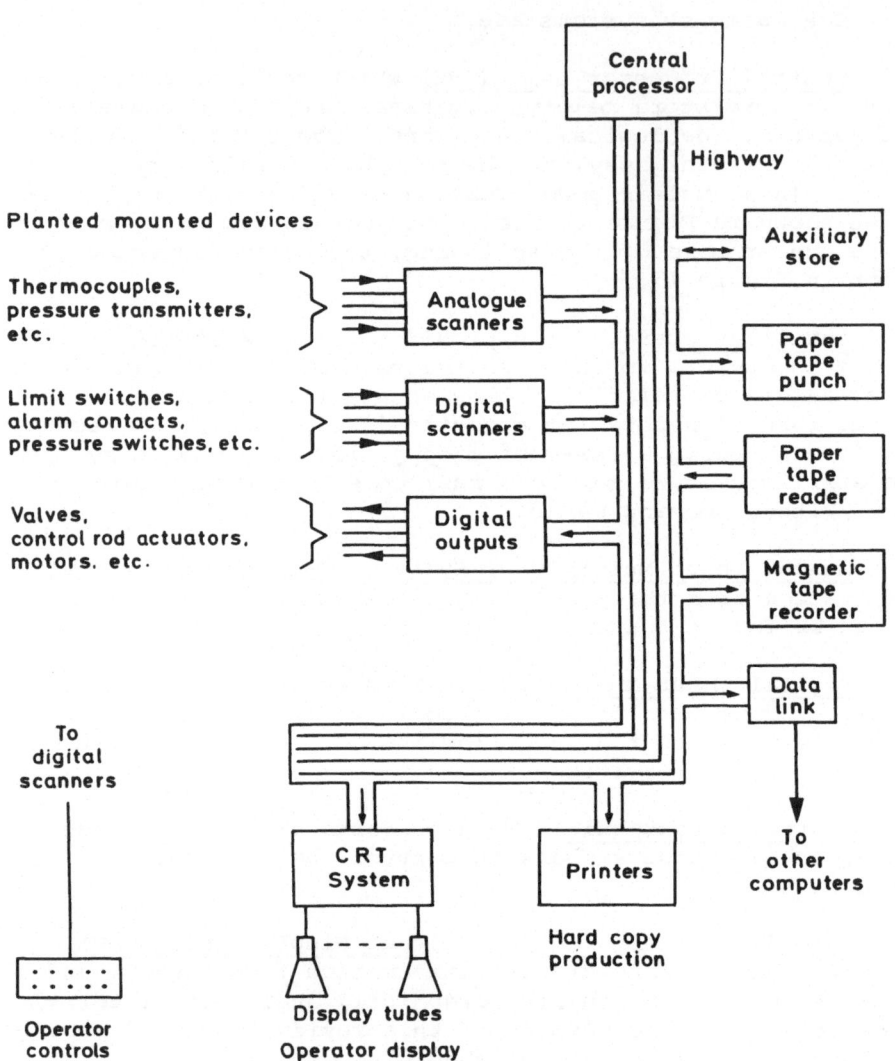

Figure 17

Basic On-Line Computer System

Paper Tape Readers. These accept punched paper tape
and feed signals to the CPU. They are used for loading programs and data into the system, e.g., when making modifications.

Magnetic Tape Store. For recording large amounts of
data via the CPU from the scanner system. This may be of
the reel-to-reel or cassette type. Magnetic tape stores are
used to load program and data and to achieve quicker reloads
than paper tape.

C.   Computer Technology

Detailed consideration of the hardware and software are
beyond the scope of this review and have been described elsewhere (1,3,4). UK systems use assembly language offered by
the computer supplier, but for the future, there is a trend
toward the use of high-level languages such as CORAL 66 (117)
and LCL (118); small systems such as those used for commissioning employ languages developed by CEGB。

Some future developments are discussed in Section XIII.

D.   Environmental Conditions and Power Supplies

Although published equipment specifications permit wide
variations in ambient temperature, some older equipment in
power plants is sensitive to temperature, rate of change of
temperature, humidity and dust. These can cause maloperation,
particularly during the early stages of installation and test,
before all civil works are complete and the air conditioning
is operational. If the surfaces of the computer room are not
properly finished and near "clean conditions" established,
dust frequently reduces reliability (37,46). Magnetic tape
equipment is sensitive to dust and humidity; these are difficult to control in the earlier operational phase.

The reliability of a computer system can be no greater
than that of its power supply. These usually take a form
similar to that used for essential C & I equipment on the
remainder of the plant; a dc/ac inverter or motor-generator
is fed from a battery that is kept charged, and an automatic
changeover to a standby covers failure of the dc/ac converter.
Post-incident recording requires the computer and the associated sensors to remain powered throughout the period involved.

A particular difficulty with inverters is their vulnerability to tripping on overload, due to the inrush current
when some computer peripherals are started.

E.   System Configurations

     For Class 4 applications, the system configuration takes
the form of a central processor unit coupled by a bus to core
store, input/output equipment, auxiliary store, and other
peripherals (37) as shown in Figure 17. In such a system, a
single fault in certain critical items will cause loss of the
whole system. However, if the failure rate is low enough and
the repair time short, an adequate system availability can be
obtained in situations in which there is extensive backup.
In some cases, the system can be split, say, with a subdivided analog multiplexer of crt subsystem does not cause
complete loss of facilities.

     Such a system can be shared between reactor units, and in
the UK Magnox plants discussed in Section VIA, the system
serves two reactor units as shown in Figure 18. While this
scheme has a lower cost than a unitized arrangement, a single
failure can cause simultaneous loss of facilities to both
reactors.

     For Class 3 applications, a lower system failure rate
is implied. This is usually obtained by providing redundancy of vital parts of the system. The degree of redundancy
must be balanced against the amount of backup provided and
the control characteristics of the plant; for example, the
time for which manual control is reasonable. Very little
quantitative information is available to the system designer,
although theoretical approaches to the problem have been
made (83,84).

     The central processor and auxiliary store are vital
parts of the system and are duplicated in many systems; for
example, at Wylfa, discussed in Section VIA and illustrated
in Figure 19. A test program is run in the system at regular intervals, and if this fails to complete, it does not
supply a signal to a watchdog timer. This causes a changeover from one CPU to the standby. The standby CPU is not
often in use and is used for off-line calculations, in particular, the processing of records. On-line duties take

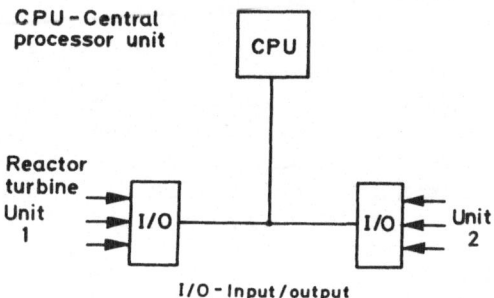

Figure 18

One Central Processor Shared Between Two Reactor Units

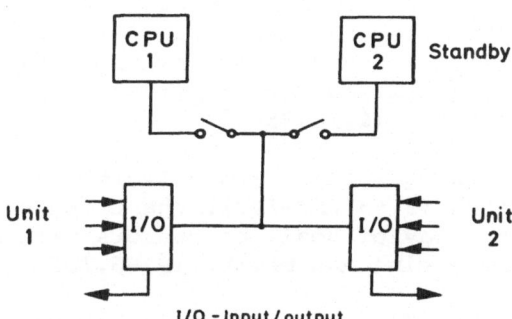

Figure 19

One Central Processor Shared Between Two Reactor Units
With Shared Standby

precedence, and the off-line work is abandoned if the processor is called upon for on-line work.

In this scheme, total failure of the system ultimately
causes loss of generation of two reactors. Such total failure can be caused by simultaneous failure of two vital elements or blockage of the common highway. The probability of
this common mode fault can be reduced by the use of a different configuration; for example, by providing one central processor for each unit with a shared standby as shown in Figure 20. Loss of service to both reactors is caused by simultaneous loss of three processors; two must fail to cause loss
of service to one reactor.

This scheme has been adopted for the UK AGRs described
in Section VIA. As in the two-processor scheme, the standby
can be used for off-line processing. This basic scheme of
Figure 20 can be further improved by splitting the input-
output and by using duplicate highways (2,120,121) as shown
in Figures 21 and 22. With such systems, the frequency and
extent of the plant having to be operated on manual control
is reduced, and the extent of manual backup control and instrumentation also can be reduced.

In Class 2 systems, such as those described in Section
VIF, it is accepted that failure of the computer system is
very likely to require shutdown of the reactor; a redundant
system is readily justified with duplicate computers provided for each reactor (8,16,85).

Many other configurations and networks are possible and
have been described (9,65,122-126), the trend being toward a
more distributed array of small processors with a flexible
allocation of processing and memory (9,65,156,157).

Class 1 systems for protection described in Section
IIIH require very special treatment of both hardware and
software in order to comply with the safety requirements.
The basic configuration necessitates both redundancy and coincidence features giving some form of "m out of n" logic;
for example, "2 and 4", shown in Figure 13, and having seg-
regation between the 'n' channels to avoid common failure
modes.

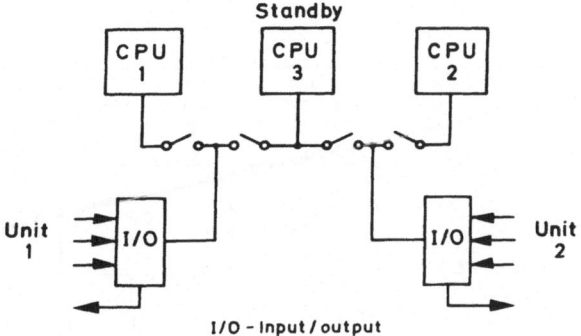

Figure 20

One Central Processor Per Reactor Unit
With Shared Standby

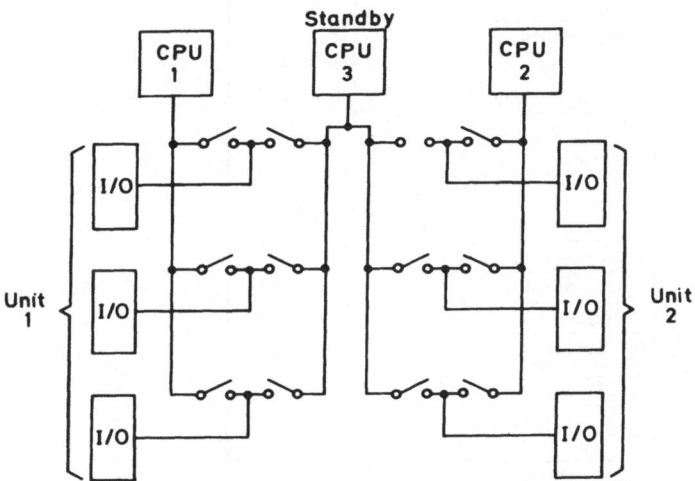

Figure 21

One Central Processor Per Reactor Unit
With Shared Standby With Duplicate Highways
and Split Blocks of Peripherals

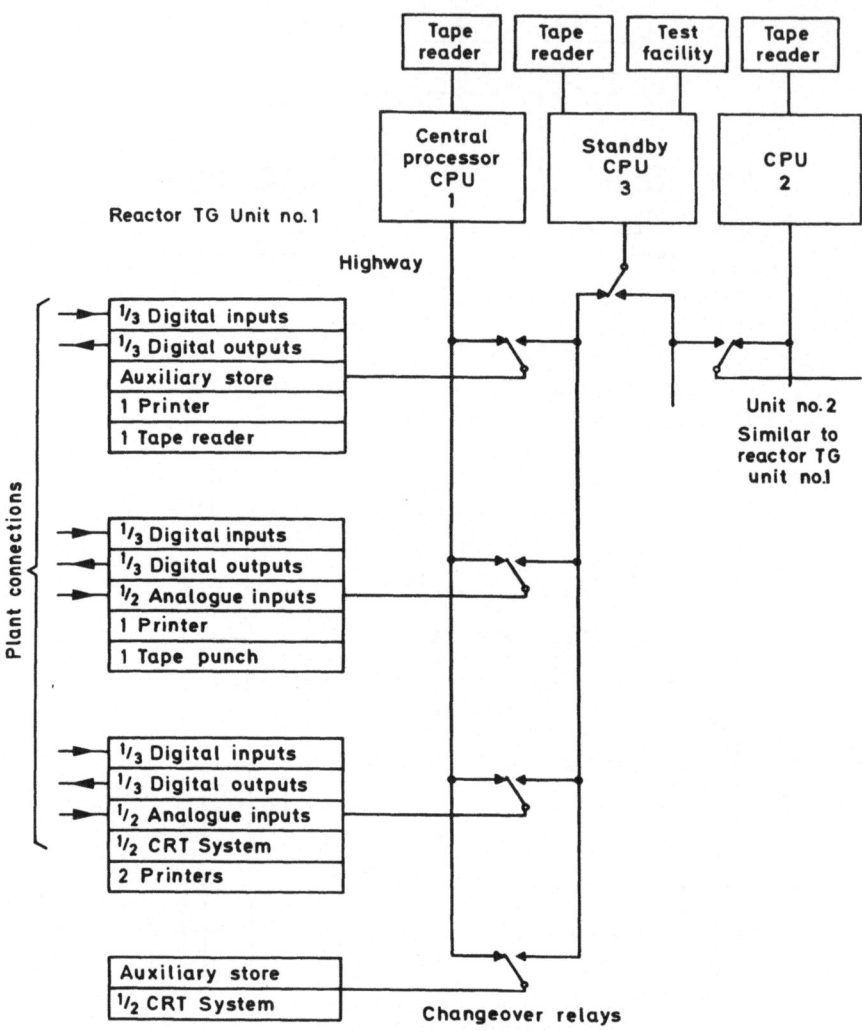

Figure 22

Distribution of Equipment in Blocks of Peripherals
In The Scheme Shown in Figure 21

VI.   EXAMPLES OF SYSTEMS IN OPERATION

A.   Plants in the United Kingdom

Oldbury. This Magnox station was the first UK CEGB nuclear plant to have a computer-based display and logging system with crt displays (97,98). A single central processor serves both reactor units as shown in Figure 18, and there is extensive backup by conventional indicators and alarm annunciators. The system was installed primarily to handle alarm analysis using the tree method described in Section IIIC.

The system now suffers from the obsolescence problems discussed in Section IX, and will be replaced by a new system that employs CAMAC (127,128), so satisfying some of the requirements of Section IX D. The new system employs more crts, but the basic concept of alarm analysis is being retained and it will be programmed, using the CORAL 66 (117) high-level, real-time language.

Bradwell. Bradwell is one of the two earliest Magnox UK commercial nuclear power plants, and was fitted with data loggers using electromechanical multiplexers, the design of which dates back to around 1958. By the early 1970's, the equipment had deficiencies caused by obsolescence, as discussed in Section IX. Furthermore, the existing equipment did not satisfactorily fulfill the functions of alerting the operator to the pattern of temperature changes in the reactor.

It has now been replaced by a computer-based system with relay multiplexers and crt display of temperatures. The crts are of the α-n type, using monoscope generation of the characters, which gives a spatial indication as discussed in Section IIIB, of the reactor channel gas outlet temperatures, as seen in Figure 5. Margins from an operator set datum are indicated and a 'drift' display alerts him to temperatures drifting away from their proper values. Additional tabular displays augment the spatial display and conventional logging is also provided. The system has proved most successful and the crt display is very acceptable to the operators.

Wylfa. This plant which has the last of the UK Magnox reactors, employs a common processor to serve both reactor units with a standby as shown in Figure 19. The drum store

has redundancy, so that a single fault does not affect the
service to both reactors. The system provides data display,
logging and automatic turbine run up, all of which have been
in operation since the plant was commissioned (99,129).

The display system employs crts with long persistence
phosphors, with a refresher rate of about 12.5 per second.
A combination of relatively high control room lighting level
and deterioration of the phosphor led to problems of flicker
and complaints by the operators. The drive circuitry has
been modified to increase the refresher rate, so reducing
the flicker.

Recently, the fault rate of the equipment which employs
early transistors, has increased, and a simultaneous fault on
two drum systems caused failure of the complete system, affecting service to both reactors. Because the system could not
be repaired and restored to service in the time permitted,
both reactors had to be shut down. Some fault rate data are
given in Reference 2. A program of progressive replacement
of the system is in hand; the first phase of improving the
displays has already been completed.

Hinkley Point 'B'. This AGR station has two reactors
that were originally rated at 660MW e each, but since have
been derated to 540MW e, at which power they have operated
during 1976. The computer systems that have been described
in considerable detail (119,120,130) perform data display,
logging, automatic reactor startup and turbine runup, and
DDC of the reactor. The latter involves 37 loops to control
spatial instabilities, the reactor core being divided into
37 zones, with reactor gas outlet temperatures in each zone
being fed back to a control rod in that zone. The control
algorithm is a discrete version of an analog double phase
lead and employs a peak seeking scheme that uses the highest
temperature measured from a group of eight thermocouples in
the control zone.

One central processor per reactor with one shared standby is arranged as shown in Figure 20. There are duplicate
drums and twin highways (2,120) and the input scanners are
subdivided with automatic isolation of defective parts.

The installation, commissioning and setting to work have
been described and some difficulties reported (119,131).

In general, the operators find the crt system acceptable,
although some overloading of the alarm system occurs when
many alarms occur within a short time.

The control room unit desk is shown in Figure 23.

Figure 23

Hinkley Point 'B'; Central Control Room Unit Desk

Insofar as DDC is concerned (131), works testing of computer function was followed by open-loop testing. This re-
vealed the presence of cyclic variations of the measured input signal due to the two halves of the interleaved scanning
system's having slightly different characteristics. This was
exacerbated by the differential term in the control algorithm.
Other limitations on the control performance were imposed by
the priority level of the control program's not being high
enough, and lack of synchronization between input scan and
output signals to the control rod actuators.

Using the technique discussed in Section IIIJ, a small
analog computer was used to simulate the reactor plant and
allow closed-loop testing of the DDC of the reactor. This
enabled more realistic operation and more effective investigation. Some off-line simulation of the computer and plant
was also used to investigate the effects on input noise on the

performance of the control system. The computer used was the same one used for data logging of the type described in Section IIIJ.

Corrective measures to make the control system more satisfactory included the introduction of dead bands and use of a single scanner. Subsequent operation has been satisfactory, the control scheme reliability being dominated by computer system failures rather than control system defects.

The computer system failures that have occurred show no individual part of the system to be particularly unreliable. During a one-year period in 1975-1976, faults on the first reactor computer system caused loss of computer service to the plant, of 16.1 hours, corresponding to an availability of 99.82%. The corresponding figures for the second reactor are 25.3 hours and 99.71%. These figures apply to the unit computer with its shared standby and include time lost due to testing, commissioning and maintenance, but exclude time lost from failure of main power supply and nonvital peripherals such as tapedecks, punches and readers.

Dungeness 'B'. This plant has two AGRs, the computer system configuration is of the type shown in Figure 20, and detailed descriptions are available (10,132,133). The reactors are not yet operational. The computers are run in a monitoring role, although a successful computer-controlled turbine runup has been achieved using steam from the Dungeness 'A' plant (133). Before this system was used on an actual plant, it was tested using a simple simulator in the manner discussed in Section III J. In addition to DDC of reactor temperature, automatic startup data display and logging, the computer provides some interlock functions. For example, movement of control rods is inhibited if the neutron flux detectors indicate an excessive doubling time, or if certain insertion and pattern constraints of the five-zone control rods are not satisfied.

The boiler feed flow trim valves, controlled by the computer, have some protection against fault situations by an analog protection system, but the computer program initiates a freeze for excessive rate of change.

Hartlepool and Heysham. Each plant has two AGRs, the computer system having a high integrity arrangement (2,121)

with subdivided input-output with a considerable fraction
of the inputs being fed to two plant-mounted multiplexers
to provide redundancy. The display system is also subdivided with duplicate discs and duplicate highways as shown
in Figure 22. The system provides data display shown in
Figure 24, with logging and DDC of reactor temperatures,
coolant mass flow and feedwater. The reactors are not yet
operational and the system is used for plant surveillance
with the control loops being checked by the closed loop
method, using a simple analog model to simulate the plant
in the way discussed in Section III J. LCL (*118*) was used
to specify the control programs, and some statistics on
faults detected during software production have been reported (*109*). So far, under test conditions, the availability
is about 99.8%, and is of a similar order to that recorded
at Hinkley Point 'B'.

B.   Other European Plants

      Reviews (*3,4*) of computer applications in Europe show
an increasing dependence on computers and of the 21 plants
listed (*4*), all provide regular data logging and alarm logging; 12 have alarm displays by crt. Recent applications
include the use of computers for safety functions. The PWR
and BWR plants have on-line programs dealing with flux power
distribution evaluation, xenon effects and fuel burnup.

C.   Canadian and other CANDU Reactors

      Although the reactors are very different, the computer
systems for the UK AGRs have some similarity to those used
on the CANDU (*11,134-143*) reactors at Douglas point, Pickering (*16,140,141*), Bruce (*66,142*) and Gentilly (*116,143*). Both
use computers that provide similar facilities for display by
crt, permanent records, automatic turbine runup and DDC, and
both employ redundant processors.

      Pickering is a Class 2 application, since it cannot run
for any length of time without the computers in operation.
This implies a high security system and twin computers are
provided. In general, the facilities for each reactor are
handled by both computers. At Pickering, the mean time between failures (mtbf) of a single computer is about 2,000
hours. In 1974, after eight unit operating years, the total
down time of the dual computer system was only five minutes
(*139*).

Figure 24

Hartlepool and Heysham
Central Control Room Unit Desk

In a further development for the Kanupp CANDU Station
in Pakistan (46,144), a similar design was adopted except
that the fueling machine control was separated and handled
by two other computers. The predicted availability of the
main computer system was 99.88%. After a period of relatively poor performance, improvements were made, so that
only 0.03% loss of production was due to the plant control
computers.

D.   Other Reactor Types

References to computer systems installed in LWRs are
given in Table II and Section IIID. In general, the application has been limited to data processing and crt display,

with little use of computers for closed loop control, a notable exception being the Halden plant (78,107).

The High Temperature Gas Cooled Reactors (HTGRs) employ computers (104,145,164) as do the LMFBRs in the UK, France and Japan, as can be seen in Tables I and II.

VII. ERGONOMIC FACTORS AND OPERATOR ACCEPTANCE

A. Control Room Design

As stated in the Introduction, an important incentive for using computers is the assistance they offer in the problems of implementing centralized control. This has been exploited in many power plants. The main influence on control room design is the use of crt displays of data and alarms as discussed in Sections IIID and IIIE.

The use of crts enables the control desk to be made smaller than conventionally instrumented desks, and the reactor can be effectively controlled by a single operator. This approach has been followed in some UK Magnox reactor plants, the AGRs as illustrated in Figures 23 and 24 and LMFBR (146), and in other countries for PWRs and BWRs (147). Table II summarizes key papers on this subject.

A major difference in the earlier designs was that only the UK used crts for alarm displays. Other countries mainly confined their use to display data, but are now introducing crts for alarm display (4,100).

The display hardware, the graphic design of display itself, and the operator selection facilities all must be of an adequate standard from an ergonomic viewpoint (163). In some earlier designs, the standard has fallen below that required with a resultant criticism from operators. It is appropriate to consider these factors and to indicate current developments in the UK.

B. Display Hardware and Printers

Particular examples of problems with crts are the irritation caused by flicker of the display presented on the crt and reflections of control room lights on the screen. Modern hardware and software enable the use of higher refresher

rates with shorter persistence phosphors. This gives an
acceptable flicker with longer tube life and greatly reduced
cost of replacement tubes (148). Reflections have been reduced by the use of filters provided integral with the tube
and by the design of the control room lighting.

Some designs, for example at Pickering (16), have printers installed in the control room. Earlier models of printers were noisy and their use was avoided, but modern designs
work on different principles and are quieter so that sound-
proofing is less important and they can be more readily integrated into control room design.

The printout can be in code or plain English, but the
former has been criticized by UK operators who much prefer
plain English, even if abbreviated.

C.   Access to Data

A major characteristic of computer-driven displays is
the limited number of displays available to the operator.
The number of values of plant measurements shown on a single
display is limited by the number of lines that can be displayed by the hardware and the number of measurements that
can be effectively presented and comprehended by the operator.
This number has not been established with confidence, but is
typically around 20.

Assuming the number of plant measurements fed to the
computer is around 2,000, one display of 20 items represents
a one percent selection from the total. If, say, five crt's
are provided, the amount of information on display represents
only a five percent "window" into the total amount of information available. Thus, if the operator wishes to examine
an appreciable sector of the plant, he must change his selection of data fairly frequently. This is not particularly
disadvantageous if the selection method is both quick and
easy, but it is bound to be less convenient than consulting
conventional meters that are always on display. If it is
not convenient, the operator will be tempted to abandon the
crt display and refer to whatever alternate instrumentation
he has available. Under these circumstances, there is a
strong argument for having a complementary system of crt
displays and conventional meters, the latter covering vital
measurements that are frequently consulted, and the crt

used for reference to information not on continuous display because of economic reasons. This approach is consistent with having the conventional meters performing some of the backup shown in Figure 2 and used in the Class 2 and Class 3 applications.

A similar situation arises in connection with alarms. crt displays of alarms have the advantage in that they have the ability to display the alarms in time order of occurrence. However, because the tube can only accommodate about 24 lines, the earlier alarms must be accommodated and made available by a paging system. The protocol of the manual paging operations can be quite complicated, and acts as a deterrent against crt alarm systems. With some systems, in a quasi-emergency situation, the operator finds the manipulation of the paging and other controls to be irksome; he would prefer immediately available alarm messages. Again, the use of conventional lamp annunciator alarms can be used for this purpose and for providing backup to the computer. This approach has been followed in the UK CEGB nuclear and fossil-fired power stations (1).

If great reliance is placed on crt displays, it is most important to make a sufficient number of crts available to the operator, displaying simultaneously, all the information he requires for the task in hand. Furthermore, if the number of chart recorders is reduced, an equivalent number of graphical trend displays (73,162) must be provided with associated hard copy facilities.

D.   Operator Selection

The design of display selection keyboard or switch array, the operational protocol, and delays in its operation are also very important and can easily turn an otherwise excellent computer system into one that is only accepted with reluctance by the operators (164).

The manipulation of the selection mechanism must not be such that it deters the operator from using the crt system. If it is not convenient, the operator will not wish to use it or will demand separate analog displays, that are always on display and can be consulted quickly without the distraction of operating a ponderous system.

E.   Graphic Design of Displays

The flexibility of the crt display can be exploited to
present the optimum combination and form of display for the
control plant under various normal conditions and abnormal
situations. This facility is important from the safety
viewpoint.

A critical time occurs during fault conditions when the
operator is presented with a lot of information, both digital
and analog. The main problem is to give the operator a good
appreciation of the situation so he can take the best action,
and this may well have safety implications. The flexibility
of the display can be exploited to give this appreciation,
but at the present time, the displays tend to be designed for
operation of the plant in the ways that can be predicted at
the design stage. However, flexibility can be built into the
computer system to allow modification as operating experience
is accumulated, so the potentiality of the crt system can be
better exploited once the plant is operational.

VIII.   RELIABILITY

In common with much other control and instrumentation
installed in early power plants, computers have not always
earned a good reputation in respect to reliability, defined
as the probability of a system successfully completing a
mission. Since World War II, the techniques of reliability
analysis and prediction have been developed,(150) and data
banks are available for recording, processing and retrieval
of failure rate data on components, subsystems and complete
systems and codes, available to predict complex system reliability.

Although not sufficient to enable systems to be designed
to meet specific reliability with great confidence, the techniques do permit a quantitative approach that has become an
essential feature of system design.(83,84) If a reliability
target is available, for example from economic considerations,
subsystem failure rates and reliability theory can be used to
indicate whether redundancy is required, and if the target
can be met. Great accuracy is not required because a decision on redundancy has a dramatic influence on the system
failure rate and mean time between failures (mtbf). A simple
example is the main and standby arrangement shown in Figure 19.

If the failure rate of each unit is λ and the time to repair a faulty unit is t, it can be easily shown that the effective failure rate is $2t\lambda^2$; e.g., if $\lambda = 10^{-3}$/h and t = 1 h (1,000 hours mtbf), the redundant configuration has an effective failure rate of 2×10^{-6}, i.e., an mtbf of 5×10^5 hours or 57 years. Thus, if the reliability target is, say, two years mtbf, great accuracy in failure rate is not required in order to make a decision on reduncancy, which changes the mtbf from one-eighth year to 57 years by providing a standby.

The predicted failure rate can be used as a guide to the expected failure rate during acceptance trials. This can be used as a basis for trials and subsequent contractual acceptance. The main difficulty is that during an acceptance trial of reasonable length, say 1,000 hours, the number of failures will be small, so the statistical confidence is suspect. Nevertheless, this procedure has been used as an acceptance method for computer systems; further development of the technique is being pursued (83).

The availability (1-outage time/running time) gives one method of indicating a "reliable" system, but it must be qualified by the number of faults and their consequent outage time. Depending on the application, a small number of relatively long outages may or may not be more serious than a large number of faults that are each cleared quickly.

Some availability figures recorded from actual service are given in Section VI.

IX.   OBSOLESCENCE AND DESIGNING FOR REPLACEABILITY

A.   General

For many years, the system designer has acknowledged the need for designing "reliability" and "maintainability". Now must be added "replaceability".

It is unlikely that the life of sophisticated electronic equipment, particularly computers (24,151), will be as long as the life of the main plant, the latter being 25 years in the case of UK CEGB power stations. It follows that the computer system probably will have to be replaced during the life of the station and arrangements made to facilitate such replacements with minimum disturbance to the normal operation of the power station.

The situation now exists in UK power plants, and several replacement schemes have been undertaken or are planned, for example, those described in Section VIA.

Replacement of a large computer system that has operator controls and indications must be carefully implemented. Furthermore, if the facilities, before or after the modification, have safety implications, the proposals will have to undergo a safety assessment and agreement of the licensing authority, as discussed in Section XII.

The replacement activity can be made easier if the need for it is recognized in the design stage; i.e., a policy of "designing for replaceability" is adopted.

B.   Reasons for Replacement

Difficulty of Maintenance Because of High Failure Rate. Some components in computer systems reach the wearout state, so the failure rate of the equipment becomes unacceptable because of the excessive maintenance effort and system downtime. In order to improve the situation, it may be better to replace the equipment rather than attempt to maintain the obsolete equipment.

Nonavailability or Excessive Cost of Spares. Some years after installation of equipment, the manufacturer introduces a new range, and although spares still may be available, the situation becomes progressively more difficult. It is often made worse by companies having financial difficulties, mergers and takeovers. These difficulties may cause the spares to be prohibitively expensive to purchase. Some items can be manufactured by the user (46), but many cannot, and it then becomes preferable to replace the system completely rather than to maintain the old one.

Desirable Operational Advantages Available with New Techniques that Were Not Available when the Plant Was First Designed. There is a time lag of some years between the equipment's being selected and its use in a working power station (1,151,152). After a few years of use, technology will have advanced to the point that new devices could give operations advantages over the original equipment. It then may be possible to justify the cost of changing the equipment to the more modern type in order to obtain such advantages.

This situation often occurs with computer peripherals such as printers and display devices.

Power Plant Operation Indicates Changes Necessary to Computer System. After some years, operation at power may reveal deficiencies in the design of the equipment and may necessitate changes to the computer system.

Sabotage, Fires, Accidental Damage (33-36,38). Recently, it has become necessary to take into account the possibility of sabotage, fires and damage caused by external hazards, and to bring the equipment into conformity with national or international (e.g., IEEE) standards. These requirements may be so extensive that it is impracticable to modify the existing design, and it may be necessary to replace the whole system or a major part of it.

C.   Factors Relevant to Replacement

The time taken for the replacement and the funds available are likely to be limited, so only partial replacement may be possible during a particular plant outage. A modular system that can be replaced in parts will be helpful in this respect.

It may be possible, and in fact, it may be insisted upon by the Licensing Authority, that the replacement system be run in parallel with the old until the replacement is proved to be satisfactory. This necessitates the allowance, in the design stage, of sufficient space to permit the installation of additional equipment at a later stage.

Rapid replacement will be easier if plugs and sockets are provided, particularly between parts that are known to be likely to become obsolete.

Use of equipment complying with standards discussed in Section VIIE will assist in replaceability.

D.   Standards

General. An ideal situation would be that in which equipment is multisourced (obtainable from many manufacturers) and conforming to a standard life comparable with

that of the power station, say 25 years. It has been said
that standards are just as replaceable as the equipment that
is designed to the standards. This is true, unfortunately,
in many cases.

To meet the multisourcing criterion and requirements
of a large-scale marketing operation, any standards must be
international. This is recognized by many organizations, and
they support efforts in this direction by membership of the
relevant IEC and ISO working groups and the ESONE committees.
The standardization concerns mechanical and electrical interconnections (153) and also software.

Mechanical Standards. At present, some standards exist
and are widely used (references to IEC standards), and employ 19" wide racks with 1 3/4" incremental heights. There
is also a DIN range, and Eurocard is available from several
manufacturers. Its use could rationalize the present situation of innumerable shapes and sizes of printed wiring and
racks.

CAMAC is a very commonly-used system that was developed
by users in the nucleonic field (127,128). It has the features required of a replaceable control system, but interconnection of CAMAC modules for an industrial environment
requires some development. Although its initial cost may be
higher than that of individual manufacturers' systems, the
overall cost of CAMAC, taking into account replaceability,
may well justify any initial cost penalty.

X. PROJECT MANAGEMENT

Nuclear power plant computer systems are major projects
involving complex hardware and software. In addition to
these technical aspects, attention must be given to the
identification of systems requirements to ensure that the
computer is properly integrated with the remainder of the
plant and is installed within the construction and commissioning program (18,152,154). This program must allow
sufficient time for analysis and specification of requirements, hardware design, manufacture, installation and test
and for software specification, production and optimization.
However, a commitment to equipment and software must not be
made so early that the plant requirements are not established
or that the hardware suffers from the obsolescence problems
discussed in Section IX.

To some extent, freely programmable computers offer considerable flexibility in the facilities they can provide with given hardware. This can be exploited to defer some software decisions until finalized information concerning the main plant and its operation is available.

XI.   PROGRAM AND DATA SECURITY

In the business computer field, there is much concern about security of data and programs. The penalty for lack of security takes the form of invasion of privacy, particularly in data banks or of stealing, in the case of bank accounts. In nuclear plants, interference with programs or data could result in maloperation of plants, leading to damage and possible danger of radioactive release, particularly where Class 1 systems are used to protect the reactor.

There is a fundamental difference between the two situations because, in the business case, the interference is due to a deliberate act, usually dishonest, while in the case of the power plant, it could be due to inadvertent error or equipment failure, although sabotage must be considered. Inadvertency is more easy to protect against by combinations of mechanical means and administrative control。

In UK nuclear power stations, with no Class 1 applications, the data and programs are classified into categories corresponding to their importance to station operation, possible safety implications and the manner in which the modifications are made. The categories, broadly, are as follows:

a.   data or program changes where neither plant
     damage, availability or safety are involved

b.   data or program changes where plant damage
     or availability could be involved but no
     direct fuel damage

c.   data or program changes where direct fuel
     damage could be involved

No control is necessary for the first category, but for the others, access to the means of making modifications is restricted by codes and locked switches, the keys being subject to administrative control. For extensive data and pro-

gram changes, access is via punched paper tape with the
processor and auxiliary store being taken off-line, and
with a comprehensive set of checks made to ensure that the
changes made have been limited to those intended and that
there are no unexpected interaction effects. Such checks
are facilitated by the use of plant simulation discussed in
Section IIIJ.

In some DDC installations for chemical plants, the importance is stressed in a facility, for easy on-line modification and development of programs, but in nuclear power
plant applications, this does not appear to be realistic,
since it introduces considerable hazard. Once the plant and
computer system have been commissioned, it is to be expected
that only minor changes will be necessary; if major changes
are required, these can be made at a plant shutdown period
with the computer off-line.

This policy minimizes the probability of unauthorized
or unintended modifications being made to the computer system that may have serious economic or safety implications.

The success of such a policy depends upon the effectiveness of the administrative control applied through the key
and access code system, and an acceptable compromise must
be maintained between the flexibility inherent in freely
programmable digital computer systems and the probability
of plant damage caused by incorrect operation as a result
of unauthorized access. In the UK, the control has proved
satisfactory.

XII.   LICENSING OF INITIAL DESIGN AND MODIFICATION DURING LIFE

The application classes discussed in Section IIB have
some safety implications, although with Class 4 they are
obviously less significant than with Class 1. The Licensing
Authority will be concerned with the operation of the computer system, and in the UK, the conditions may be referred
to in the Site License; e.g., the length of time that records on paper tapes must be kept. As indicated in Section
VIIF, certain modifications can be implemented by the operators without reference outside the plant, and once recognized and defined, no further action is necessary; however,
some modifications have more serious consequences, and a
rigid control procedure is adopted.

In the UK there are not yet any Class 1 applications.
For the other classes, the Site License defines conditions
of operation through the Station Operating Rules. These
define conditions under which the reactor can be operated;
for example, they may call for a reactor shutdown if certain
computer facilities are lost for certain lengths of time.
For example, in a gas-cooled reactor, there is a time limit
for the operation of a reactor without temperature scans
being available. The implications of this leads to decisions
regarding the necessary computer system reliability relative
to the cost of forced outages, due to computer system failure.

Some experience in the United States has been reported
by the Nuclear Regulatory Commission (60) on considerations
of software design and qualification with special reference
to Class 1 applications.

It is clear that the regulations for computer systems
will be more difficult to formulate and apply than for hard-
wired systems. There is a great deal of work to be done before it will be so closely defined as in the case of hard-
wired systems.

XIII.   FUTURE DEVELOPMENT

A.   Results and Management Centers

One way to avoid burdening the control desk operator
with excessive information is to process it and arrange to
give him only that which is strictly necessary. Though this
can be applied to some extent, experience in the UK with
alarm analysis indicates that this is difficult to apply.

An alternate, simpler approach is to reduce the volume
of data passed to the operator by routing some of it to a
separate result center (31) and/or a management center where
it can be consulted by others; for example, reactor physicists, maintenance and planning engineers. These centers
would be equipped with printout devices and crts, enabling
ready access to all reactor data without interference with
the desk operator, and processing for particular purposes,
for example, reactor optimization, maintenance strategy or
planning of refueling (155). Such an arrangement implies a
common data base for the complete plant and a system of computers that can access the data, process it and produce displays and hard copy.

B.   Multicomputer Systems

In all but the simplest Class 4 applications, the
arrangement with a single processor serving more than one
reactor (Figure 18) makes it vulnerable to reactor outage
due to a single processor failure.

The main and standby systems adopted in the United
Kingdom and in Canada, discussed in Section VI, provide a
solution but require "main frame" computers implying complex software systems. An alternative is to use a multiplicity of processors, memory and input/output units in an
interconnected array (9,65,156,157). Such a scheme has been
pioneered at Halden (3), the whole array giving sufficient
computing power to serve the whole reactor; but failure of
a single part does cause the whole to fail to the point in
which a reactor shutdown is enforced.

A further alternative is to split the computer functions
and to allocate them to more discrete computers (processor and
local memory) with less interconnection than in the Halden
scheme. Each computer works autonomously, and although there
is some interconnection, it is not essential, and the system
continues to operate suboptimally if the interconnection
fails. Failures now have a limited and local effect, and
if manual control is possible until it is repaired and the
loop reinstated, there is no need for redundancy, although
this can be provided on particularly critical loops if required. This approach has been made cost-effective because
of the reduced cost of computers using the latest semiconductor technology, and is termed a "distributed" system.

C.   Color Displays and Alarm Analysis

Although many papers describe color crt systems and
claim advantages for them, these appear to be marginal relative to well-designed monochrome systems. However, as techniques improve, these may become better established, and
color crts will be common for many applications. Alarm
analysis has not yet proved itself to be justified, but there
is a continuing interest in the concept (158) and the work on
disturbance analysis (43,44) and software-selected displays
(89) may well lead to new development in computer-aided displays to the operator.

D. Control Room Design

Some experiments have been conducted with reactor control using only crts and computer interface controls (*19*) but current designs of control rooms retain a large amount of conventional indications, controls and alarm annunciators. These are likely to remain, for the reasons discussed in Section III, although they may be driven by microprocessors (*159*) and the form of display may be different, for example, light-emitting diodes or plasma panels rather than lamp annuciators and crts.

E. Computer Technology

As stated in Section VIIIB, semiconductor technology development and packaging techniques have made computer hardware cheaper and also more reliable (*159,160*). In large systems, the auxiliary store, a rotating magnetic drum or disc, make a large contribution to the system failure rate. Solid state mass stores are now becoming available, and these should improve this situation (*161*). Magnetic core stores have a significant failure rate and will be replaced by semiconductor memories.

At present, analog inputs are multiplexed using relays. These give the high rejection of common mode and series mode interference that is important with existing techniques. However, with the availability of new techniques, inexpensive, high-performance amplifiers with the necessary interface rejection capability, solid-state multiplexers, individual analog-to-digital converters, one for each input may be cost-effective. These may well be located near the plant sensor with digital transmission, as discussed in Section IIIH and Reference 104.

XIV. CONCLUDING REMARKS

The progress of nuclear power would not have been so rapid if digital computers had not been available for design calculations. Nowadays, efficient operation of nuclear power plants also relies heavily on digital computers, and in general, the demands of the plant operators have been met by the computer designers and suppliers. Furthermore, much development potential remains, and computers will continue to be employed on an increasing scale.

Many notable successes have been accompanied by a number of problem areas that have evaded completely satisfactory solution. Among these are the use of crts and alarm analysis; both occur at the operator interface.

The existing ways of exploting crts constrain the flow of information through a relatively small number of channels and the operator must continually select and combine these to obtain an adequate indication of plant conditions. This is an area requiring investigation and improvement. A particular aspect is the establishment of a methodology for analyzing and selecting plant data into significant groups that can be displayed through a relatively small number of channels. Another is the provision of improved mechanisms for the operators to select the information they wish to consult. The work at Halden and Risø, and feedback from systems in operation will contribute to making better systems available in the future.

Computers are well-suited to solve the problems of implementing the logical processing of alarm data that is required for alarm analysis, and it is somewhat surprising that it has not been more successful. In fact, it has not been applied in the United Kingdom since the AGR stations were designed, but the more recent work at Halden and Munich on disturbance analysis may give a lead to better systems.

These examples emphasize the fact that major factors in the exploitation of computers are the identification and specification of the task they have to perform. Many problem areas, identified as computer problems, are, in fact, due more to inadequate attention being given to operational research and reliability targets than to computer system deficiencies. The dramatic rise in computing power and drop in memory cost make computers attractive in many areas. For example, the trend toward the acceptance of computers for protection is largely a result of economic factors. The application of computers to LMFBR protection is perhaps the ultimate challenge. It has acted as a stimulus to some recent developments, and although computers may not be used in this role, the hardware and software techniques will be relevant to other control and protection applications of computers in nuclear power plants.

ACKNOWLEDGEMENTS

This chapter is published by permission of the Central
Electricity Generating Board. Thanks are due to Nuclear
Power Company Limited for agreement to publish Figures 23
and 24.

REFERENCES*

1. Jervis, M. W.,"Online Computers in Power Stations,"
    Proceedings IEE Reviews, 119 8R, pp 1052-1076, August,
    1972.

2. Welch, B. R., "Computer Systems in CEGB Nuclear Power
    Stations," Nuclear Eng. Int. Technical Survey No. 10,
    Computer Applications in Power Plant Control, pp 25-30,
    January, 1973.

3. Lunde, J. E.,"Overview of Current Trends in Applications
    of Process-control Computers to Nuclear Power Plants,"
    Proceedings of Conference on Mathematical Models and
    Computational Techniques for Analysis of Nuclear Systems,
    Ann Arbor, Michigan, April, 1973. Washington: USAEC
    Publication Conf-730414-P2, pp VIII, 1-19, 1973.

4. Hoermann, H.,"Review of Online Computer Applications in
    Nuclear Power Stateions in Europe." Ibid, pp 36-50.

5. "Application of Online Computers to Nuclear Reactors,"
    Proceedings of ENEA/OECD Conference, Sandefjord, Norway,
    1968.

6. Jervis, M. W., "Online Computers in CEGB Nuclear Power
    Stations. Ibid, pp 51-78.

7. Pekarek, H., and Veras, A. F., "The Online Computer at
    the Dresden Nuclear Power Stationand Its Nuclear Fuel
    Management. Ibid, pp 793-817.

---

*   In the case of papers presented at the same meeting, the
    full source reference is quoted in the first reference
    only and subsequent ones are marked "Ibid".

8. McAffer, N. T. C., "The Computer Instrumentation of the Prototype Fast Reactor." Ibid, pp 351-380.

9. Basten, R. G., "Impact of Nuclear Reactor Control on the Structure of Computer Systems." Ibid, pp 517-535.

10. Cameron, A. R., "The Online Digital Computer System for the Dungeness 'B' Nuclear Power Station." Ibid, pp 273-300.

11. Pearson, A., "Computer Control on Canadian Nuclear Reactors." Ibid, pp 123-144.

12. "Online Computers for Nuclear Reactors," Nuclear Eng. Int. 14, pp 950-952, 1968.

13. Bullock, J. B., Seminar on the Application of Online Computers to Nuclear Reactors, Nuclear Safety 10, pp 148-154, 1969.

14. Langlade, R., and Leroy, R., "Level of Automation and Use of Computers at the French Nuclear Power Stations Currently in Operation," IAEA Symposium SM/168, January, 1973.

15. Freymeyer, Ph., "Prozessrechner für Kernkraftwerke," Atomwirtshaft 10, pp 524-528, October 1971.

16. Key, F. J., "Station Control," Nuclear Engineering Int. 16, pp 511-513, 1970.

17. "Application of Digital Computers to Nuclear Reactor Control and Instrumentation," International Electrotechnical Commission Draft Standard 45A, Secretariat 37, June, 1975.

18. Jones, R. T., "The Nuclear Control Room Experience Directs the Future," International Atomic Energy Agency Specialist Meeting on Plant Control Room Design, San Francisco, June, 1975. IEEE 75 CH 1065 - 2, Paper 1-7.

19. Lunde, J. E., and Netland, K., "Experimental Operation of the Halden Reactor, Utilizing a Computer and Color Display-based Control Room," Ibid, Paper 1-1.

20. Dowling, E. F. and Castanes, J. A., "Efficient Plant
    Operation Through Condensed Information Display." Ibid,
    Paper 4-6.

21. Brookes, J. G., and Meijer, C. H., "Safety System-
    Operator Interfaces during Abnormal Conditions." Ibid,
    Paper 3-5.

22. Malmskog, S., and Versluis, R., "Colored Graphic Design
    Display - An Aid in Reactor Core Supervision and Con-
    trol." Ibid, Paper 1-2.

23. Fukuzaki, T., Kishi, S., Kiyokawa, K. and Serizawa, H.,
    "Man-Machine Studies of a Computer-Based Console for
    BWR Plant Operation." Ibid, Paper 3-1.

24. Darke, R. S. and Coley, W. A., "Operator Interface Re-
    quirements for Nuclear Generating Stations - A Review
    and Analysis." Ibid, Paper 3-6.

25. Niki, Y., Iida, H. and Inaba, G., "Importance of Graphic
    Display for Nuclear Power Plant Operation." Ibid, Paper
    4-2.

26. Gotoh, S., Aoki, R., Makino, M., Kawahara, H. and Satch,
    T., "An Application of the Process Computer and CRT Dis-
    play System in BWR Nuclear Power Station." Ibid, Paper
    4-3.

27. Shukla, J. N. and Wong, R. H., "Nuclenet Control Complex."
    Ibid, Paper 4-5.

28. Mazzetti, M. S., "Designing an Operator-oriented Nuclear
    Plant Control Station." Ibid, Paper 3-2.

29. Smith, J. E., Ashwell, R. E. and West, R. G., "Control
    Panel Evolution in the Canadian Nuclear Power Plant
    Program." Ibid, Paper 1-4.

30. Fenton, E. and Mahon, H., "Design Aspects of Multi-Unit
    Control Centers for Ontario Hydro's Nuclear Generating
    Stations." Ibid, Paper 1-6.

31. Frezel, W. J. and Lilly, G. M., "Application of Modern
    Control Room Principles to Westinghouse Nuclear Steam
    Supply Systems." Ibid, Paper 5-7.

32. Owens, G. R., "Applying Computers and Related Technology
    Progressively Toward TVA's First Advanced Control Room
    Designs." Ibid, Paper 4-7.

33. Spurgin, A. J., "IEEE Standards Relating to Control
    Room Design." Ibid, Paper 5-1.

34. Luna, S. F. and Williams, S. J., "A Multi-purpose Class
    IE Cabinet Design." Ibid, Paper 5-2.

35. Reisch, F., "Swedish Standards Require Separation to
    Prevent Spreading of Cable Fires." Ibid, Paper 5-3.

36. Söderman, E., and Myren, I., "Lighting and Fire Pro-
    tection in a Twin BWR Power Station." Ibid, Paper 5-4.

37. Shultz, T. S. and Brown, E. M., "Power Plant Computer
    System Reliability and Availability." Ibid, Paper 5-6.

38. Corcoran, P. J., "Arrangement of Control Building Com-
    plex." Ibid, Paper 5-5.

39. Goodstein, L. P. and Pedersen, L. P., "Do Control Room
    Designers Have Adequate Bases for Computer Displays?"
    Ibid, Paper 4-1.

40. Worth, G. A. and Patterson, J. R., "Computer Applica-
    tions for the Fast Flux Test Facility." International
    Atomic Energy Agency Specialists' Meeting on Use of
    Computers for Protection Systems and Automatic Control,
    Munich, May, 1976, Paper 5-6.

41. Netland, K. and Øvreeide, M., "A Computerized Super-
    vision and Control Concept for Nuclear Power Plants
    Based on Experience Obtained from Operation of the
    Halden Reactor." Ibid, Paper 4-4.

42. Amiel, J. and Guesnier, G., and Chambrette, M., Presen-
    tation of the Complementary Information Processing by
    Computer before Equipping the 900 MW Plants in Construc-
    tion." Ibid, Paper 4-5.

43. Grumbach, R. and Hoermann, H., "Plant Disturbance Anal-
    ysis by Process Computer - Basic Development and Experi-
    mental Tests." Ibid, Paper 1-6.

44. Dahll, G., Grumbach, R. and Netland, K., "Plant Distur-
    bance Analysis by Process Computer-on-line Operation
    and Operator Communication." Ibid, Paper 1-7.

45. Zwingelstein, G., "Supervision and Diagnosis of the
    Operation of Nuclear Reactors by Real-time Identifi-
    cation Techniques." Ibid, Paper 1-1.

46. Hashmi, J., Iqleem, J., Jafri, M. N. and Siddiqui, Z.
    H., "Experience in the Operation and Maintenance of a
    Digital Computer Control System of a Nuclear Power
    Plant." Ibid, Paper 3-2.

47. Hanke, W., "A Computerized Strategy for Controlling
    Xenon Oscillations." Ibid, Paper 2-4.

48. Karppinen, J., Blomsnes, B. and Verluis, R. M., "Two
    Optimal Control Methods for PWR Core Control." Ibid,
    Paper 2-5.

49. Lawrence, L. A. J., Page, N. and Wells, F., "A Study of
    Failure-Survival Closed-Loop Plant Autocontrol Systems
    Employing Microprocessors." Ibid, Paper 4-3.

50. Oeder, K., "The Formation of Safety Parameters with the
    Digital Calculating Module." Ibid, Paper 1-3.

51. Johnstone, W. and Pooley, D., "The Determination and
    Use of Sensor Status Information in Computer-based
    Monitoring and Protection Systems." Ibid, Paper 5-2.

52. Grisollet, J. and Quenee, R., "The Computers Processing
    the Temperatures in the Core of PHENIX." Ibid, Paper
    5-7.

53. Jeannot, A., Grisollet, J., Gourdon, M. and Paiziaud,
    A., "Computers in the Core Supervision and Protection
    System at Creys Malville (SUPER-PHENIX)." Ibid, Paper
    5-8.

54. McAffer, N. T., "Microprocessors for Plant Protection."
    Ibid, Paper 5-3.

55. Keats, A. B., "A Failsafe Computer-based Reactor Pro-
    tection System." Ibid, Paper 5-4.

56. Thomasson, R. K. and Kneller, B., "Design Study For a
     Computer-based System for Fast Reactor Protection."
     Ibid, Paper 5-5.

57. Jungst, U., "Design Features of the Fuel Element Com-
     puterized Protection System BESSY." Ibid, Paper 6-3.

58. Edelmann, M., "Two On-line Methods for Routine Test-
     ing of Neutron and Temperature Instrumentation of
     Power Reactors." Ibid, Paper 5-1.

59. Gallagher, J. M. and Lilly, G. M., "System Architecture
     for Microprocessor-based Protection System." Ibid,
     Paper 4-1.

60. Beltracchi, L. and Bullock, J. B., "Safety Evaluation
     Experience with Digital Computer Software." Ibid, Paper
     6-1; alsim Nuclear Safety 176, pp 693-700, Nov., 1976.

61. Voges, U., "Aspects of the Design and Verification of
     Software for a Computerized Reactor Protection System."
     Ibid, Paper 6-2.

62. Volkmann, K. P., Hoermann, H. and Ehrenberger, W.,
     "Statistical Test Data Selection for Reliability Eval-
     uation of Process Computer Software." Ibid, Paper 6-4.

63. Simon, A., "Computer-aided Commissioning of the Reactor
     Control System." Ibid, Paper 3-3.

64. Champiot, G. and Varaldi, G., "Study of the Analog Low-
     Level Data Acquisition System Used in the 900 MW
     Plants." Ibid, Paper 4-6.

65. Hassan, M. A., Ibrahim, M. K., Madkour, M. A. and
     Ghonaimy, M. A. R., "A Micro/Mini Computer Network for
     Applications in Nuclear Station Instrumentation and
     Control." Ibid, Paper 4-2.

66. Chou, Q. B. and Stokes, H. W., "Computer Functions in
     Overall Plant Control of CANDU Generating Stations."
     Ibid, Paper 3-1.

67. Cundall, C. M., "CRT Display Systems," _Electronics and Power_ 14, pp 115-120, 1968.

68. Groner, G. F., "A Guide to Display Terminals that Enhance Man-Computer Communications," Santa Monica: _Rand Corporation Report R-1183_, 93 pages, January, 1973.

69. Umbers, I. G., "CRT/TV Displays in the Control of Process Plants: A Review of Applications and Human Factors Design Criteria," UK Department of Industry, _Warren Springs Laboratory Report, LR242(CON)_, 1976.

70. Cropper, A. G. and Evans, S. I. W., "Ergonomics and Computer Display Design," _The Computer Bulletin_, July, 1968, pp 94-98.

71. Rouse, W. B., "Design of Man-Computer Interfaces for On-line Interactive Systems," Proceedings IEEE 63, No. 6, pp 847-857, June, 1975.

72. Jervis, M. W. and Butterfield, R. J., "The Use of Alphanumeric CRT Displays in CEGB Power Stations," IEE Conference on _Man-Computer Interaction,_ September, 1970.

73. Jervis, M. W., "CRT Displays of Graphs in CEGB Power Stations," IEE _Displays_ Conference, September, 1971.

74. Goodstein, L. P., "Man-Machine Studies at AEK, Riso. OECD Halden Reactor Project Enlarged Halden Program Group Meeting on Computer Control, Leon, Norway, May-June, 1972, Report _HPR 161_, 1973, Paper 28.

75. Netland, K., Hveding, T., Augustin, J., Strøm, S., "A Computer-based Operator/Process Communication System," Ibid, Paper 29.

76. Andersen, A. B., Hernes, T., Hol, J. Ø., Netland, K., Petterøe, A., Sletvold, B., Thomassen, B. B., Øhra, G., "Application of a Computer-based Operator Communication System in the Operation of the Halden Reactor." Ibid, Paper 30.

77. Lindgren, P., Andersson, S. A., "The Use of Process Computers in ASEA-ATOM BWR Power Plants." Ibid, Paper 5.

78. Blomsnes, B., Karlsson, R., Sato, K and Stoffelsma, A., "A Computer Control Concept for Load-Follow Control of a Nuclear Plant." Ibid, Paper 8.

79. Böhm, B., Schrüfer, E., "Application of Process Computers in a Reactor Protection System." Ibid, Paper 32.

80. Schüller, H., "Self-checking Features of a Process Computer." Ibid, Paper 34.

81. Jörgensen, U. Scot, Korpås, T. H., Ueberall, J., "Software Development for a REactor Safety System Based upon Computers." Ibid, Paper 33.

82. Ehrenberger, W., Soklic, M., "A Hybrid Method for Testing Process Computer Performance." Ibid, Paper 35.

83. Nevalainen, M., "A Quantitative Method for Determining the Role of the Computer in Nuclear Power Stations." Ibid, Paper 7.

84. Dahll, G., "Reliability Analysis for Multicomputer Systems." Ibid, Paper 22.

85. Bock, H. W., Corpus, W., and Profos, D., "Application of a Dual Processor Concept to the Reactor Power Control at Stade Nuclear Power Station." Ibid, Paper 2.

86. Sigurdson, E. L., Endresen, E., Kjeldstad, J. K. and Lunde, J. E., "DEMP-A Multiple Computer System Built for High Reliability." Ibid, Paper 18.

87. Laakso, O. and Koponen, P., "Availability Test Specification for the Computer System of Nuclear Plant Loviisa." Ibid, Paper 3.

88. Øvreeide, M., Palmgren, T., Kvallheim, O., and Berge, G. K., "System Software Organization for the Multi-computer System DEMP." Ibid, Paper 19.

89. Kishi, S., Nagaoka, Y., Yoneda, Y, Fukuzaki, T., Kiyokawa, K., Serizawa, M., Nigawara, S., "Plant Monitoring by Color CRT Displays for Boiling Water Reactor," Hitachi Review 25 (8), pp 265-270, 1976.

90. Preliminary Study of the Advantages to be Gained by the Use of Color Coded Information on Electronic Data and Radar Displays for Air Traffic Control. Procurement Executive Ministry of Defence Report JFA/0181, Dec.,1972.

91. Christ, R. E. and Teichner, W. H., "Color Research for Visual Displays (U)," AD-767 066, New Mexico State University, Department of Psychology F/G 5/5. Contract N00014-70-A-0147-0003, Unclassified. MMSU-JANAIR-FR-73-1, JANAIR 730703, July, 1973.

92. De Mars, S. E., "Human Factors Considerations for the Use of Color in Display Systems," North America Space Administration (NASA) Technical Report (TM-X-72196, TR-1329), Page 36, January 5, 1975.

93. "Performance Requirements for Electrical Alarm Annunciator Systems," London: Engineering Equipment Users Association, EEUA, Document 45D, 1972.

94. "Specification and Guides for the Use of General Purpose Annunciators," Instrument Society of America, Document ISA-RP18.1, 1965.

95. Popp, H. and Schwarz, H. G., "Acquisition, Processing and Evaluation of Signals in Thermal Power Stations," Siemens Review, 36, PP 95-100, 1969.

96. Andreiev, N., "Annunciators Hold Ground Against the CRT," Control Engineering, PP 46-48, March, 1976.

97. Patterson, D., "Application of a Computerized Alarm-Analysis System to a Nuclear Power Station," Proceedings IEE 115, (12), PP 1858-1864, 1968.

98. Kay, P. C. M. and Heywood, P. W., "Alarm Analysis and Indication at Oldbury Nuclear Power Station," in Automatic Control in Electricity Supply, IEE Conference Publication 16, Part 1, PP 295-317, 1966.

99. Welbourne, D., "Alarm Analysis and Display at Wylfa Nuclear Power Station," Proceedings IEE 115, (11), PP 1726-1732, 1968.

100. Graf, G., "Alarm Design Concept with Data Display Units
     Using Computers," Presented at the IAEA NPPCI Special-
     ist Meeting, Brussels, Belgium, Paper PL481/9, 1971.

101. Turner, G., "On-line Programs from Control Logic Dia-
     grams," Control Engineering 15, PP 102-104, 1968.

102. Ulman, R. P. and Vetter, J., "Computerized Radiation
     Monitoring," Power Engineering, 80.7, PP 55-57, July,
     1976.

103. Dawson, R. E. B. and Jervis, M. W., "Instrumentation
     at Berkeley Nuclear Power Station," J. British IRE,
     PP 17-33, 1 January 1962.

104. Union, D. C., "Nuclear Power Plant Computer System
     With Remote Multiplexing," American Power Conference,
     Chicago, April/May, 1974.

105. Corbett, B. L., "Fuel Meltdown at St. Laurent 1,"
     Nuclear Safety, 12, PP 35-39, 1971.

106. Cummings, J. D. and Butterfield, M. A., "Application of
     Modern Control Theory in Nuclear Power," IAEA Symposium
     SM/168, January, 1973.

107. Grumbach, R. and Blomsness, B., "Development and Appli-
     cation of Advanced Concepts for Nuclear Plant and Core
     Control," IAEA Symposium Symposium SM/168, January, 1973.

108. Jolly, M. E. and Wreathall, J., "Principles and Practice
     of Reactor Safety Systems," Nuclear Eng. Int., PP 42-
     45, February, 1976.

109. Welbourne, D., "Computers for Reactor Safety Systems,"
     Nuclear Eng. Int., PP 945-949, November, 1974.

110. IEC 2131A. Supplement to IEC 231, General Principles
     of Nuclear Reactor Instrumentation, 1969.

111. US NRC Document 10CFR, Part 50, "Licensing of Production
     and Utilization Facilities," Section 50.55(a).

112. "Criteria for Nuclear Power Plant Protection Systems,"
     IEEE Std 279-1971: ANSI N42.7, 1972.

113. "Safety Criteria for Nuclear Power Stations," Ministry of the Interior, Reactor Safety Institute of the Technische Überwachungs - Verieine eV, ANS N41.2, 1972.

114. Damon, D. L., "Application of Minicomputers to Testing Safety Systems," IEEE Paper 73/94.

115. Jacquin, et al. "Commissioning of Control and Monitoring Equipment Using Analog Simulation," Automatism 13, 1968.

116. Craik, N. G., Cloutier, P. M., and Duchesne, A. J., "Testing of the Computer Control of the Reactor Power System at the Gentilly Nuclear Power Station," J. Br. Nuclear Energy Soc. 12, 1, PP 85-94, 1973.

117. "Official Definition of CORAL 66," HMSO, 1970.

118. Welbourne, D., "A Limited Control Language for DDC," IEE Conference Publication 102, Software for Control, July, 1973.

119. Taylor, J. G., "The Site Construction Phase with Particular Reference to Hinkley Point 'B' Power Station," Proceedings Second Conference on Trends in On-line Computer Control Systems, University of Sheffield, 1975, IEE Conference Publication No. 127, PP 58-65, 1975.

120. Shirra, J. M. and Bull, M. G., "A High Security Computer System for Nuclear Power Plant Monitoring and Control," Nuclex 72, Paper 10/17, 1972.

121. Welbourne, D. and Graham, G. V., "Hartlepool-AGR Survey-Computer Control Applications," Nuclear Eng. Int. 15, PP 988-990, 1969.

122. Fergus, P. J. B. and Taylor, J. M., "High Integrity Systems Atomic Energy REsearch Establishment Report," No. R6940, 1971.

123. Reid, J. B. and Selmeczy, J. C., "Computers in Nuclear Plants Beyond 1985," IEEE PES Winter Meeting January, Paper C75 156-5, 1975.

124. Morita, K., Takuma, Y., Kitanosono, H. and Koyama, T.,
     "Progress of Applications and Technology of Hitachi
     Control Computer," Hitachi Review 256, PP 197-204, 1976.

125. Kitanosono, H. and Nigawara, S., "Computer Control Sys-
     tems," Hitachi Review 256, PP 205-212, 1976.

126. Kuwabara, H., Ide, J. and Hirai, K., "Hitachi Control
     Computer 80 System," HItachi Review 256, 213-222, 1976.

127. Bisby, H., "The CAMAC Interface and Some Applications,"
     Radio & Electronics Eng. 41, PP 527-537, 1971.

128. Peatfield, A. C., "The Potential of a Standard Highway-
     Interface (CAMAC) for Computer Systems in Real-time
     Applications," in Trends in On-line Computer Control
     Systems, IEE Conference Publication 85, PP 142-146,
     1972.

129. Welbourne, D., "Data Processing and Control by a Com-
     puter at Wylfa Nuclear Power Station," Proceedings
     Institute Mechanical Engineering 179, PP 131-140, 1965.

130. Makin, J. E. and Shirra, J. M., "Digital Computer
     Application to AGR Systems," Proceedings of the Nuclex
     Conference, Basel, Switzerland, PP 1-16, 1969.

131. Morrish, M. F. G., "Experience with DDC at Hinkley
     Point 'B' AGR Power Station," IEE Colloquium-Operating
     Experience with DDC Systems, May, 1977.

132. Williams, J. R., "The Design and Implementation of a
     Fail Safe On-line Computer Control System for Dunge-
     ness 'B' Power Station," in Trends in On-line Computer
     Control Systems, IEE Conference, Publication 85,
     pp 121-126, 1972.

133. Williams, J. R., Simm, K. J. and Hinze, P., "The
     Application of Computers in Nuclear Power Stations,"
     J. Institute Nuclear Engineers, 18, 3, pp 74-80, May-
     June, 1977.

134. Lennox, C. G. and Pearson, A., "High-speed Monitor for
     Closed Loop Control," Nucleonics 20, PP 73-74, 1962.

135. Pearson, A., "Control and Instrumentation on Canadian Nuclear Power Plants," IAEA NPPCI First Working Group Paper PL431/2, 1971.

136. Smith, J. E. and Morris, D. I., "Prospects for Computer Control in Nuclear Power Plants," Nuclex 69 Technical Meeting Report 6/6, 1969.

137. Davis, H. L., "Instrumentation and Control for Canadian Nuclear Reactors in Instrumentation for Nuclear Power Plant Control, 119, IAEA, 1970.

138.. Davis, H. L., "Toward Closed Loop Control in Nuclear Plants," Nucleonics, 20, PP 71-72, 1962.

139. Koekebakker, J., "Canadian Experience in DDC of Nuclear Power Stations," Power Engineering, May, PP 42-44, 1975.

140. Mahood, T. B., "Computer Control in Pickering Nuclear Power Station Nuclex 72, Paper 10/8, 1972.

141. Mahood, T. B., "Computer Control at Pickering," Nuclear Eng. Int., pp 31-36, January, 1973.

142. Morris, D. I., "Computer Control at Bruce Nuclear Generating Station," Nuclear Safety 15, 6, PP 691-701, November/December, 1974.

143. Cooper, W. R. and Whittall, W. R., "Direct Digital Control of the Gentilly Nuclear Power Station," Proceedings of the 1973 Conference American Nuclear Society, Ann Arbor, Michigan, VIII-20 - 35, 1973.

144. Luopa, J. A. and Anderton, R. D., "Direct Digital Control of a Nuclear Power Reactor," IEEE Transactions Nuclear Science, NS17, Part 1, PP 586-593, February, 1970.

145. Nowicki, S. J., Hubbard, D. G. and Jones, R. T., "On-line Data Acquisition and Processing for the HTGR," IEE Transactions on Nuclear Science NS-20, Paper 73/84, PP 717-723, February, 1973.

146. Mummery, P. W., Evans, A. D. and Barclay, F. J., "The PFR Station Control System," Paper 10/7, Nuclex 72, Basle, Page 8, 1972.

147. Rippon, S. E., "Taking the Complexity Out of the Control Complex," Nuclear Eng. Int., PP 36-39, January, 1973.

148. Bryden, J. E., "Some Notes on Measuring Performance of Phosphors Used in CRT Displays," 7th National Symposium Information Display, Page 83.

149. Patalon, W., "Sequoyah Nuclear Power Plant," Nuclear Eng. Int. 17, PP 845-869, 1971.

150. Green, A. E. and Bourne, A. J., Reliability Technology, John Wiley & Sons Ltd, London, 1972.

151. Jervis, M. W., "Power Station Computer Systems - Inquiry Specification and Tender Assessment," Electronics and Power (IEE), PP 180-182, February, 1975.

152. Welch, B. R., "Planning of Power Station Computer Systems," Organization and Management of Computer-based Control and Automation Projects, IEE, London, October, 1973.

153. Riley, W. B., "Interface Pact Gains Momentum," Electronics, PP 115-116, October 26, 1970.

154. Williams, W. D., "Initiating a Computer Control Project," Control Engineering, PP 87-90, September, 1974.

155. Nail, J. H., "Quick Access to Nuclear Data Bank," Power Engineering, PP 44-47, December, 1973.

156. L'Archeveque, J. V. R. and Yan, G., "On the Selection of Architectures for Distributed Computer Systems in Real-time Applications," IEEE Nuclear Science Symposium, New Orleans, October, 1976.

157. Capel, A. C. and Yan, G., "An Experimental Distributed System Development Facility," IEEE Nuclear Science Symposium, New Orleans, October, 1976.

158. Bristol, E. H., and Wade, H. L., "Alarm Analysis Can Diagnose System Faults," Control Engineering, PP 47-49, February, 1976.

159. Aspinall, D., "Microprocessors - New Components for the Electronics Engineer," Electronics and Power, PP 437-443, July, 1976.

160. Vacroux, A. G., "Microcomputers," Scientific American, PP 32-40, 1975.

161. Feth, G. C., "Memories: Smaller, Faster, Cheaper," IEEE Spectrum, PP 37-43, June, 1976.

162. Dallimonti, R., "Human Factors in Control Center Design," Instrumentation Technology 23, PP 39-44, May, 1976.

163. McGowan, T. E. and Bruno, S. J., "Ergonomic Considerations in the Design of Power Plant Control Rooms," Joint IEEE/ASME/ASCE Power Generation Technical Conference, Portland, Paper PG 75 635-3, 1975.

164. Bear, D. E., "Plant Operator's Computer Interface," Instrumentation Technology 22, PP 29-34, October, 1975.

FUEL FOR THE SGHWR

D. O. Pickman, J. H. Gittus and K. M. Rose

UKAEA, Reactor Fuel Element Laboratories
Springfields, Preston, Lancashire, United Kingdom

I. INTRODUCTION

The Steam Generating Heavy Water Reactor system (SGHWR) was selected for development in the United Kingdom as an alternative to the Commercial Advanced Gas-Cooled Reactor (CAGR) in the late 1950's. As an alternative, it was required to have diverse technical features such as coolant, moderator and pressure containment.

Contstruction of a prototype 100 MW(e) reactor was started in May, 1963, and full power was attained at the end of 1967. This reactor is situated at Winfrith Heath in Dorset, England. Construction of six 660 MW(e) commercial reactors was authorized in August, 1974 (1) four of which are to be built at Sizewell in Suffolk for the CEGB and two at Torness in East Lothian, Scotland, for SSEB.

The engineering design of the Winfrith SGHWR has been described by Bradley et al. (2). Some changes have been made to the commercial design (CSGHWR) as a result of experience and of the more stringent safety criteria now prevailing, but these are changes of detail, or sometimes materials, and the original design principles are retained throughout (3).

The SGHWR system is a pressure tube boiling light water reactor with steam drums from which the separated steam

passes directly to the turbine. The condensate is returned
directly to the steam drums via a feed train with full flow
ion exchange purification to limit the buildup of impurities
in the primary circuit and to prevent excessive crud deposition on the fuel elements.

The primary circuit in the Winfrith SGHWR is austenitic
stainless steel, but will be ferritic in the commercial reactors on grounds of technical preference. Key parts of the
feed system are in austenitic stainless steel; copper alloys
are minized because of deleterious effects on fuel element
crud deposits. No additives are made to the coolant and the
system is self-pressurizing as steam is generated, so that
hydrogen overpressure is not required. As a result, the
oxygen content of the water at core entry is relatively high,
0.05-0.08 NTP cm^3/kg, which influences corrosion behavior of
the fuel element cladding.

The heavy water moderator is contained in a separate
unpressurized calandria tank with zirconium alloy internals
and stainless steel end plates and barrel. The nuclear design parameters of importance are H_2O/UO_2 and D_2O/UO_2 volume
ratios and enrichment. Their choice affects system economics,
control and safety. Control is effected by boric acid additions to the D_2O and by small moderator height adjustments.
In larger reactors, some interlattice tubes are used for power shaping by liquid absorbers. Shutdown is by rapid injection of liquid absorbers in other interlattice tubes with
a secondary D_2O dump. There are, therefore, no movable control rods. The axial flux shape is a very symmetrical chopped
cosine with only minor perturbations from the small D_2O level
changes. Power changing to accommodate grid frequency changes
or system demand does not distort the cosine distribution,
thus avoiding local power peaking, and in steady operation,
no power peaks sweep along the fuel from control rod movement. The effect is to minimize damaging fuel/clad interaction in the fuel pins and also to minimize fission gas release.

The SGHWR operates with either a four-batch or eight-
batch fuel cycle, giving all fuel a four-year residence time
in reactor. Refueling is off load at six or twelve month intervals and initial cores require some burnable poison
(UO_2/Gd_2O_3) to hold down reactivity. Fuel is differentially
enriched, and two different levels of feed enrichment are

used in outer and inner zones. There is no fuel shuffling,
and new fuel peaking factors are in the region of 10% or
less, which imposes no significant power ramp on partially
irradiated fuel.

The SGHWR has engineered safety systems to prevent excessive release of activity under any credible accident condition. Blowdown studies show that with critical size pipe
rupture in particular positions, bidirectional flow to the
breach may result in coolant flow stagnation in some fuel
channels. Although such stagnation conditions do not persist,
and are likely to be accompanied by some convective cooling,
it is a requirement that fuel temperature rise must be controlled by an independent emergency cooling system. The
system adopted is to provide a separate supply of cooling
water into each fuel channel known as the emergency cooling
water supply (ECW). This supply is piped via the standpipe
and hanger bar into the center of each fuel element, and
flow is initiated in the event of a breach in the primary
circuit. The quantity is in the range of 5% - 10% of normal full channel flow.

II. FUEL DESIGN

The design of nuclear fuel elements for SGHWR, as for
other water-cooled reactors, has two major components, the
fuel rod that contains the fissile material, and the structure that retains the array of fuel rods in an optimum configuration for physics and thermal-hydraulic purposes. There
are certain design criteria that are set to ensure safe and
reliable performance over the four-year dwell. A high degree
of integrity is necessary to avoid the need to prematurely
discharge leaking fuel, despite the fact that in the SGHWR
system, such fuel can be easily and positively identified
by the individual channel detection system and readily removed within a very short shutdown period.

In recent years, fuel element design has been increasingly dominated by the absolute necessity of minimizing release of fission products in any conceivable accident. The
criteria that ensure this become essentially fuel design
criteria.

In this account of the SGHWR fuel element, a short, general description is followed by more detailed accounts of

the design principles of the fuel rod and the structure.
Attention is concentrated on the well-proven, 36-rod fuel
element, but some reference is made to the 60-rod fuel element currently under development for the commercial SGHWR's.

A.   General Design Description

     The circular pressure tubes of the SGHWR require a cir-
cular array of fuel rods, and the fuel-handling arrangement
makes it convenient to hang the fuel elements from the individual channel seal plugs. A neutron shield plug is an integral part of the hanger bar which also serves to supply
the emergency cooling water to the fuel element.

     The fuel element is subdivided into a 37-rod bundle,
the central position being occupied by a structural tube
known as a sparge tube which serves also to distribute the
ECW throughout the element via a series of carefully-
positioned spray holes. The fuel rods are arranged in
three rings of 6, 12 and 18, with an additional outer row
of 24 in the 60-rod version. All rods are bolted to a
strong, stainless steel top support grid with auxiliary
holes for coolant passage, as shown in Figure 1; they are
not required to serve as axial support members. At the
bottom end, the fuel rods are free to move axially, but are
supported in their designed spatial locations by a spring-
type support grid that forms a part of the nose unit, as
shown in Figure 2. This includes an arrestor plate to retain parts of broken fuel rods and a split ring to stabilize
the fuel element in the pressure tube. Anti-bowing support
is provided by intermediate spacer grids that may be of two
types, both of which have been extensively tested in the reactor. The ring spacer used in most 36-rod elements is a
simple, lightweight grid made up of stainless steel ferrules
brazed together and with an extended central ferrule by which
it is riveted to the central structural tube, as shown in
Figure 3. Fuel rods are centered in the ferrules with a
small clearance by three equispaced pads, resistance-welded
to the fuel cladding, as is shown in Figure 4. The spring
spacer used in some 36-rod elements and proposed for the
commercial 60-rod element is generally similar in construction (see Figure 5), but has extended support members carrying double-sided leaf springs. The ferrules have elongated
dimples that act as fixed supports for the fuel pin. There
are one spring and two fixed supports per rod at each inter-

mediate grid, the contacts being at the same axial level.

   In the 60-rod design, additional emergency cooling
water is supplied by six outer sparge tubes in interlattice
positions between the two outer rows of rods. These are
fed from a distributor attached to the top support grid,
as shown in Figure 6. In addition, fuel rod length has been
increased by 160 mm to accommodate a larger gas plenum.

   General views of the two types of fuel element are
shown in Figure 7, and the leading dimensions in Table I.

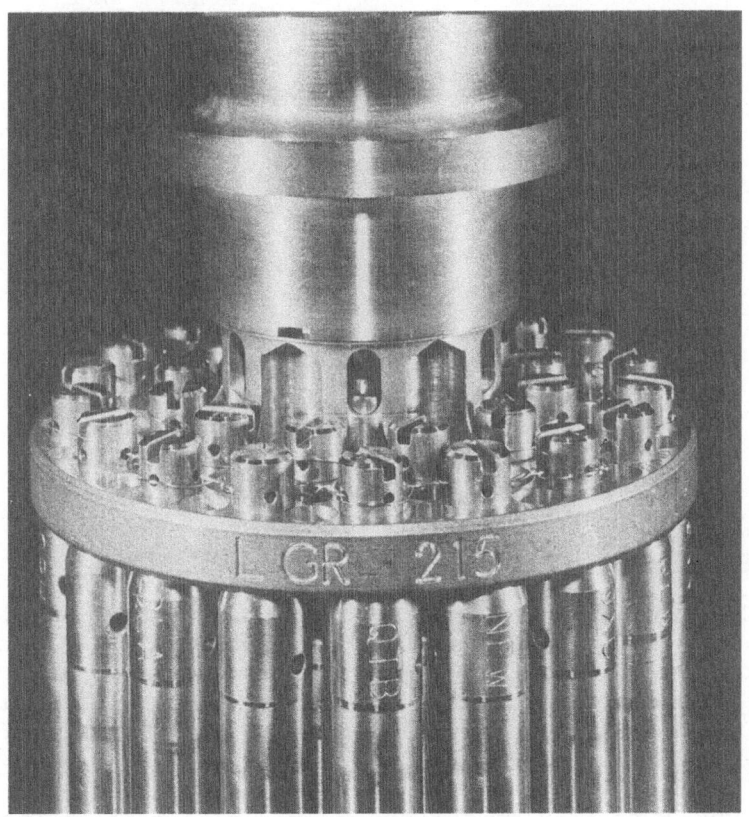

Figure 1

Top Support Grid (36-Rod Element)

Figure 2

Nose Unit (36-Rod Element)

B.   Fuel Rod Design

All SGHWR fuel rods have from the beginning been designed with safety under accident conditions in mind.   The
cladding material is cold-worked, stress-relieved, Zircaloy-
2, but experience with fully annealed material is being
accumulated and the preferred condition for CSGHWR will be
decided later. Fuel rod diameter was fixed at 16 mm for the
36-rod design as a result of a cost optimization, but has
been reduced to 12.2 mm for the 60-rod design to increase
margins in a loss-of-coolant accident (LOCA). The cladding
is designed to be free-standing under the 67-bar coolant
pressure and to have a low creepdown rate under normal oper-

Figure 4

Detail of 36-Rod Element at Spacer Grid
Showing Projection-Welded Spacer Pads

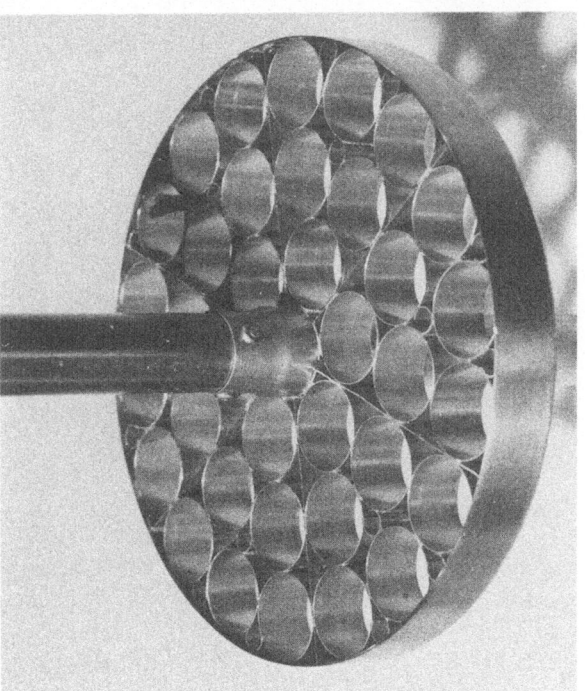

Figure 3

Ring Spacer Grid (36-Rod Element)

ating conditions (0.025 mm/year) so that the fuel rods will
perform well under conditions of large fast power changes
(power ramps). Irradiation creep effects and corrosion
losses are taken into account in fixing the thicknesses at
0.635 mm (36-rod) and 0.60 mm (60-rod).

The UO_2 is in the form of sintered pellets with a length-
diameter ratio close to unity and with double end-dishes to
accommodate differential thermal expansion and some of the
UO_2 swelling. UO_2 density is 10.6 Mg/m^3 nominal (97% theoretical). At this density, and with the method of manufacture
adopted, there is virtually no open porosity, considerably
easing the problem of controlling moisture content within the
tight limits necessary to avoid hydride defects in operation.
Structural factors, including grain size, can influence fission
gas release and a minimum of 5 μm is specified, together with
a control on type and size of pores. The stoichiometry is
also important in influencing thermal conductivity and
UO_2/H_2O compatibility and is held within limits of 2.000-
2.003.

Figure 5

Spring Spacer Grid

Fuel-to-clad gap is 0.166 mm (36-rod) and 0.125 mm
(60-rod). The 36-rod design therefore has a positive fuel-
clad diametral interaction at ratings toward the top end of
the spectrum and for gaps in the lower half of the tolerance
band. In the 60-rod design there is no initial mechanical
interaction, but gaps are very small at power. A maximum
start of life diametral strain of 0.5% is permitted. This
close gap helps appreciably to minimize fission gas release
and stored thermal energy.

Fuel rods are sealed by resistance welds with an applied
upset force and the resultant external flash is machined off.
Atmospheric pressure filling is with helium, together with a
small proportion of Kr85. The helium gives a good gap conductance in the early stages and the active tracer permits
a check on gas filling and on leak tightness at any stage
after the initial mass spectrometer leak test. The gas
plenum for storage of released fission gases is at the top
of the fuel rod and contains a light, stainless steel helical
spring to prevent gross pellet movement during transport and
handling. It has no designed operational function, but does
continue to apply some force to the top of the UO_2 stack
through most of the rod life.

Fuel cladding is not pickled or autoclaved, either inside or out, and any residual fluorine from manufacturing
operations is removed by grit blasting the bore and belt
grinding the outer surface.

The design criteria for the fuel rods are:

1. The cladding must be elastically stable under
   coolant pressure throughout the irradiation
   life. The critical thickness/diameter ratio
   is 0.035, which gives minimum wall thickness
   values of 0.56 mm (36-rod) and 0.43 mm (60-rod).

2. Cladding start-of-life hoop strain is not to
   exceed 0.5%, and end-of-life (or peak in life)
   hoop strain is not to exceed 1%.

3. Peak-in-time center UO_2 temperature is not to
   exceed 2500°C to provide an adequate margin to
   center melting at 2800°C.

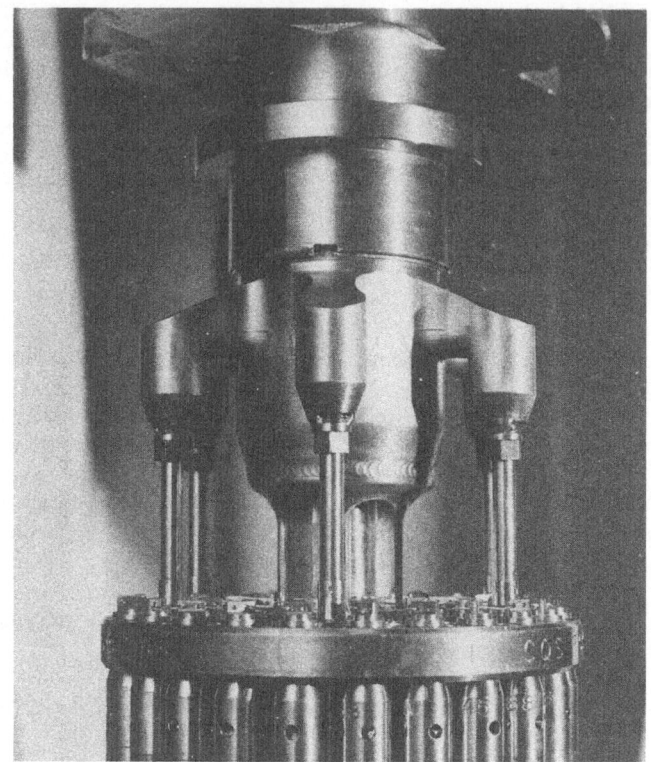

Figure 6

Top Support Grid of 60-Rod Element
Showing Projection-welded Spacer Pads

4. Peak internal gas pressure in operation is not
    to exceed 5 MPa.

There are, however, additional operational and safety
criteria that may and sometimes do override the basic design
criteria. The operational criteria are:

1. Local cladding wall thinning as a result of power
    cycling not to exceed 10%.

2. Peak cladding stress as a result of power ramps
    not to exceed 480 MPa.

a   b

Figure 7

General View of 36-Rod and 60-Rod Elements

TABLE I

SGHWR FUEL ELEMENT DESIGN PARAMETERS

	36-Rod	60-Rod
Rod distribution	3 rows, 6,12,18	4 rows, 6,12,18,24
Sparge tubes	1 central	1 central,6 outer
Spacer grids	Ring with spacer pads(11)	Spring (9)
Distance bet. grids	305 mm	394.5 mm
Fuel element length	4.162 m	4.315 m
Fuel element weight	0.3 Mg	0.250 Mg
Weight of UO_2	0.1939 Mg	0.1774 Mg
Fuel rod length	3.878 m	3.966 m
Pellet stack length	3.582 m	3.582 m
Gas plenum volume	42 cm^3	31 cm^3
Filling gas	He/Kr at 1 atm	He/Kr at 1 atm
Weight of fuel rod.	7.000 kg	4.005 kg
Weight of U	5.385 kg	2.957 kg
Fuel cladding	CW/SR Zircaloy-2	CW/SR Zircaloy-2
Clad O. D.	16 mm	12.2 mm
Clad wall thickness(min)	0.635 mm	0.597 mm
Fuel/clad gap (dia.)	0.165 mm	0.125 mm
UO_2	Enriched	Enriched
UO_2 density	10.6 Mg/m^3	10.6 Mg/m^3
Pellet length	15.24 mm	12.7 mm
Pellet diameter	14.53 mm	10.75 mm
Pellet dish volume	2.9%	2.77%
Number of pellets	235	282

The safety criteria in a worst credible accident are:

1. Peak cladding temperature not to exceed 1200°C.

2. Peak cladding hoop strain not to exceed 5%.

3. Cladding embrittlement not to result in cracking
    on quenching at any stage of the transient.

C. Fuel Element Structural Design

Many factors are involved in the structural design of
fuel elements. These have been reviewed in the context of
the SGHWR by Pickman (4).

The structure of the SGHWR fuel element has been designed to provide adequate support to the fuel rods in reactor and during transport and handling, including the important charge/discharge operations. Experience in reactor
and special trials have shown that provided there is no maloperation and transverse acceleration is restricted below
5 g_n, both the ring grid/spacer pad and spring grid designs
are satisfactory. The nose unit has a chamfer to provide a
lead-in to the charge machine and fuel channel, and the
outer bands of spacer grids are rolled over to prevent snag-
ging. The 36-rod element is fairly flexible and the 60-rod
element much stiffer, but both have shown satisfactory loading and general handling behavior.

The fuel stringer must have a clearance within the reactor channel, which is nominally 1.5 mm, and stability is
essential to prevent any fretting damage to the fuel and to
the pressure tubes and standpipes. There are a number of
sources of destabilizing forces including turbulent flow at
entry and within the fuel element, with cross-flow components, flow past the neutron shield plug, and flow across the
hanger bar at the riser pipe position, where the two-phase
mixture is led from the channel to the steam drum. The nose
unit is stabilized within the pressure tube by an interference fit split ring, seen in Figure 2. Development tests
and reactor experience show that there is some degree of low
frequency movement that is not damaging, but further endorsement testing is in progress to confirm the behavior of the
longer hanger bars to be used in the commercial SGHWR's. It
is feasible to introduce additional stabilizers should this
prove necessary.

In the ring grid/spacer pad design, freedom from fretting up to a limiting channel mass velocity of about 5
Mg/m^2s is achieved as a result of the hydraulic conditions
in the rod/ferrule annulus. Subject to satisfactory control
of oxidation and crud deposition, this is combined with freedom from axial interactions between fuel rods and spacer
grids that would lead to stressing of the ECW sparge tubes.
This is particularly important for the 60-rod design because
the six outer sparge tubes are in stainless steel and the
ECW water supply drops in temperature during a LOCA, generating differential thermal contraction. In the spring grid
designs, the axial interactions resulting from the rod/spring
contact must be taken into account in sizing the sparge tubes
and the various couplings to avoid unacceptable stress levels.
Fuel rod stability is a function of spring force applied and
the number of spacers. It is found that a lesser number of
spacers is required with springs than with spacer pads.

Fuel rods increase in length with irradiation and in
both length and diameter during a LOCA. Spacer pads are
sized and positioned asymmetrically, as shown in Figure 4,
to retain location in the grids at all times and rod end
clearance within the nose unit of 62 mm accommodates the
axial length changes without end interference.

III. OPERATIONAL EXPERIENCE

Typical operating parameters for SGHWR fuel elements
are given in Table II. It can be seen that the change from
36-rod to 60-rod fuel reduces peak linear rating from 44 to
34 kW/m, and peak center UO_2 temperature from 1480°C to
1256°C. This has a substantial effect in reducing fission
gas release and stored thermal energy. The flow parameters
for the 60-rod design are essentially the same as for the
36-rod. The spread of flow conditions between high power
channels at start of life and low power channels at end of
life is wide, and it is important that the extremes and
associated fuel element aging effects are covered by the
prototype reactor irradiation program and in laboratory loop
tests.

A.   Rating and Burnup

Some 382 fuel elements have been irradiated in the 104-
channel Winfrith SGHWR (WSGHWR) since the start of operation

TABLE II

SGHWR FUEL ELEMENT PERFORMANCE PARAMETERS

	36-Rod	60-Rod
Channel power, heat to coolant (best estimate)	4.43 MW	4.43 MW
Peak linear rating (best est.)	43.7 kW/m	33.8 kW/m
Nominal UO_2 center temperature	1480°C	1256°C
Maximum surface heat flux	0.89 MW/m^2	0.87 MW/m^2
Maximum outlet steam quality	16%	16%
Coolant pressure (core exit)	6.73 MPa	6.64 MPa
Coolant exit flow*	16.2 kg/s	
Goolant exit mass velocity*	1760 kg/m^{2}s	
Coolant inlet velocity*	4.53 m/s	
Coolant exit velocity*	15.60 m/s	
Coolant exit flow**	28 kg/s	
Coolant exit mass velocity**	4770 kg/m^{2}s	
Coolant inlet velocity**	6.23 m/s	
Coolant exit velocity **	7.70 m/s	
Design burnup	20 MWd/kg U	21.5 MWd/kg U

*   Peak best estimate, start of life.

**   Minimum channel, end of life.

in January, 1968. Fuel element performance has been discussed by Pickman et al. (5). A variety of fuel element
types have been irradiated, but there has been a general
progression from a Mk I to a Mk II 36-rod design, and more
recently, the Mk III 60-rod design in a spacer pad embodi-
ment. Important variants have been 36-rod spring spacer
elements and elements with fully annealed Zircaloy-2

cladding. The spring grid 60-rod design is to be introduced as the standard replacement fuel at the end of 1976.
Irradiations have covered ratings up to 80 kW/m, with a
burnup of 8 MWd/kg U, but are more normally restricted by
safety requirements to about 55 kW/m (36-rod) and 42 kW/m
(60-rod). Burnup has been restricted by limitations imposed
by design features of some of the early elements, and by the
low overall load factor (51%) of WSGHWR because of its large
experimental program; however, the design burnup of 20
MWd/kg U has now been exceeded with 36-rod fuel and 14
MWd/kg U has been reached with 60-rod fuel. Substantial
numbers of 36-rod fuel elements are close to the design
burnup and there are plans to take a number to at least
25 MWd/kg U.

B.   Coolant Flow and Steam Quality

A range of flow conditions is obtained in WSGHWR as a
result of power variations, but extreme high flow conditions
for CSGHWR are not normally attained; so four channels have
been provided with additional flow from an auxiliary pump
and have been used to endorse fuel stringer stability and
freedom from fretting. The normal peak channel exit steam
quality in WSGHWR is less than the 16% of CSGHWR, although
a limited number of high power channels have reached this
value. A number of experiments, therefore, have been performed with channel exit qualities of 20% and 30% to confirm
fuel behavior at and beyond peak random channel conditions.
In these experiments, flow has been adjusted to the appropriate values by gagging in standard channels and by flow
control valves in a full-size loop.

C.   Surveillance Program

In establishing the suitability of fuel elements for
service in large power reactors, there are a number of technical areas in which design assumptions and results of development work must be supported by full-scale reactor endorsement in the correct environmental conditions. By design, CSGHWR takes the identical fuel element as WSGHWR and
has the same system pressure; WSGHWR irradiation therefore
gives a definitive endorsement in these important areas of
behavior. One of the major roles of the WSGHWR supported
by post-irradiation examination (PIE) facilities at Wind-
scale, Winfrith and Harwell is to irradiate fuel elements

for a surveillance program. The position in areas of interest is summarized below.

Fretting. Fretting of spacer pads on fuel rods has not
been found in loop tests at coolant mass flow values up to
5 Mg/m^2s. However, in early designs without the bottom end
spring grid, severe fretting was observed at about 4.1
Mg/m^2s. These findings have been confirmed by reactor experience at similar flows. Aging effects of importance are
increasing overhang of the bottom ends of fuel pins upstream
of the bottom spring grid, and thermal and irradiation re-
laxation of the Nimonic PE16 springs. Freedom from fretting is well established to beyond 10 MWd/kg U as a result
of the surveillance program. Measurements of spring relax-
ation suggest that the minimum load of about 0.7 kg, established from loop tests, will be available with some margin
at full design burnup.

No fretting damage has been found in any spring spacer
elements irradiated in WSGHWR. There are slight wear marks
only at spring contact positions.

The type of fretting damage that occurs if a spring
load is insufficient is shown in Figure 8. The fretting
occurs mainly at the position of the fixed supports in the
spacer grid.

Dimensional Changes. Dimensional changes in fuel rods
are important in relation to axial clearance and fuel rod-
spacer grid interactions. All fuel rods in WSGHWR have been
shown to increase in length under irradiation with a consistent pattern of behavior, inner row rods (lowest rating,
highest fast flux) increasing the most and outer row rods
(highest rating, lowest fast flux) least. The maximum rate
of rod growth is around 1 mm per MWd/kg U, with an initial
increment of about 1.5 mm caused by the fuel/clad mechanical interaction. The mechanism is a combination of irradiation growth and ratcheting, contributing approximately
equally as judged by measurements taken from the unfueled
central sparge tube. The larger growth of the low-rated
fuel rods is attributed to a larger ratcheting contribution
as a result of more complete pellet relocation.

There is a reduction in UO_2 pellet stack length under
irradiation, partly as a result of UO_2 densification, partly

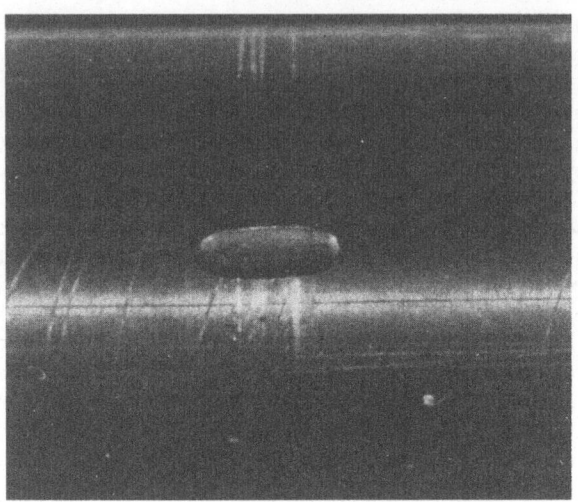

Figure 8

Fretting Damage at Position of Fixed Support in
Spring Spacer Grid Resulting from Insufficient Spring Load

from compressive creep, the magnitude of which varies, depending on the extent of interaction, but which may be as
much as twice the fuel rod length increase. Despite this
pellet stack shrinkage, no significant axial gaps (i.e.,
> 1 mm) have been observed. This is attributed to the rela-
tively small amount of densification shown by the high-
density UO_2 and the restricted pellet relocation because of
the interacting design.

Fuel rod diameter reduces steadily with irradiation,
certainly up to 15 MWd/kg U, showing a minimum that varies
in position from the bottom end to the upper third of the
rod. The minimum diameter position generally shows the
largest circumferential ridging and greatest local UO_2 stack
length decrease, probably attributable to locally greater
ratcheting. Maximum ridge heights are about 0.02 mm.

Fuel/Clad Gap. Changes in fuel/clad gap at a very
early stage of irradiation are of interest in the safety
context. Assessment of gap changes from PIE measurements

is difficult, but measurements have been made on fuel rods
irradiated in the range of 20-40 days, which show that the
maximum gap increase over a wide range of ratings is about
0.025 mm, and that at ratings above 30 W/g (42 kW/m on 36-
rod fuel) there is no increase. At these higher ratings,
some UO_2 swelling caused by grain boundary fission gas
bubbles occurs as rapidly as the densification, shown in
Figure 9, and thus effectively offsets it. There is also
some axial fuel stack compression that contributes to reducing any gap increase. No evidence for any radial relocation of pellet fragments has been obtained because observed fragment misorientations on cut sections cannot be
unambiguously interpreted.

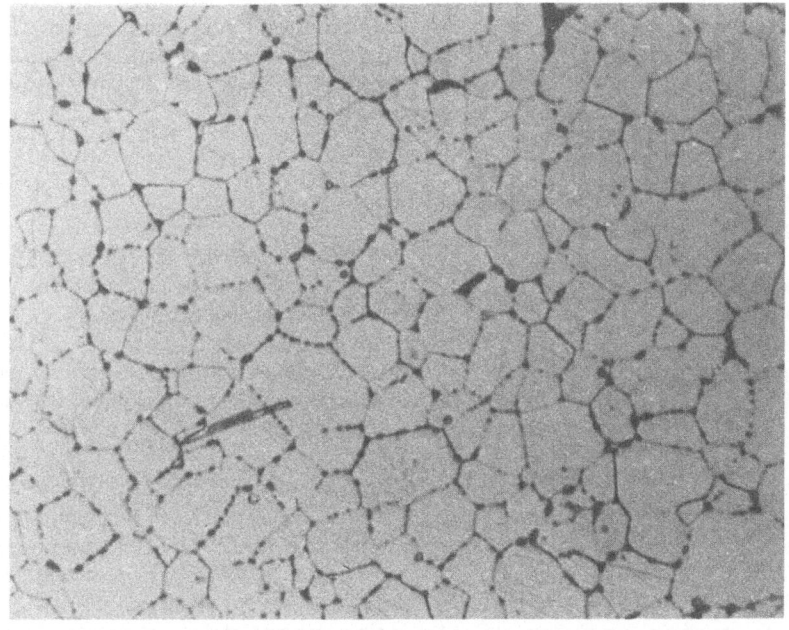

Figure 9

Grain Boundary Fission Gas Bubbles in UO_2
Irradiated in WSGHWR at 28 W/g for 24 Days

Fission Gas Release. While fission gas release from irradiated UO_2 is primarily temperature dependent, there is also a burnup dependence in the release from fuel below about 1300°C. The fission gas release in fuel rods operated at the peak linear ratings proposed for the CSGHWR's is in the range of 1% - 2%. However, gas release increases rapidly with rating above about 40 kW/m, so that the need to stay within acceptable gas pressure limits in a LOCA for peak random rating has been a major factor in the decision to develop the 60-rod element.

The MINIPAT computer code for prediction of fission gas release (6,7) has been adapted for use with SGHWR fuel rods. This code is based on a well-established physical model and has been shown to give a good best estimate prediction of a large number of PIE measurements made on WSGHWR rods. Some irradiation experiments have been carried out with gas pressure transducers on up to 12 rods for on-line measurement and MINIPAT has consistently overpredicted the measured pressures. These experiments have also demonstrated the substantial advantage obtained with the small fuel/clad gap in SGHWR rods. Figure 10 shows the effect of gap size in two rods from one such experiment.

UO_2 Swelling. For many years, UO_2 fission product swelling was a bogey that led designers to adopt UO_2 densities as low as 10 Mg/m^3. In the SGHWR rod design, consistent with the priority on safety, such densities were not acceptable, and it was anticipated that there would be some increase in rod diameter at the design burnup. However, UO_2 swelling has been proven to be lower than expected and to be accommodated within the UO_2 voidage and pellet dimples without any increase in rod diameter.

Corrosion and Deposition. Corrosion of the Zircaloy-2 cladding was expected to be small, but experience has shown that a new mechanism occurs under the neutral boiling coolant conditions of SGHWR, leading to relatively heavy corrosion in certain areas (8,9). This mechanism is referred to as nodular corrosion because of the characteristic appearance as shown in Figure 11. The nodules form after an incubation period and eventually coalesce into a uniform layer of oxide. In WSGHWR, this phenomenon has been much more severe on fuel cladding in the region of the spacer grids. It appears to bear little relationship to steam quality or flow velocity

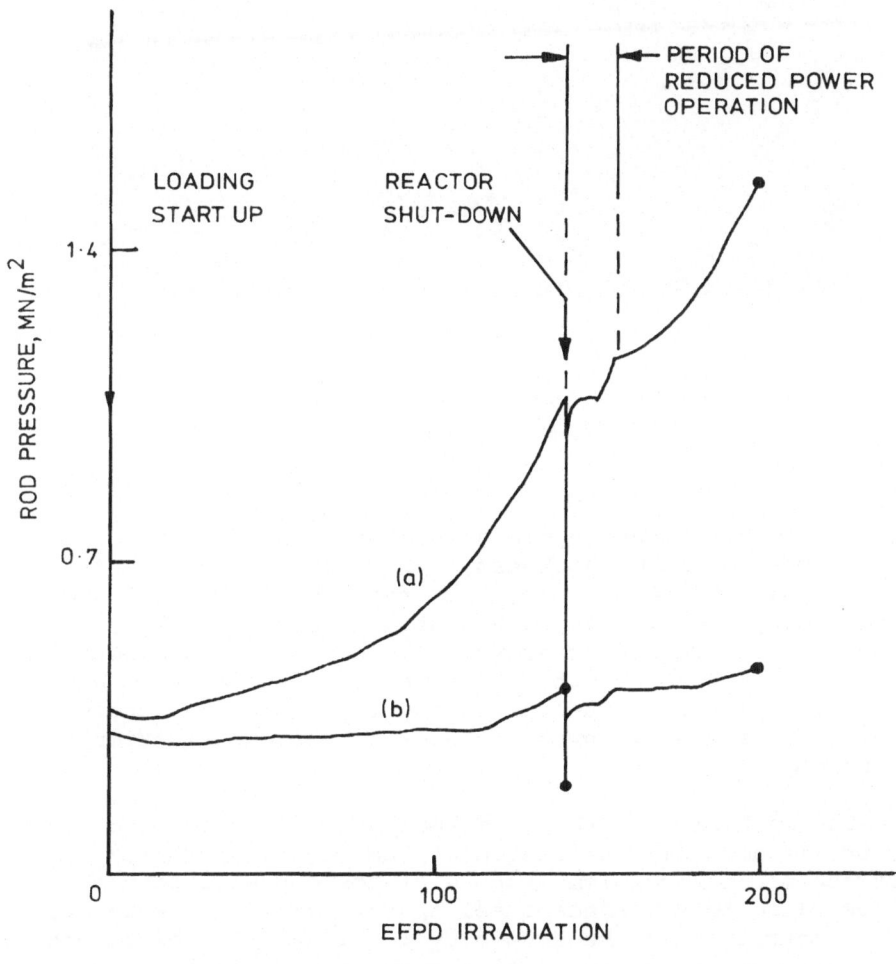

Figure 10

Internal Gas Pressure Variation with Burnup for
Two Fuel Pins from the Same Fuel Element
Fuel/Clad Gap is 0.35 mm (a), 0.18 mm (b)

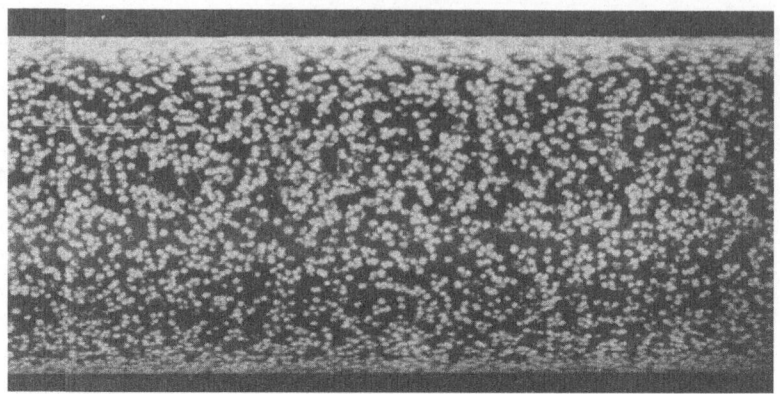

Figure 11

Nodular Corrosion on Central Sparge Tube
of 36-Rod Element

as judged by distribution along the length of fuel rods.
The predicted maximum thickness at the maximum fast neutron
dose in CSGHWR is in the range of from 150-200 μm, although
average thickness over any substantial area is unlikely to
exceed 150 μm. This represents a loss of metal thickness in
these local areas of about 100 μm which is unlikely to have
any adverse effect on fuel performance. Certainly, there
are no indications of any defects arising from oxidation at
high burnup.

The mechanism of this phenomenon is of scientific interest, but is not fully understood. One might hypothesize an
electrochemical mechanism because of the concentration of
nodules close to stainless steel spacer grids, but equally,
the concentration may be caused by the local flow disturbance.
A mechanism analogous to pitting corrosion in iron, with
intermetallic particles giving anodically active sites is also
under examination. This topic is discussed in more detail
by Trowse et al. (8) and by Tyzack and Sheppard (9).

Hydrogen pickup from the corrosion reaction has followed the expected trends and has not been found to increase
in line with the heavier nodular corrosion in the region of

spacer pads. The maximum concentration is not expected to exceed 100 ppm.

Crud deposition gave unexpected early problems (10) due to copper infiltration as a result of faults in the ion exchange cleanup plant in the feed train. Since that time, impurity levels of Fe and Cu have averaged about 10 and 7.5 ppb, respectively, in the feed water and 50 and 15 ppb in the primary circuit; crud has been of a low density, iron oxide type with a maximum thickness of 25 μm. This appears to be an equilibrium level, attained in a relatively short period, and there is evidence from activation studies that there is a constant interchange of material at all levels within the crud layer. Such crud layers have been shown to have excellent heat transfer properties and result in very low temperature drops.

Mechanical Properties of Irradiated Cladding. The continued integrity of the UO_2 containment depends upon the ability of the Zircaloy-2 cladding tubes to accommodate the stresses and strains imposed on them without fracture. It is known that fast neutron damage results in increased yield stress and reduced ductility, and particularly, in a loss of uniform elongation, thus leading to the early onset of plastic instability.

Mechanical tests carried out on irradiated SGHWR cladding have confirmed these trends. They have also shown a marked dependence of ductility on the initial condition of the cladding, decreasing with increased levels of cold work. Provided the tensile strength of the unirradiated cladding does not exceed about 675 MPa, then the residual ductility is unlikely to be below 3%, expressed as total circumferential elongation in a closed-end burst test. Tests have shown annealed Zircaloy-2 tubing to retain higher residual ductility in such a test, values falling in the range of from 5 - 10%. In both cases, the nonuniform necking strain is little affected by irradiation and values around 20% are common. The ductility changes are substantially complete at a fast neutron dose of 10^{24} neutrons/m^2, and the residual ductility should be quite adequate for reliable service, provided there are no stress-corrosion agents present when high stress levels are induced in the cladding. This is a special situation that can arise under certain operational conditions and is discussed in the next section.

D.   Power Cycling and Power Ramps

Operational requirements dictate that power reactors
will not always operate at steady power levels. There are
two characteristic types of operation to which fuel responds
very differently, regular short-term power changes of large
or small magnitude and infrequent power increases after
long periods of substantially steady operation. These
characteristically different modes of operation are referred
to as power cycling and power ramping, respectively.

In both modes of operation, there are power increases
that may be expected to increase the stress level in the
fuel cladding. The essential difference is the timescale
of events and the extent to which the fuel rod reaches an
equilibrium condition (clad creep, UO_2 swelling or relocation)
at the lower power level. In power cycling, the extent of
interaction for a given power increment is less than in power
ramping and the damaging effects will also be less. The
theoretical aspects of these processes and the models that
have been set up to understand and predict their effects
are more fully discussed in a later section.

Tests have been made in WSGHWR to study both types of
operation. The reactor was cycled 205 times on a daily
basis between full power and 70% power, spending 6 h each
day at the lower level. The expected mode of damage is a
progressive clad wall thinning over radial pellet cracks.
No evidence for such thinning was observed in PIE of fuel
elements that experienced this cycling, confirming that the
allowable thickness reduction of 10% would not occur in less
than 2,000 such daily cycles. No other adverse effects, including enhanced gas release or defects were observed as a
result of this experiment.

Power ramps can occur in SGHWRs after prolonged low
power operation for any reason or as a result of loading
new fuel. There are no other systematic means short of
accident conditions since SGHWR does not contain movable
control rods that can generate local ramps as in BWRs.
Generally, the refueling ramps will be less than 10%, but in
WSGHWR, because of its extensive experimental program, there
have been a number within the range of 10 - 15%. None of
these have resulted in defects. In order to study ramp limits,
special shuffling experiments have been performed (11).

These have shown that fuel defects certainly can be produced given the right conditions, and a clad stress criterion of 480 MPa is the best indicator. The experiments
have shown very consistent behavior, but also show that it
is difficult to quantify a safe power ramp magnitude because of the dependence on so many other factors such as
pre-ramp history, starting power level, burnup and ramp rate.
There are obviously fuel rod design factors that influence
ramp behavior such as type and thickness of cladding,
initial fuel/clad gap and UO_2 pellet design and density.
SGHWR fuel rod design has not been strongly influenced by
power ramp considerations since it is well able to survive
all normal system demands without defects. To give some
idea of the range of ramp upratings and liability to defects,
fuel rods have survived ramps of up to 60% power increase
without failure, and some have defected at around 40% power
increase. However, these varying results can be rational-
ized on the basis of the critical clad stress for stress-
corrosion cracking. The fractures that result are always
substantially brittle, and there seems little doubt that

Figure 12

Typical Longitudinal Crack in Cladding
Caused by Power Ramp

this is an example of stress-corrosion cracking with iodine
as the active agent (12). A typical crack resulting from a
power ramp is shown in Figure 12.

E.   Defects

There have been 52 fuel element defects in WSGHWR in
the nine years since first full power was attained at the
end of 1967. Of these, 33 occurred during the first year
of operation and were caused by the abnormal crud deposition
referred to earlier. Of the other defects, eight were in
experiments designed to go to failure, such as power ramps,
or in instrumented elements that eventually defected as a
result of leaks in instrument cables. Five were in a small
batch of elements manufactured with pickled and autoclaved
bores in the cladding tubes, and have been attributed to
this feature. The mechanism was bore hydriding and the
supposition is that the presence of the oxide layer enhanced
the susceptibility to local hydriding, although this is not
a confirmed explanation since fuel rod hydrogen content at
about 1 mg per rod was well below the limit normally expected to cause problems.

Nothing systematic about the remaining six fuel element
defects suggests a common cause. All except one, however,
defected very early in life, which suggests materials or
fabrication defects. It is most probable that at least two
were caused by defective cladding.

WSGHWR, therefore, has given no evidence of defects
that are related to design, rating or burnup. In view of
the results of the surveillance program, this is not surprising, as the only factor that might conceivably result
in the onset of systematic defects is the loss of metal
from the cladding by corrosion, which would not be expected
on a significant scale until beyond the design burnup level.

The defect rate of fuel rods in WSGHWR can be assessed
in different ways, depending on which of the special experiments is included and whether more weight is given to recent
experience with the latest fuel specification. The rate is
between 1 in 10^3 and 1 in 2 x 10^3 over the most recent five
years, ignoring the 'high risk' experiments.

No special action has been taken to remove defects prior
to a refueling shutdown except when there was a requirement

to remove the element for PIE. Some defects have operated
for about 250 days after the leak was first detected, and
there is no real evidence for serious deterioration. Experience is being accumulated on the correlation between activity
release and defect severity to act as a guide to when an element should be discharged.

IV. PERFORMANCE MODELS: THE SEER-SLEUTH COMPUTER CODE

In the early days of nuclear fuel element design and
development there was little real understanding of the important design parameters and materials properties. Only the
advent of performance models has helped to produce a logical
basis for design and performance prediction. The essence of
these models is that they are dynamic, and attempt to predict the way in which changes in the rating, temperature
and burnup will influence the state of stress and strain in
components of the fuel element; the likelihood of clad fracture is also calculated in certain cases.

Broadly speaking, performance models can be divided
into two classes: those that concern themselves with performance under ordinary day-to-day conditions of reactor
operation and those that deal with performance under fault
conditions such as a loss of coolant accident (LOCA). Consideration of the latter is deferred to a later section and
this section is confined to a brief appraisal of the SEER-
SLEUTH computer code, the performance model developed at RFL
Springfields, and applied there not only to SGHWR fuel but
also to the fuel for AGR and the LMFBR (13). Since it is
applied to such a wide range of metal-clad oxide fuel, the
SEER-SLEUTH code has arguably received a wider validation,
by comparison with the results of fuel element irradiation,
than any other in the world today.

In this code, for any arbitrary history of rating and
clad-temperature, the time dependence of clad stress, strain
and fracture is calculated. The pellet is modeled as indicated in Figure 13 (14); it comprises an outer region of
blocks that are separated from one another by radial cracks.
The crack-tips lie on an isotherm, the bridging isotherm.
An increase in reactor power causes the material in which
the bridging isotherm is situated to expand, and this forces
the radial cracks to open more; if the clad is in contact
with the pellet, the radial pellet cracks may propagate into

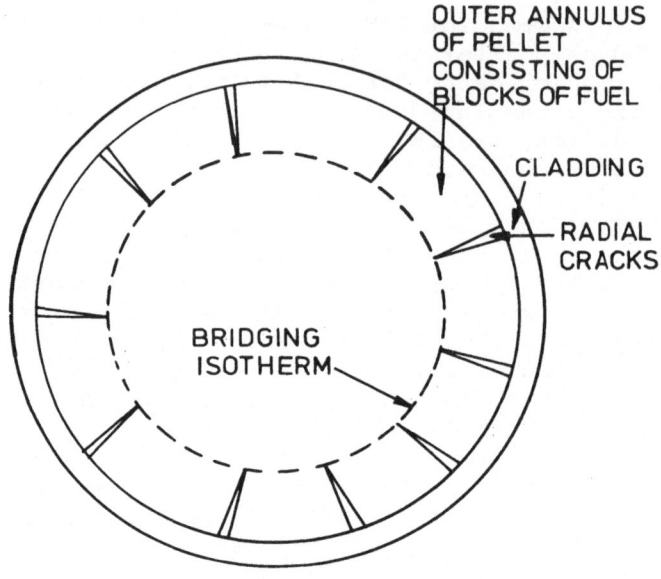

OUTER ANNULUS
OF PELLET
CONSISTING OF
BLOCKS OF FUEL

CLADDING

RADIAL
CRACKS

BRIDGING
ISOTHERM

Figure 13

Model of a Ceramic Pellet
Used in the SEER-SLEUTH Computer Code

it.   Two things oppose this type of clad failure, the ductility of the cladding and the sliding at the interface;
thus, if the clad were brittle and the interface welded,
then the crack would certainly propagate into the clad.
If the interface has zero friction coefficient, then the
whole of the clad-circumference is stretched during the uprate and so the strain that any region of the clad must sustain is reduced.   If the clad has enough ductility, then it
will not fail.   The friction coefficient and the number of
radial pellet cracks are therefore important parameters in
the SEER-SLEUTH model (15).

Important also is the position of the bridging isotherm.
The nearer to the surface of the pellet this isotherm lies,
the smaller the extent to which its temperature will rise
and the smaller the associated crack opening displacement
for a given uprate.   The position of the bridging isotherm

depends on the interplay of irradiation creep in the pellet
exterior, thermal creep in its interior and gas-bubble swelling in the UO_2, all of which are related to the burnup and to
the restraint that the clad and coolant pressure impose on
the pellet. The SLEUTH portion of the code calculates the
position of the bridging isotherm, taking account of these
mechanisms as a function of time. It uses a simplified
model of the cladding and of the fuel-clad interaction (zero
interfacial friction, instantaneous, recoverable primary
creep). Then the SEER portion of the code repeats the calculation. Its input is the time-dependence of rating, clad
temperature and bridging-isotherm radius. It uses the
simple pellet model of Figure 13. All the sophistication
of pellet behavior that went into the SLEUTH part of the
calculation comes across into the SEER via the time dependence of the bridging-isotherm radius.

The SEER part of the code then has a simple pellet
model (albeit mirroring, via the bridging isotherm radius,
all the complexity of the prior SLEUTH prediction of
pellet behavior). The SEER part of the code, however, has
a very sophisticated representation of the clad and of the
fuel-clad interaction. It calculates, if necessary, for
nonaxisymmetric temperature conditions, the way in which
the strain in the cladding varies near a radial pellet crack.
It can deal with a variable number of pellet cracks and with
any value of the friction coefficient. The cladding is cred-
ited with the capacity to deform in all of the observed ways:
elastically, in a time-dependent plastic manner, in primary
creep (recoverable and nonrecoverable components are included)
and in secondary creep. The effect of irradiation on the
yield point and on the creep rates (16) is also taken into
account. Finally, four clad-fracture parameters are calculated. These include:

1. the fraction of ductility that has been exhausted
    by monotonic straining; every power increase tends
    to thin the clad by more than it thickens during
    subsequent collapse at low power, a consequence
    of the strain-concentration over pellet cracks.

2. the fraction of the low-cycle fatigue endurance
    that has been exhausted - another power cycling
    effect.

3. the depth to which time-dependent, stress-corrosion
    cracks have penetrated the bore of the clad (*17*).

4. the peak stress in the clad.

This last parameter is a crude version of the stress-
corrosion cracking parameter and it is observed that for
SGHWR, when the peak clad stress exceeds a threshold value,
the clad fails, apparently by the unstable extension of an
iodine stress-corrosion crack.

Table III shows the predictions that were obtained
from one of the early versions of the performance model and
compares them with what has happened to SGHWR. Figures 14
and 15 show some of the tabulated information on axes of
initial rating and percentage increase. If the code pre-
dicts a peak clad stress in excess of 504 MPa, then the observed failure-rate is 9 in 11. If, on the other hand, the
stress never exceeds 483 MPa, then the observed failure rate
is zero in more than 10,000 fuel rods. Significantly, laboratory experiments show that 500 MPa is about the threshold
stress for iodine stress corrosion cracking in this type of
Zircaloy (*12*).

TABLE III

The relationship between the peak stress, calculated to be
produced in SGHWR fuel cladding by an uprate, and the number
of fuel rods that have failed. The failure probability has
been calculated from the observed failure rate.

Predicted Peak Stress (MPa)	No. Rods At Risk	No. Fuel Rods That Failed	Calculated Probability of Failure (95% Confidence)
> 504	11	9	0.55-1.0
483-504	36	2	0.00-0.14
< 483	>10,000	0	0.0

The code has also been used to predict the effects of daily power cycling in SGHWR. Here the stress is always well below the threshold for iodine stress corrosion, but now the gradual thinning of the clad over the pellet cracks, as the latter open and close during the cycling, can eventually exhaust its ductility and lead to failure. The code indicated that more than 2,000 of a certain, representative type of daily power cycle would be needed to produce failure. A SGHWR experiment reached 205 such cycles without causing failure.

The SEER-SLEUTH code has also produced excellent agreement with the results of LMFBR experiments in which the power level has been altered in ways designed to produce failure. The code was the basic tool in the redesign of the AGR fuel(18) and in the selection of an improved clad variant for that fuel system.

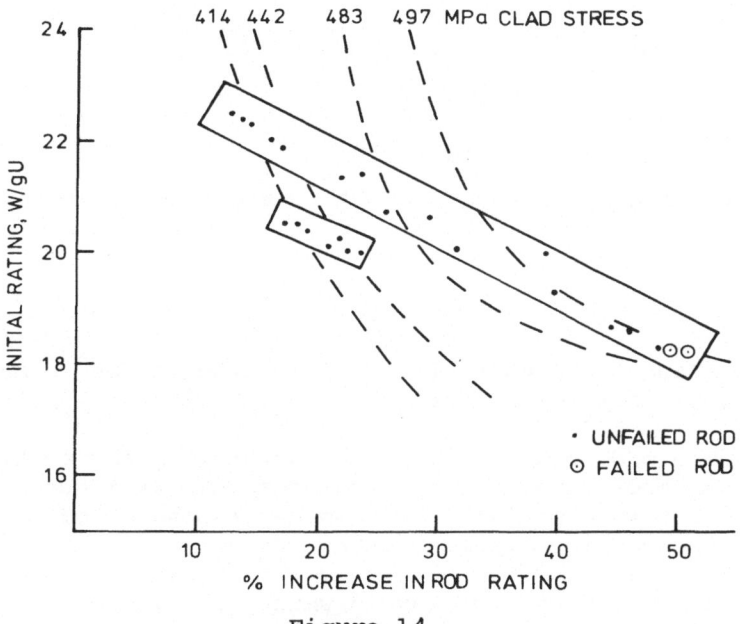

Figure 14

WSGHWR Power-Ramp Test Results
Showing Relationship to Calculated Clad Stresses

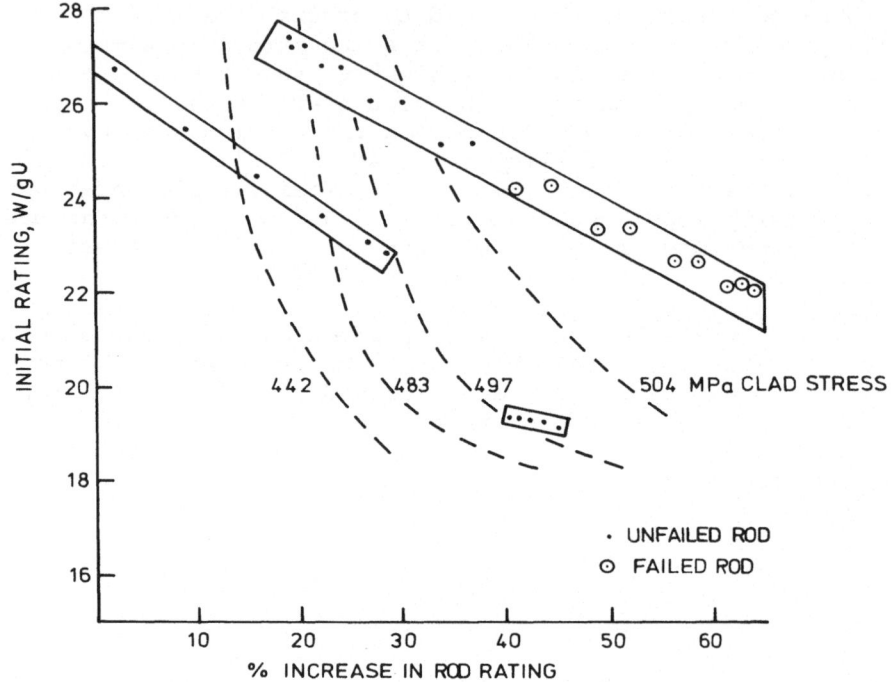

Figure 15

Same as Figure 14 except for Other Experiments in SGHWR

The SEER-SLEUTH code differs from other codes in many ways, notably the following:

1.   The code can be run, at comparatively low cost, to model the whole course of a complex reactor history comprising perhaps may thousands of changes in temperature and rating.

2.   It calculates the way in which the clad will fail, taking into account the mechanical interaction between pellet and clad in a very general manner.

3. It uses physically sound models of both gas bubble swelling and pore sintering in the pellet and takes into account the effects of pellet fracture on both the behavior of the clad and that of the pellet.

4. It models the deformation and fracture of the cladding with great realism.

5. It runs quickly and comparatively cheaply (although admittedly a large storage facility is needed in the computer). Its quickness stems from the division of the model into two parts, linked via the bridging-isotherm radius calculation. Some precision is lost because the pellet calculations are done for a simple clad model, but for most purposes, this does not reduce the accuracy significantly.

V.   BEHAVIOR IN ACCIDENT CONDITIONS

A.   Origins of Major Accidents

The diversity of the control systems of the SGHWR ensures that the reactor shuts down in the event of a coolant circuit rupture. The large amount of energy in the fuel and the coolant circuit is then most likely to be dissipated in a safe way by boiling the coolant, and the escape of steam as the reactor returns to the cold condition. However, some circuit ruptures create conditions in which it is not at first obvious that no hazard arises. These fall into two groups: ruptures that lead to coolant stagnation in one or more channels and rupture near the fuel. Coolant stagnation leads to high fuel temperatures but small mechanical forces applied to the element. Ruptures close to the fuel keep it cool as the coolant discharges over it, but subject it to high drag forces. These accidents require special consideration.

B.   Stagnation Accidents

Coolant stagnation occurs when the reactor channel depressurizes by discharging equal amounts from both ends. The stored heat in the fuel and the heat from decay of fission products boil the water in contact with the rods

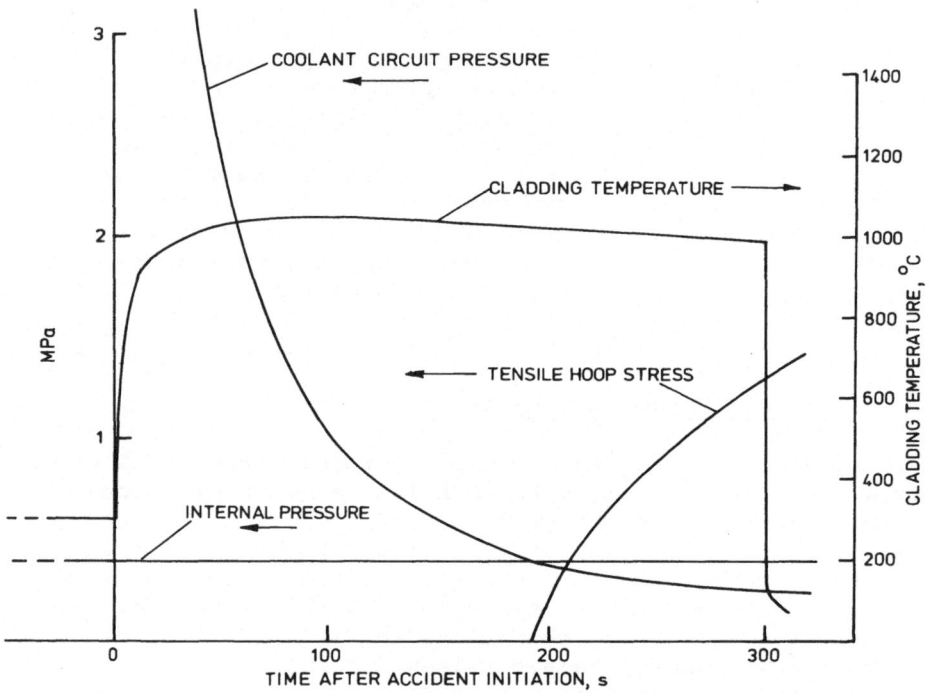

Figure 16

Typical Temperature and Pressure Transients Following a
Stagnation LOCA and Associated Clad Hoop Stress

which become dry and suffer a large decrease in heat transfer coefficient. The subsequent rise in the Zircaloy cladding temperature may be about 100 K/s, and it is this that
if not controlled, will lead to a hazardous situation in
which the cladding disintegrates; the fuel pellets can drop
to the bottom of the channel and the mass can overheat and
melt the reactor structure. Alternately, the hydrogen generated during oxidation of the cladding by steam may form
an explosive mixture capable of disrupting the containment.
In SGHWR, these situations are avoided by sparging the fuel
rods with an emergency supply of cooling water (19). This
ensures that the nuclear heat, the stored heat, and the
heat from the reaction of the Zircaloy cladding with steam
are all lost by radiation to the pressure tube which is
kept wet by the emergency water supply and by evaporation
of the sparge water. The cladding temperatures are kept
within acceptable limits in this way, and finally, the
cooling water rewets the rods. This process may be assisted
by and is eventually overtaken by cooling from water that is
supplied to reflood the channels.

The performance of the emergency cooling water system
naturally has been the subject of much study. The behavior
of the fuel during a stagnation accident must be such as to
allow this system to operate as designed; that is, the geometry of the fuel element must be maintained during a LOCA
should it occur at any time in the fuel element life.

C.   Maintenance of Coolable Geometry in New Fuel

Clad Ballooning. When the coolant pressure falls below
the gas pressure within the rod, there is a hoop stress tending to increase the cladding diameter. A general increase of
18% would close the gaps between the outer row rods, prevent
the radial flow of sparge water at certain axial locations
and hence, reduce the heat lost by radiation to the cooled
pressure tube. To maintain a coolable fuel element geometry,
this increase is limited to 5% which has been shown in fuel
element spray cooling tests to have no significant effect.
Figure 16 shows a typical circuit depressurization curve and
the associated cladding hoop stress. A typical cladding
temperature transient is also shown, and this, in combination
with the net internal pressure, must not produce clad swelling in excess of 5%.

The means of estimating the cladding diameter increase for combinations of temperature and pressure transients is the simulation of accident conditions in a specially designed test rig known as PROPAT (Programmed Pressure and Temperature), shown in Figure 17. A short length of new cladding (generally fully annealed) is made to follow the temperature transient by passing an electric current through it, controlled by a thermocouple. The inside of the cladding specimen is filled with argon at a rising pressure equal to the difference between the coolant and rod internal pressures and the whole is surrounded by steam at atmospheric pressure. The cladding diameter is recorded photographically every ten seconds from which the rod diameter at the time of rewetting is deduced. This method has the advantage that the complex metallurgical and chemical factors that influence cladding behavior are automatically included. This is particularly helpful where the cladding remains for a long time around the α/β transition where the strength and ductility are sensitive functions of strain rate and temperature. Repeated tests on suitable samples produce an estimate of the variations within a bulk supply of material. The effect of small changes in accident conditions can be obtained by interpolation from the large amount of available data. An alternate means of estimating cladding diameter increase is the CANSWEL computer code described later.

These methods are used to estimate the rod internal pressure just before the accident, which would develop 5% cladding diameter increase. The difference between this pressure and the expected pressure in service is a measure of the operating margin, often expressed in terms of a rating margin.

Reduction of Cladding Ductility. The absorption of oxygen by zirconium above about 900°C reduces the cladding ductility, and in high temperature transients it falls below 5%. The limit on rod internal gas pressure is then determined by the avoidance of cladding rupture. At higher temperatures still the cladding fails by thermal shock caused by the relatively cold water fed from the emergency cooling system. Extensive cracking is possible with the loss of the well-defined fuel element geometry. The PROPAT rig is used for assessment of transients in this way also, but an alternative is to specify a limiting oxygen uptake. For LWR fuel rods, 17 mole % oxygen has been used (20) and this is consistent with measurements made on SGHWR cladding.

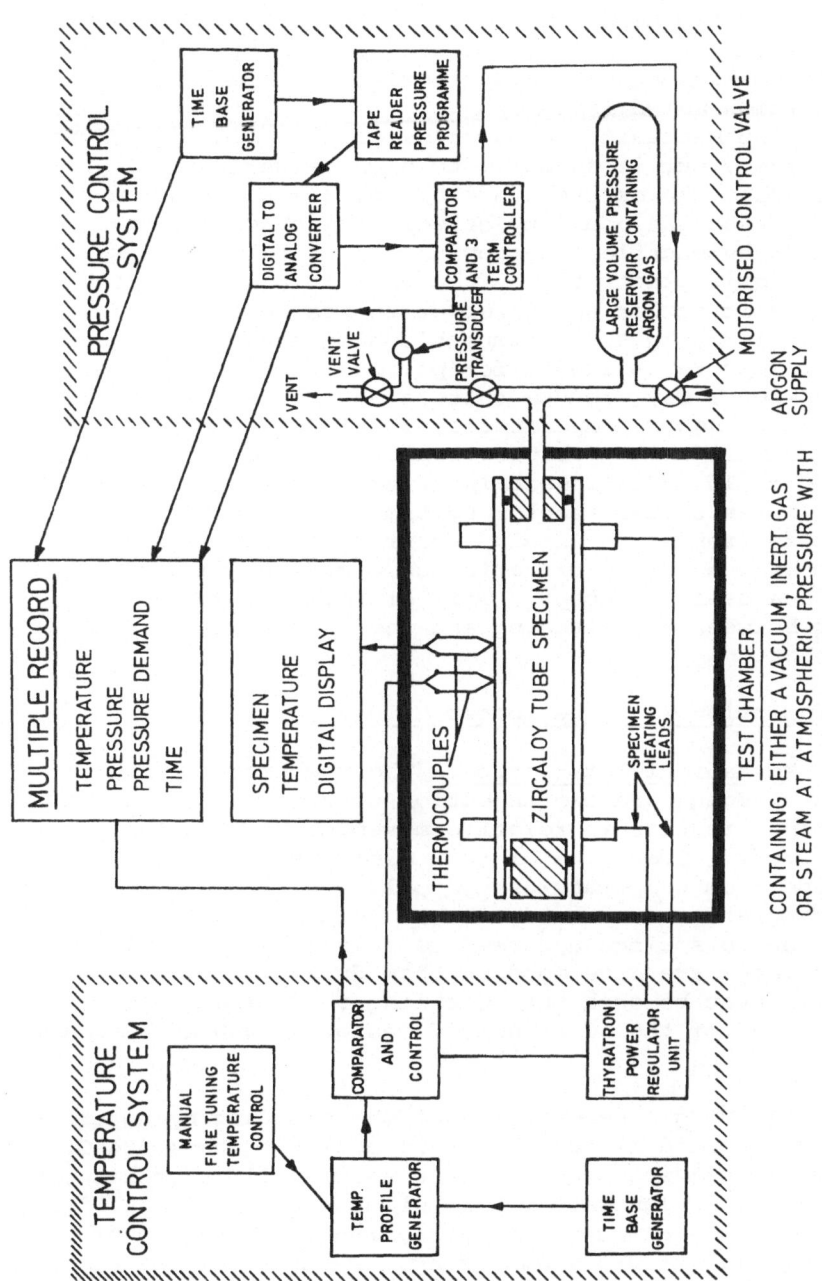

Figure 17

Control System for the PROPAT Fuel Rod Testing Rig

Throughout a LOCA transient, the layers of zirconium oxide and stabilized α phase on the surface of the cladding have uniform although steadily increasing thicknesses. In this condition which is dependent on a steep radial oxygen gradient, the mechanical properties of the cladding can be characterized, but if exposure to high temperature is prolonged so that the oxygen tends toward its equilibrium distribution, then the boundary between the phases becomes ragged and the ductile β zirconium that forms the bulk of the cladding is penetrated by the brittle α phase. Such material is irregular and inferior in its behavior and is avoided by limiting the maximum cladding temperature to 1200°C. Exposure to 1200°C for periods much longer than LOCA transients may produce these alpha incursions and short exposures to higher temperatures, say 1300°C, do not produce them. The maximum temperature limit is therefore time dependent, and could be expressed in alternate ways; however, the basic features of the reactor system constrain the cladding temperature transients within a fairly narrow range of types, making a safety criterion of a maximum temperature of 1200°C a useful concept. This criterion also avoids alloying between the cladding and the spacer grids, discussed next.

Interaction with the Spacer Grids

a. <u>Mechanical Interaction.</u> The temperature rise in the fuel cladding and the injection of cold water into the sparge pipes results in relative movement between the grids and fuel rods. Quenching of fuel elements with plain ferrule grids from temperatures up to 1200°C has demonstrated rod sliding and shown that coolable geometry will be maintained. Where springs are used to stabilize the rods, compressive loads are developed in them during quenching and the design must be such that buckling is avoided. The rod buckling load of the SGHWR 60-rod design at 900°C is about 200N.

b. <u>Alloying.</u> Penetration of the cladding by an iron-zirconium eutectic is theoretically possible above 934°C, and by a nickel-zirconium eutectic above 961°C, but has not been observed in laboratory LOCA simulations below 1275°C. Oxidation of the zirconium by steam and the presence of crud have a significant delaying effect. Puncture of the can and release of gaseous fission products from a fraction of the

rods is not a problem, but extensive interaction breaking
the rod into two or more pieces is unacceptable. Since
penetration by molten phases is very rapid, the PROPAT
rig is used to establish the absence of alloying for anticipated SGHWR transients.

D.   Modification to Fuel Behavior During Irradiation

        Fission gas release increases the pressure within the
pin, the release in the most highly-rated elements exceeding the filling gas volume. Calculation of fuel temperature
change following an accident and pellet-heating experiments
indicate that an insignificant further release of fission
gas would occur during a LOCA. A similar conclusion has
been reached by USNRC (21). The consequent decrease in
the operating margin is in part offset by the falling rating
of the fuel that corresponds to lower fuel temperatures,
lower stored heat and less decay heat. Fuel temperature is
also influenced by the thermal resistance of the fuel/clad
gap that is slowly closed by cladding creep but is initially
increased by fuel densification. Densification can have a
significant effect on stored heat and hence, cladding temperature during a LOCA, but is rapidly offset in high rating
regions by grain boundary gas bubble swelling and by axial
interactions between fuel and cladding in the SGHWR design.

        The strength of Zircaloy-2 is increased by irradiation
hardening during normal reactor service. Annealing studies
have shown, however, that the properties are restored in less
than one second at 800°C. The creep strength of cladding
from a fuel element irradiated to 11 MWd/kg is shown in
Figure 18, and is compared with unirradiated material. Cladding also becomes oxidized on the bore and outer surface
while in the reactor. There are small effects due to the
loss of metal and decrease in thermal conductivity. When
substantial oxidation has occurred prior to a LOCA, the surface oxide dissolves in the metal during LOCA transients,
approximately compensating for the reduction in the rate of
strengthening from direct oxygen absorption from steam by
unoxidized cladding during a transient.

E.   High Drag Force Accidents

        Failure of the pressure tube is extremely improbable,
but it is desirable to know the effect on the fuel element

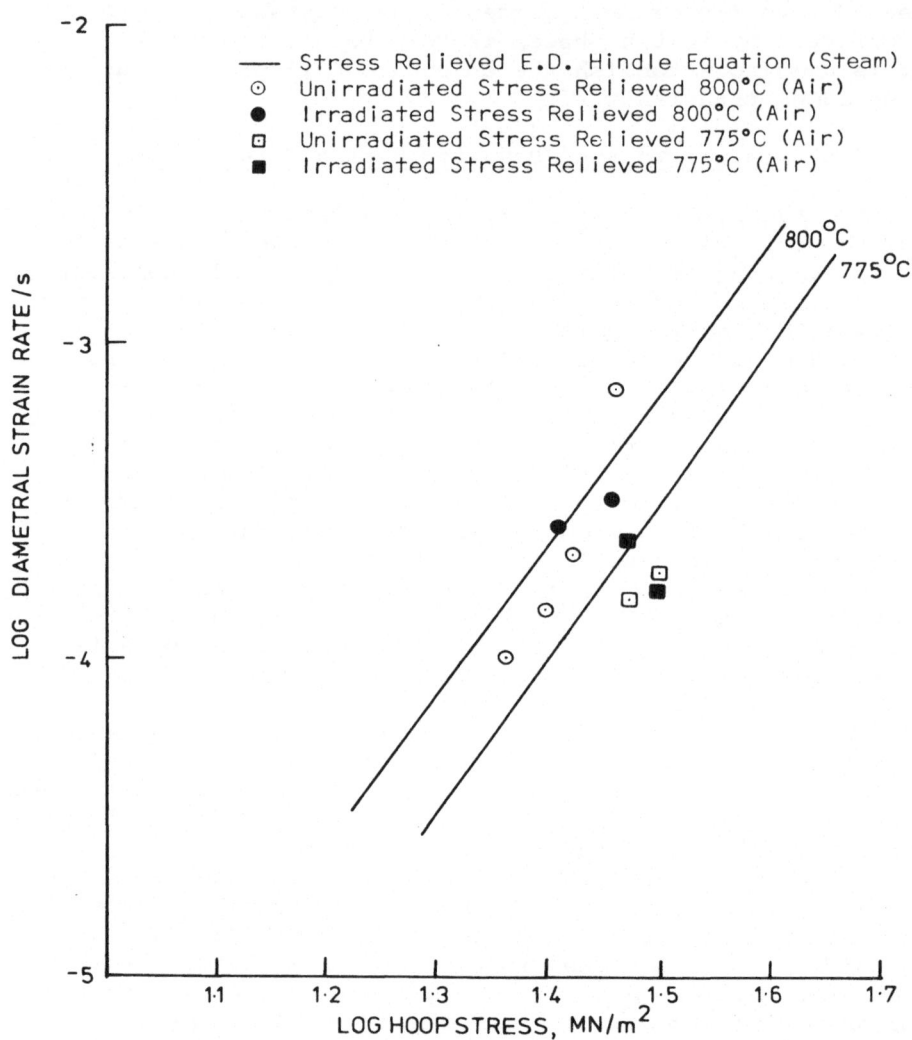

Figure 18

Minimal Diametral Creep Rate of SGHWR
Irradiated Fuel Cladding Compared with Unirradiated
Zircaloy-2

of a longitudinal split in the pressure tube or a double-
ended break in the pipework either just above or below the
fuel element. In either event, the contents of the coolant
circuit are discharged over the fuel which sustains the
major pressure drop. The behavior of the fuel in these
situations has been determined in a blowdown facility known
as the "BLUSTER" rig, shown in Figure 19.

A channel break below the fuel element initiates a
pressure front that sweeps rapidly over the fuel element
before bulk motion of the coolant is affected. The design
of the fuel element is such that the impulsive loads applied
to the structure cause no deformation. When the coolant discharge starts, the rods, being held at the upstream end,
suffer a tensile force that they are designed to withstand.
A break above the fuel loads the rods in compression and by
themselves they are unstable, so that without the spacer
grids, the bundle of rods would twist together like a stranded
rope. It has been demonstrated in the BLUSTER rig that
this mode of collapse is prevented by the assembly of grids
and sparge tubes. The circuit is filled with water and
raised to the reactor operating pressure. When the temperature in the drum reaches 280°C, the discharge over the
fuel is initiated. Initially, no instrumentation was
attached to the fuel since the test is required to reveal
only major damage; instrumented tests are to follow.

Longitudinal pressure tube splits have been studied in
a rig that is designed to rupture the pressure tube as the
pressure reaches its normal operating value. Drag forces
on the rods cause marked radial displacement, but the spacer
grids are strong enough both to retain the rods and maintain
a separation between them. In tests to date, no rods have
broken.

VI.   TRANSIENT MODELING

In a previous section, an account of the SEER-SLEUTH
computer code has been given, showing how it correlates the
performance of SGHWR fuel under ordinary day-to-day conditions of operation in the reactor. If a LOCA occurs, then
there is a need to be able to predict the fate of the fuel
with just as much certainty. In particular, the need is to
know whether it will swell to an unacceptable extent. A
computer code known as CANSWEL 1 has been developed at RFL

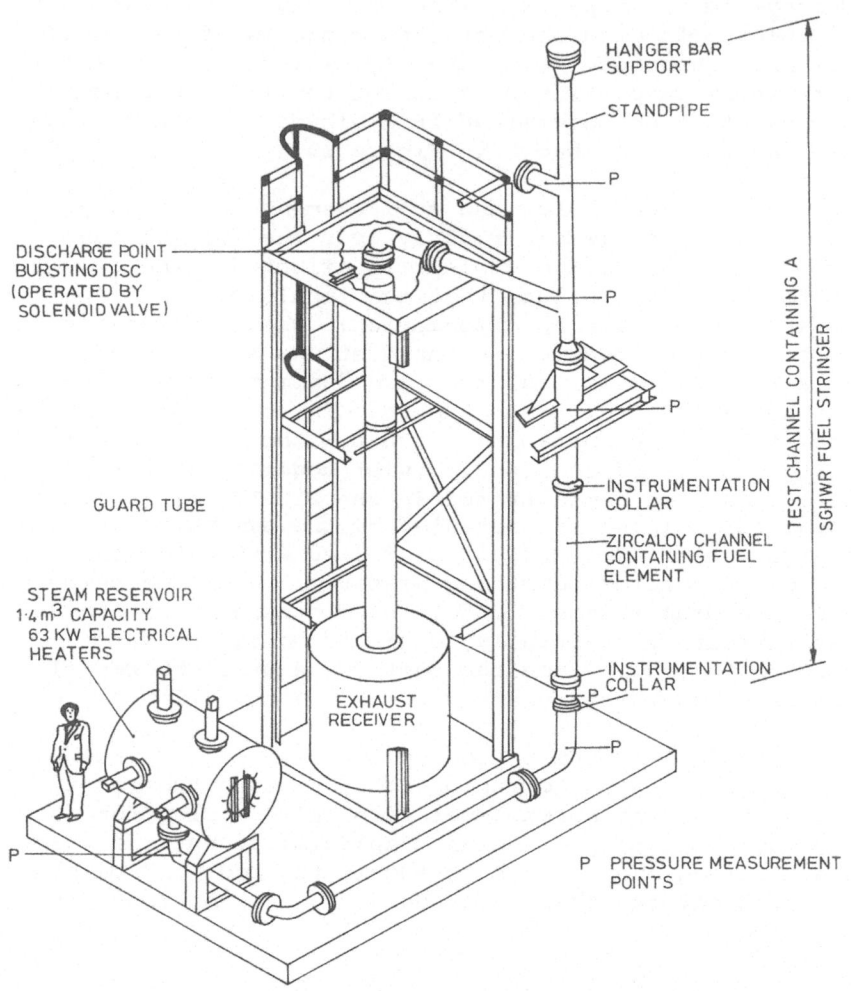

Figure 19

The "BLUSTER BLOWDOWN RIG" for Exposing Fuel Elements to the Full Force of Coolant Discharge During an Accident. Alternately, the discharge point may be at the bottom end of the fuel channel.

Springfields for this purpose. Its more sophisticated
successor, CANSWEL 2, is under development currently.

A.   CANSWEL 1

     This model calculates the straining of the clad that
occurs during a LOCA. It assumes that the clad is in an
inert atmosphere.

     The clad is treated as a single axi-symmetric shell
under plane strain conditions, and all strain is assumed to
be elastic or creep. The imposed stresses can be chosen
from any combination of tube differential pressure, axial
load and equibiaxial stresses. In place of the axial load,
an imposed axial strain applied via a spring can be included
to simulate the effect of interference between the fuel pin
and the support structure. For efficiency of computation,
the time steps chosen for the interpretation are calculated
automatically to give predetermined strain increments.
Stress and temperature histories during the transient are
divided into periods in which they can be assumed to change
linearly with time. Approximate functions allow the effect
of the changing conditions during a time step to be taken
into account. This enables creep strain increments equal to
several elastic strains to be used, thus reducing the number of time steps required. For example, tube failure in
a temperature ramp test can be calculated with about 50
steps.

     Constitutive Equations. The constitutive equations
used for the creep of Zircaloy in the alpha phase field are
similar to those used for the thermal creep strains in SEER-
SLEUTH. Additional data obtained at higher temperatures
were also taken into account (22-25).

     At temperatures between approximately 820°C and 980°C,
where both alpha and beta phases are present, the creep properties change markedly. Several workers have reported superplastic properties, but minima in the ductility-temperature
curve are also found. Further experimental work has confirmed this showing high ductility and marked uniformity of
strain at low strain rates but low ductilities at high strain
rates. This work has also confirmed an observation of
Hindle (22) that at temperatures above about 900°C, the maximum strain at low stress occurred at the cooler ends of the

specimen. At higher stresses, the maximum strain occurred
in the hotter central region. Analysis of the results
showed that the behavior was consistent with two competing
creep processes, one of which has a high stress sensitivity
and dominates at high stresses, the other with stress expon-
ent approaching unity dominating at low stress. This latter
is a superplastic mechanism (26) and has a maximum contribution at about 900°C. The first process can be accounted
for by conventional creep of the alpha and beta materials
in their respective volume fractions.

The superplastic process is modeled by equations that
take account of the manner in which the relative volumes of
alpha and beta change with changing temperature.

Use of the Model. The resultant behavior of the model
is shown in Figure 20 which gives the stress required for
failure, or for a particular strain in a given time as a
function of temperature for isothermal, constant pressure,
tests. The point symbols are interpolations from experimental data for comparison, and are in reasonable accord
with prediction. The abscissa is a reciprocal absolute
temperature.

The model has been tested against a wide range of data
reported for inert atmosphere tests relevant to a LOCA.
Much of these data are in the form of failure temperature
in constant heating rate, constant pressure tests. Experimental results from several authors (24,27,28,29) are compared with the code predictions in Figure 21. In general,
the agreement is seen to be reasonable. Simulated PWR
transients with various temperature and pressure histories
have been reported by Hindle (27). The experimental data
here are in the form of failure times and these are also
predicted reasonably as Figure 22 shows.

For SGHWR, the prime concern is the 5% strain limit.
Of the many simulated transient tests done at RFL, only a
few can be used to test the current version of the code,
since the majority have been done under conditions giving
oxidation strengthening that has not yet been modeled. Up
to about 850°C close agreement is obtained between observed
and calculated strains, but errors become significant at
higher temperature.

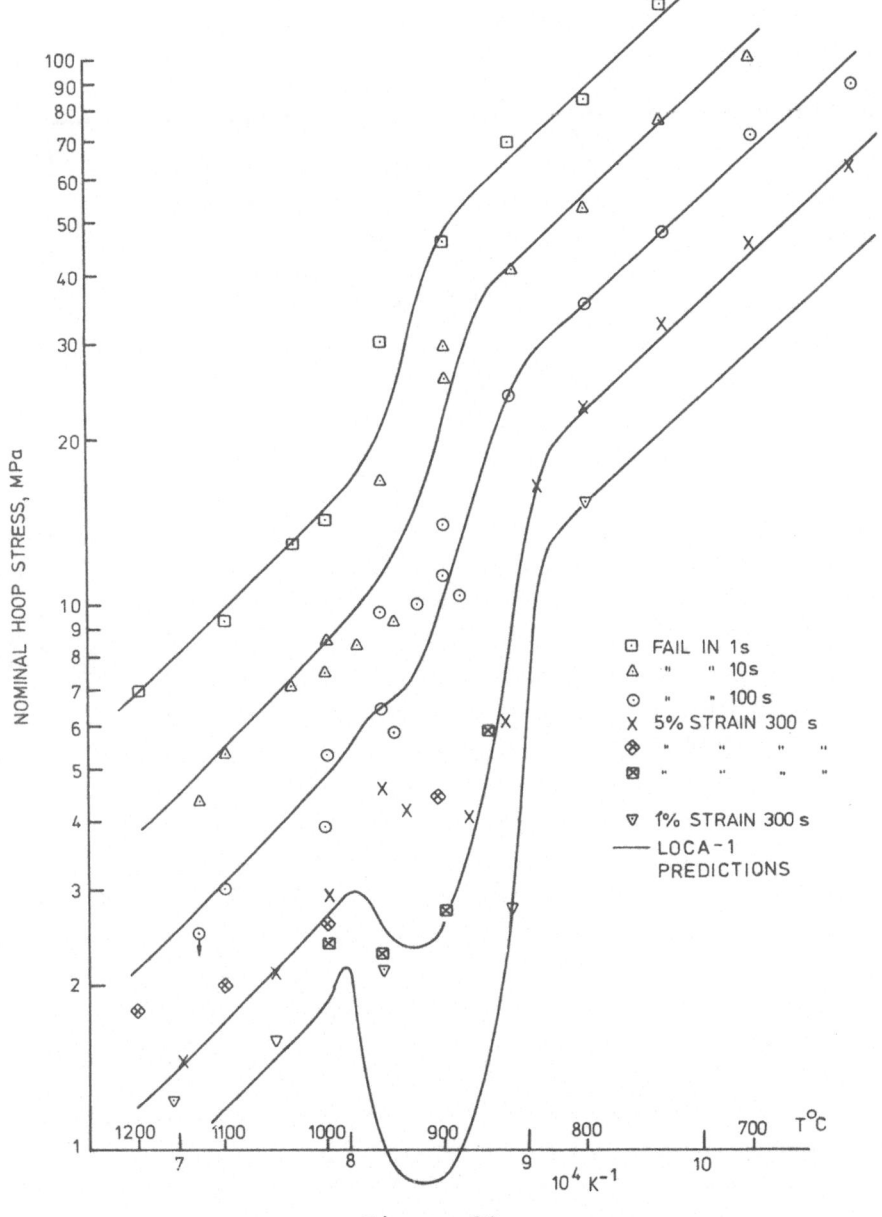

Figure 20

Comparison of CANSWEL 1 Model
With Isothermal Creep and Strss-rupture Data

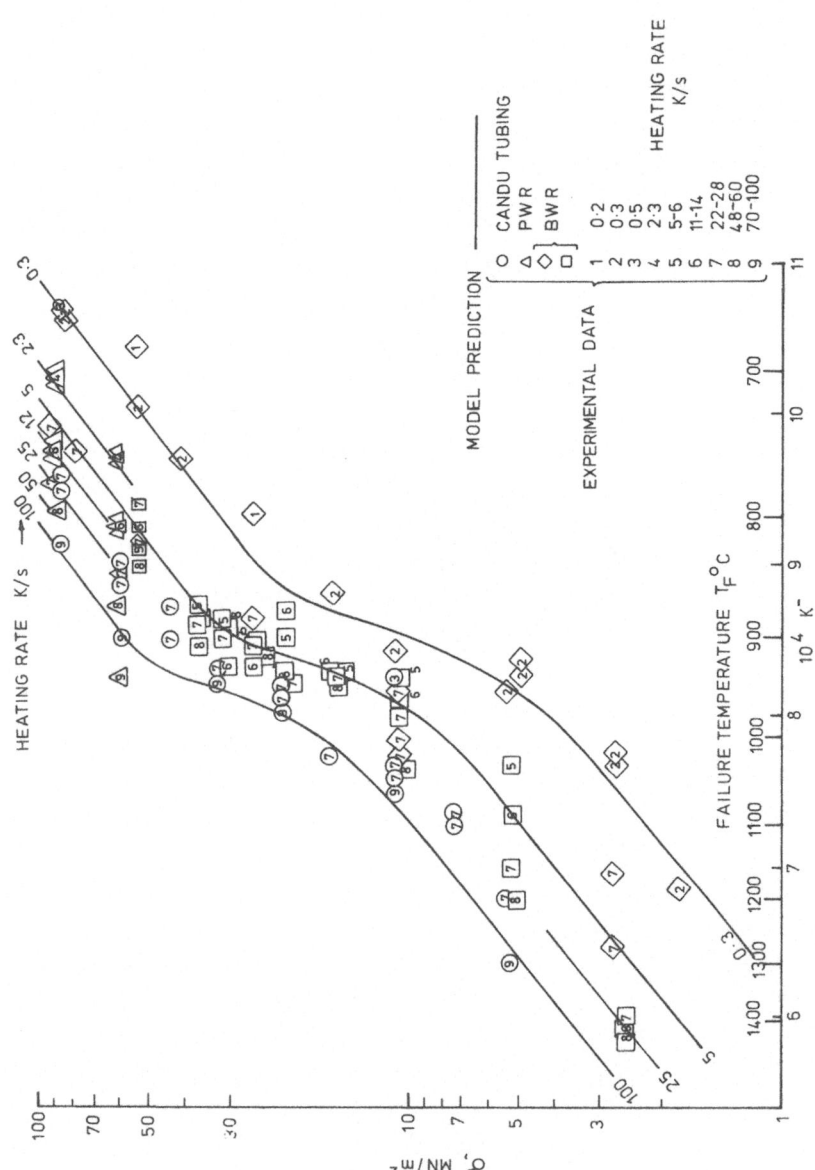

Figure 21

Comparison of CANSWEL 1 Model Predictions with Experimental Failure Times
In PWR Transients

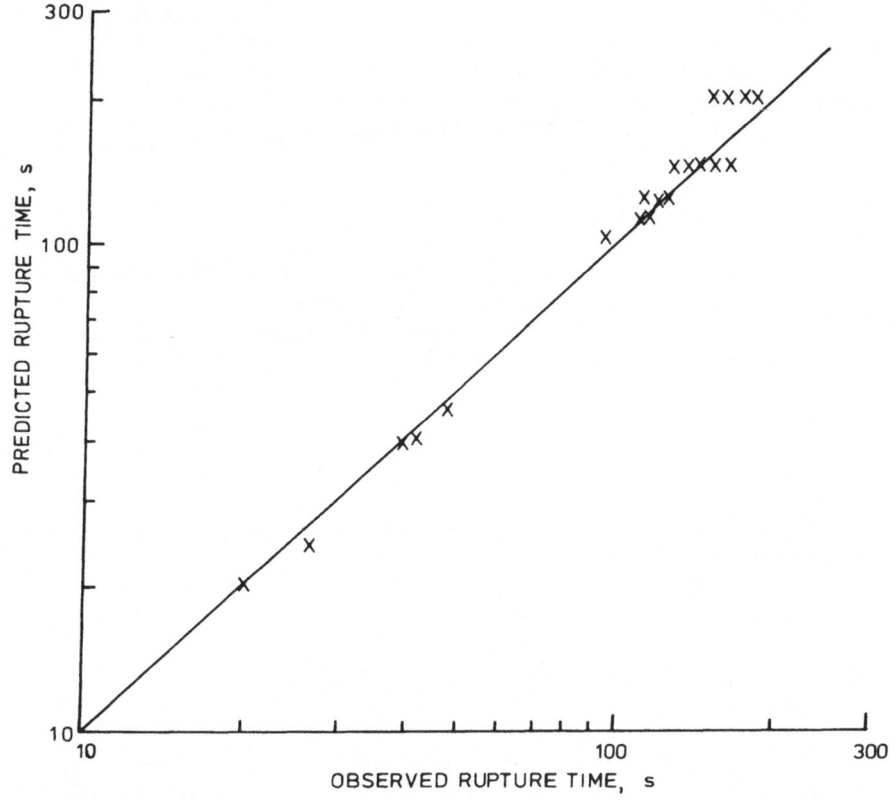

Figure 22

Comparison of CANSWEL 1 Model Predictions
With Experimental Failure Times in PWR Transients

B.   CANSWEL 2

The reason that the CANSWEL 1 code gives an increasingly poor fit to data obtained from tests in steam at the
higher temperatures is primarily because of the strengthening effect of dissolved oxygen that tends to offset the
weakening associated with temperature rise.   In CANSWEL 2,
this effect is taken into account.

It seems, in fact, that a shortcut may be acceptable in a proportion of cases. Thus, the equation that theoretically determines the rate of creep ($\dot{\epsilon}$) at stress σ, due to dislocations when their glide is opposed by solute drag, has been calculated by Bittus (*30*) and is of the form:

$$\dot{\epsilon} = \{C/(1 + C)\}^n \exp (E_1 - [Q_1/RT])$$

Here, C is to a first approximation proportionate to the concentration of dissolved solute (oxygen). As the latter is, at least at equilibrium, an Ahrrenius function of temperature, the following equation can be justified:

$$C = \exp \left(E_2 - [Q_2/RT]\right)$$

Suppose now that $Q_2 \approx Q_1$:

Then the creep rate becomes independent of temperature above a temperature T_c K defined by

$$T \geq T_c : C \geq 1$$

so $\exp (E_2 - Q_2/RT_c) = 1$; i.e., $E_2 = Q_2/RT_c$.

This class of behavior has in fact often been observed, and the creep data for tests done in steam can be used to define the values of the parameters in the above creep equation.

The development of this code is under way and it is expected to be in use early in 1977 when its validity will be first checked.

VII. FUTURE DEVELOPMENT

It will be clear from the foregoing parts of this report on SGHWR fuel that not many problem areas remain to be studied; however, two areas identified for further work are:

a. fuel cladding corrosion

b. incidence of defects

An intensive program of work is under way to ascertain the mechanism of nodular corrosion and to find ways of reducing its severity. This includes consideration of the use of alternate cladding alloys, although a change of alloy would not be entered on lightly because of the magnitude of the proving work involved.

The defect incidence of less than 1 in 10^3 fuel pins is an irritant only, but does involve expense and must be looked at in the context of increasingly strict limits on activity release from nuclear power station sites. In view of the conclusions of the review of causes of defects, it is apparent that the defect rate can be reduced by attention to component specifications and quality assurance procedures. Both are under review and have already shown benefits. It remains to be seen where the economic limit lies, but hopefully, it will be in the region of 1 in 10^4 fuel pin defects or better.

Defects can be produced by power ramps, and some operational guidelines are required to avoid them. Further refinement of performance models and more extensive validation will be part of an ongoing development program.

Although not a problem against the current reactor design specification, the power restrictions imposed by accident conditions are such that if they were removed, ratings could be almost doubled before other restrictions were met. In the ongoing development program for future CSGHWR's, changes that will go some way to reducing this gap will have high priority. Emphasis will be placed on improving blowdown heat transfer and reducing fission product gas release and stored thermal energy. It is unlikely that there will be further changes in fuel rod diameter, but fuel element structural changes may be made, particularly in the 60-rod element.

REFERENCES

1. Nuclear Reactor Systems for Electricity Generation, HMSO, London, July, 1974.

2. Bradley, N., Dawson, D. J. and Johnson, F. G., Engineering design of SGHWR's, Proceedings BNES Conference, "Steam Generating and Other Heavy Water Reactors," London, 1968, Page 11.

3.   Smith, D. R. and Phillips, J. L., "The SGHWR – Design
      and Operational Experience," Nuclex 1975, Technical
      Meeting 3/13.

4.   Pickman, D. O., "Design of Fuel Elements," Nuclear
      Engineering Design 21, 1972, Page 303.

5.   Pickman, D. O., Willey, D. H. and Eldred, V. W.,
      Proceedings BNES Conference, "Nuclear Fuel Performance,"
      Paper 51, London, 1973.

6.   Collins, D. A., Hargreaves, R. and Hughes, H., Proc.
      BNES Conference, "Nuclear Fuel Performance," Paper 49,
      London, 1973.

7.   Hargreaves, R., BNES Journal, to be published, 1976.

8.   Trowse, F. W., Garlick, A. and Sumerling, R., "Nodular
      Corrosion of Zircaloy-2 and Some Other Zirconium
      Alloys in SGHWR and Related Environments," Paper to be
      presented at ASTM Symposium, Zirconium in the Nuclear
      Industry, Quebec City, August, 1976.

9.   Tyzack, C. and Sheppard, M. F., Paper to be presented
      at ASTM Symposium, Zirconium in the Nuclear Industry,
      Quebec City, August, 1976.

10.   Pickman, D. O., "Fuel Performance in the Prototype SGHWR
      Power Station," UKAEA TRG Report 1943(S), HMSO, London
      December, 1969.

11.   Bond, G. G., Cordall, D., Cornell, R. M., Fox, W. N.,
      Garlick, A. and Howl, D. A.,"SGHWR Fuel Performance
      Under Power Ramp Conditions," BNES, 1976.

12.   Garlick, A., J. Nuclear Materials 49, Page 209, 1973.

13.   Gittus, J. H., Howl, D. A. and Hughes, H., "Theoretical
      Modeling of Nuclear Fuel Performance," Report of Work
      at the UKAEA Springfields Laboratories, TRG Report
      2743(S), 1975.

14.   Gittus, J. H., Howl, D. A. and Hughes, H., Nuclear
      Applied Tech. 9, PP 40-46, 1970.

15.   Gittus, J. H., Nuclear Engineering Des. 18, PP 69-82,
      1972.

16. Gittus, J. H., <u>Creep Viscoelasticity and Creep-Fracture in Solids</u>, London: Elseviers Applied Science Publishers Ltd; New York: Halstead, Division of Wiley, 1975.

17. Gittus, J. H., <u>ORNL TM 2857</u>. See also, Gittus, J. H., Nuclear Engineering Design <u>28</u>, PP 252-256, 1974.

18. Gittus, J. H., Proceedings International Conference Metallurgy of Reactor Fuel Elements, CEGB, Berkeley Nuclear Laboratories, Gloucestershire, England, PP 369-373, 1973.

19. Smith, D. R., Bradley, N., Hicks, D. and Cowan, A., "SGHWR Loss-of-coolant Accidents," Nuclex 1975, Technical Meeting 5/09.

20. Scatena, G. J., "Fuel Cladding Embrittlement During a Loss-of-coolant Accident," <u>NEDO 10674</u>, October, 1972.

21. USNRC, Technical Information Division, "The Role of Fission Gas Release in Reactor Licensing," <u>NUREG-75/077</u>.

22. Hindle, E. D., UKAEA RFL Springfields Internal Document.

23. Clay, B. D., Redding, G. B., "Creep Properties of Alpha-phase Zircaloy-2 Cladding Relevant to the Loss-of-Coolant Accident," CEGB Report <u>RD/B/N3187</u>, March, 1975.

24. Hardy, D. G., "High Temperature Expansion and Rupture Behavior of Zircaloy Tubing," ANS Topical Meeting on Water Reactor Safety, Page 254, 1973.

25. Clendenning, W. R., "Primary and Secondary Creep Properties for Zircaloy Cladding at Elevated Temperatures of Interest in Accident Analyses," Third International Conference Str. Mech. Reactor Tech. <u>C2/6</u>, London, September, 1975.

26. Gittus, J. H., "Creep of Two-phase Material," Presented at the Royal Society's Discussion Meeting on Creep in Engineering Materials and the Earth, London, The Royal Society, 1977.

27. Hindle, E. D., UKAEA RFL Springfields, internal document.

28. Emmerich, K. M., Juenke, E. F. and White, J. F.,
     "Failure of Pressureized Zircaloy Tubes during Thermal
     Excursions in Steam in Inert Atmospheres," ASTM STP
     458, PP 252-268, 1969.

29. Hobson, D. O., Osborne, M. E. and Parker, G. W.,
     "Comparison of Rupture Data from Irradiated Fuel Rods
     and Unirradiated Cladding," Nuclear Technology, Vol. 11,
     Page 479, August, 1971.

30. Gittus, J. H., Acta Met. 22, Page 1179 (Equation 12),
     1974.

THE NUCLEAR SAFETY RESEARCH REACTOR (NSRR)
IN JAPAN

Michio Ishikawa and Teruo Inabe

Japan Atomic Energy Research Institute
Tokai Research Establishment
Tokai-Mura, Ibaraki-Ken, Japan

I.   INTRODUCTION

A.   Background

Safety has been an important consideration from the very beginning of the development of nuclear reactors. Especially in recent years, the significance of reactor safety research has been emphasized so much that the research work in this field is entering a new stage.

The recent requirement for reactor safety research seems to have been brought about by the fact that the accident consequences are increasing as plant capacity becomes large, although today's plants are well-provided against accidents, and that the afety margin in reactor design, once applied excessively, is being reduced to a practical and economical level through advances in reactor technology and accumulation of experience in reactor operation. This situation requires precise information to better resolve reactor behavior problems under off-normal operating conditions.

Another factor that encourages reactor safety research is reconsideration by modern technology and industry as a result of side effects that have caused undesirable changes of nature and environmental pollution. Fortunately, the

nuclear industry has so far been less harmful in this respect than other industries. But continuous efforts must
be made to secure safety since the nuclear technology is a
typical big and complex modern science and has the role of
providing a major portion of the world's energy in the future.

Based on this situation, many reactor safety research
programs are under way or are being planned in various
countries, and technical collaboration including exchange
of information among these programs is being initiated.

The Nuclear Safety Research Reactor (NSRR) program to be
presented in this paper is one of the experimental research
programs related to reactivity-initiated accidents (RIAs)
of light water power reactors (LWRs) and fast breeder reactors (FBRs) being carried out in the Japan Atomic Energy
Research Institute.

Since an RIA results from the loss of appropriate control of nuclear fission and is an accident unique to nuclear
reactors, the research on RIAs was started at the very beginning of nuclear reactor development. Early research
work in this field was on reactor kinetics in super-prompt
critical states represented by the BORAX (1) and SPERT (2)
experiments in the United States.

In spite of these efforts, however, more than a dozen
RIAs took place in the history of nuclear reactor development due to the lack of practical knowledge and experience.
The most serious of these cases was the SL-1 accident (3) in
1961, which was caused by manual withdrawal of a control rod
from the reactor core during maintenance work. The reactor
core was completely destroyed and the pressure vessel suff-
ered damage showing that violent mechanical energy was generated in the reactor. Three workers were killed in this
accident.

After that incident, major research work on RIAs shifted
to reactor destructive experiments as demonstrated in the
SPERT-1 (4) and SNAPTRAN (5) in the United States. From these
tests, it was found that mechanical energy is generated only
when fuel failure occurs. It was then understood that hot
molten fuel dispersed into the surrounding coolant following the fuel failure brings about very rapid vapor formation

and this can result in the generation of violent mechanical energy like steam explosion.

Since the consequences of RIAs were made clear to such an extent through these experiments, further expensive reactor destructive tests may not be required. More economical and smaller-scaled experiments to investigate fuel failure phenomena in a confined environment, a capsule, or a loop, under a power excursion state were believed to be more appropriate. In the fuel failure experiments with capsules or loops, tests can be repeated as many times as necessary to obtain statistically reliable data, and detailed information on fuel behavior and on the consequences of fuel failure can be obtained by attaching adequate instrumentation.

The first trial of fuel failure experiments was started in the SPERT-CDC (6) in the United States in the late 1960's. These experiments made clear the fundamental behavior of LWR fuel rods under overpower transients with cold startup conditions, and thus made a significant advance in safety research on RIAs, and contributed to establishing the safety criteria for reactor design and operation. Because of facility limitations, however, the experiments could not well simulate practical reactor operating conditions.

As discussed at the beginning of this section, today's nuclear power reactors do not have excessive safety margins, and hence, require the information to be as precise as possible to provide assurance of reactor safety. The NSRR program has been planned in the light of this background.

The NSRR is a pulsed reactor that can generate very rapid overpower transients to irradiate clustered fuel rods as well as a single fuel rod under realistic system conditions with respect to temperature, pressure, flow rate, and other properties of the coolant surrounding the test fuel rods. We expect that the NSRR experiments will provide practical data on fuel behavior during RIAs, both for regulating use and design use.

B.   Objectives and History of the NSRR Project

The primary objective of the NSRR project is to clarify fuel behavior during rapid overpower transients that corres-

pond to reactivity-initiated accidents and thus to help
toward securing greater safety of nuclear reactors. The
major concerns on fuel behavior from the safety point of
view are:

   1.   What are the fuel failure thresholds in RIAs?

   2.   Can mechanical energy generated following fuel
        failure give any damage to the pressure boundary?

     In finding answers to these questions, fundamental
knowledge concerning fuel failure mechanisms is required.
There can be several patterns of fuel failure with different
thresholds, such as failure caused by melting of cladding,
melting of fuel pellets, increase in internal pressure,
pellet-cladding mechanical interaction, and so forth. If
the mechanism of fuel failure is different, the mechanical
energy generated by fuel failure may work in a different
form.

     It follows, then, that the primary objective of the
NSRR experiments is to ascertain fuel failure mechanisms
and their thresholds. Another objective is to ascertain
the relation between fuel failure mechanisms and the behavior of mechanical energy. If fuel failure mechanisms
are clarified and factors affecting the mechanisms are
distinguished, it will help toward improving the design of
reactor fuel.

     The difficulties in attaining these ends are that fuel
failure is a rather irregular event with a certain variation in threshold, and that fuel failure mechanisms can
be too complicated to identify in a single test. Therefore,
experiments should be repeated as many times as necessary to
provide reliable data from a statistical point of view and
to enable detailed examination of failure mechanisms with
changing parameters pertaining to fuel design.

     Fortunately, test fuel rods do not become so radio-
active in an irradiation with pulsing power in the NSRR and
they can be handled in a small glove box instead of a large,
hot cell; this makes it possible to carry out numerous experiments. Furthermore, capsules mainly are used rather
than loops as experimental rigs in the NSRR experiments, because capsules are easy to handle and facilitate the experi-

ments. Specifically, experiments with capsules with ordinary system conditions are frequently carried out to obtain
fundamental data on fuel behavior and to study the effects
of many fuel design parameters. Experiments with capsules
with system conditions of high pressure and high temperature, or with loops that simulate the operating condition
of power reactors will be carried out selectively when required to ascertain the effects of pressure, temperature
or flow rate of coolant on fuel behavior. As discussed
above, numerous tests to obtain reliable data are the significant consideration in the NSRR program.

The planning of the research program and investigation
of the driver core were initiated in 1968. Several types
of pulsed reactors were examined, and the TRIGA-Annular
Core Pulse Reactor (ACPR) (7) was finally chosen in 1970
because of its sufficient pulsing capability and its having
a large experiment cavity. The construction of the facility
was started in 1973, and it was completed in 1975. The
reactor reached its first criticality on June 30, 1975.
After three months of reactor performance tests, the first
experiment on fuel behavior was carried out in October, 1975.
Since then, a total of 37 experiments have been performed
as of the end of June, 1976, when this paper was prepared.

II. NSRR FACILITY AND EXPERIMENTAL CAPABILITY

A. Description of the NSRR Facility

The major facility of the NSRR consists of the reactor
building, reactor control building, test fuel handling
building, sodium handling building, and mechanical equipment building as shown in Figure 1.

The reactor building houses the reactor and also includes facilities necessary to carry out primary inspections
on test fuel rods after the transient tests. The reactor
control building is located at about 30 meters away from the
reactor building. Also included in this building is a data
acquisition and processing system to record the transient
data of the experiments. The mechanical equipment building
which contains water control, electric power, and ventilation equipment, is located at right angles to the control
building direction so that electric cables from the mechanical equipment building will not give any perturbation to

the signals from the instrumentation. In the test fuel
handling building, test fuel rods are fitted into capsules
or loops with necessary instrumentation and undergo inspection before transient tests. Special inspection of test
fuel rods after transient tests can also be performed in
this building. The sodium handling building has facilities
necessary to handle sodium for the experiments with FBR fuels.

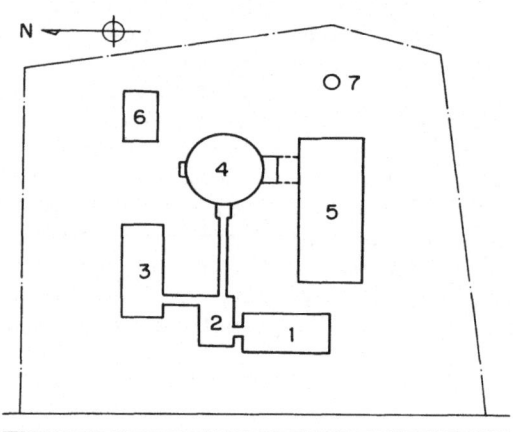

1 . Laboratory

2 . Reactor Control Bldg.

3 . Test Fuel Handling Bldg.

4 . Reactor Bldg.

5 . Mechanical Equipment Bldg.

6 . Sodium Handling Bldg.

7 . Ventilation Stack

Figure 1

Layout of the Facilities

The driver core of the NSRR is a modified TRIGA-
Annular Core Pulse Reactor whose features are: (1) the use
of uranium-zirconium hydride (U-ZrH) fuel-moderator elements
with a large prompt negative temperature coefficient of reactivity, which allows insertion of positive reactivity of up
to 3.43% Δk/k ($4.70) for pulsing operation; and (2) the
existence of a large dry experiment cavity (220 mm in inner
diameter) in the central region of the core, where test fuel
rods contained in a capsule or a loop are installed and exposed to a high pulse power.   The core structure is mounted
on the bottom of a 9 m deep, open-top swimming pool as shown
in Figure 2.

The top of the core is approximately 7 m below the
water surface.   The experiment tube extends from the top of
the pool, passing through the core, to the subpile room be-
neath the pool bottom.   The experiment tube consists of
three parts: (1) a test space in the core region; (2) a
vertical loading tube; and (3) an offset loading tube.   The
outer wall is made of aluminum and has a hexagonal shape to
match the lattice pattern of the fuel-moderator elements
with small clearance to achieve high fast neutron fluence
in the test space.   The inner wall is a cylindrical stainless steel tube, and works as a thermal neutron absorber.
It provides nuclear decoupling between the test space and
the driver core to reduce perturbation given by an experimental capsule or a loop to the pulsing performance.

Access to the test space is provided by the vertical
loading tube and the offset loading tube that join in a "Y"
fitting at the top of the test space.   The offset loading
tube is designed for experimental capsules of the maximum
size of approximately 0.2 m in diameter by 1.2 m in length.
Because of the offset, no radiation-shielding plug is required in the offset tube; this allows easy insertion and
removal of a capsule from the tube and thus facilitates
numerous experiments.   The vertical loading tube contains a
shielding plug suspended inside the tube to prevent radiation
from streaming up the tube.   The vertical loading tube is
intended for experimental capsules or loops too long to
handle through the offset loading tube.

Inside the vertical tube, a capsule hold-down device
is provided which extends downward until it rests on top of
a capsule to prevent the capsule from jumping up, due to the

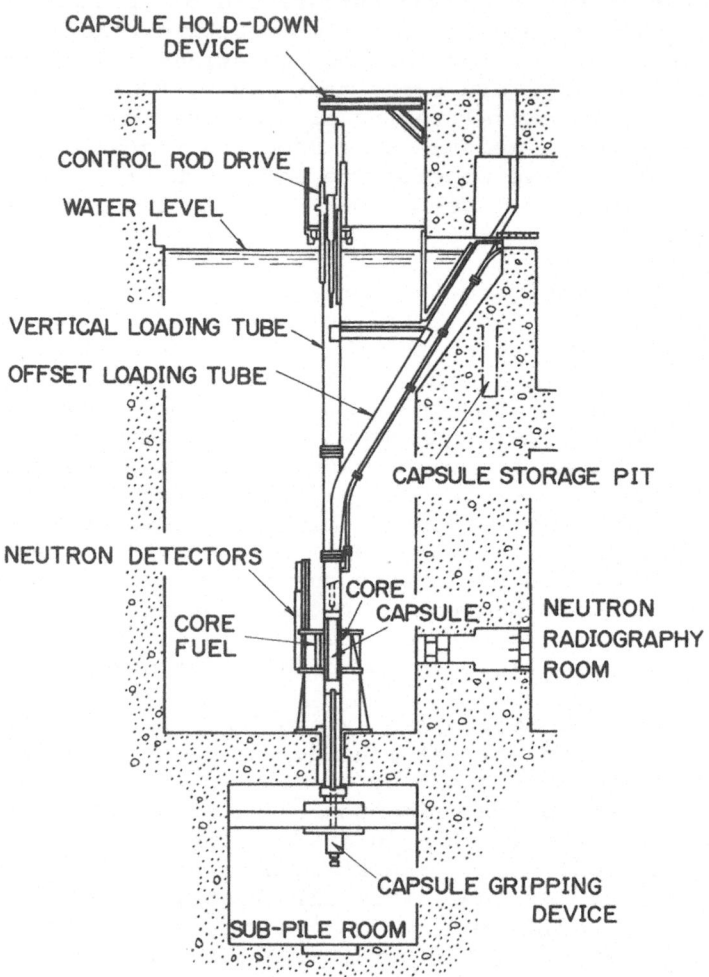

Figure 2

Vertical Cross-section of the NSRR

mechanical energy generated inside the capsule during the
tests. In addition, a capsule gripping device is provided
inside the lower section of the experiment tube as another
mechanism to hold a capsule firmly. The gripping device is
fixed to two 300 mm I-beams whose ends are embedded in the
walls of the subpile room, and it can withstand a shock
load of up to 700 kN. The subpile room is also intended to
accommodate a loop supporting structure and other equipment.

The driver core contains U–ZrH fuel-moderator elements
in which the ZrH moderator is homogeneously combined with
enriched uranium fuel. The active section of this fuel-
moderator element is 381 mm in length by 35.6 mm in diameter
and contains 12 weight percent uranium and 88 weight percent
ZrH. The uranium is enriched to 20 percent in ^{235}U and the
hydrogen-to-zirconium atom ratio of the ZrH moderator is 1.6.
Graphite cylinders 87 mm in length act as top and bottom reflectors. The fuel meat and the top and bottom graphite
cylinders are contained in a stainless-steel cladding 0.5 mm
thick. The cladding is provided with dimples that act as
spacers to ensure a thermal gap between the fuel meat and
the cladding.

The nuclear feature of this core is to have the large
prompt negative temperature coefficient of reactivity
(-9.3×10^{-5} $\Delta k/k$. This temperature coefficient arises
primarily from a change in the disadvantage factor resulting
from the neutron-spectrum hardening effect in the fuel-
moderator elements due to the temperature rise in the moderator (8). This effect is prompt because the fuel is intimately mixed with the moderator; thus, the fuel and moderator temperatures rise simultaneously. In addition to
this feedback effect, Doppler broadening of ^{238}U resonances
also contributes to the prompt shutdown process as the fuel
is heated.

The fuel-moderator elements are arranged on a triang-
ular spacing grid to form a hexagonal pattern surrounding
the experiment tube. The operational core loading of the
reactor comprises 149 fuel-moderator elements, 8 fuel-
followed control rods, and three transient rods, as shown
in Figure 3. The transient rods are boron-carbide poison
rods with air-followers, and the pulsing operation is performed by quick withdrawal of the transient rods from the
core using pneumatic drive systems. One of these rods, the

adjustable transient rod, also can be moved by a rack-pinion drive system to adjust reactivity so that pulses of any size may be obtained. The total amount of reactivity worth of the three transient rods is approximately 3.7% $\Delta k/k$ ($5).

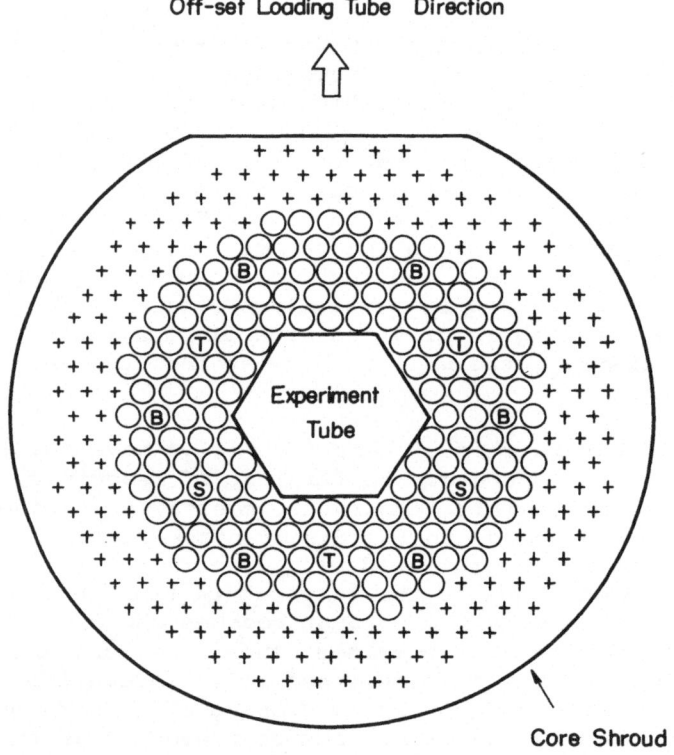

Figure 3

Cross-Section Plane View of the NSRR Reactor Core

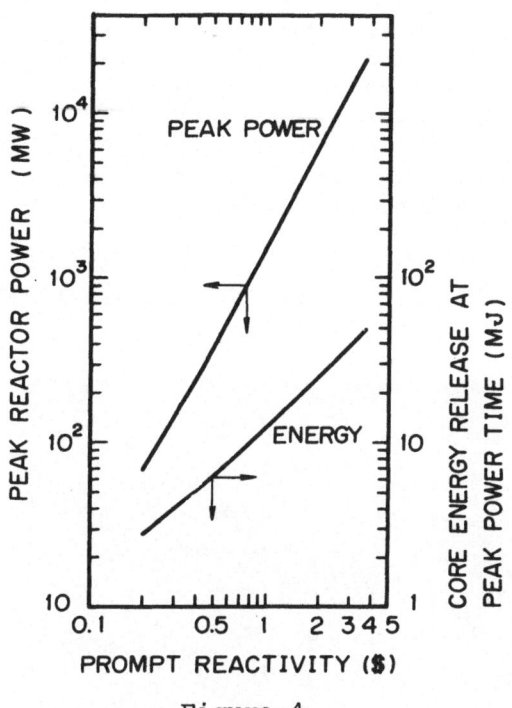

Figure 4

Peak Reactor Power and Core Energy Release at Peak Power
Time as a Function of Prompt Reactivity Insertion.*

B.   Reactor Performance and Experimental Capability (9,10)

The maximum reactivity insertion allowed for pulsing operation of the NSRR is 3.43% $\Delta k/k$ ($4.70), and the actual re-
active stroke of the transient rods can be withdrawn within
50 ms; thus, the maximum reactivity insertion rate is
approximately $100/s. This reactivity insertion rate is
larger than any other one in credible reactivity-initiated
accidents.

Figure 4 shows the measured peak reactor power and
core energy release at peak power time as a function of the
prompt reactivity insertion,* $\delta\rho$. As shown in this figure,

---

*   Defined as the portion that is above critical.

the peak powers for pulses vary approximately as the second power of $\delta\rho$, and the energies released from pulses vary approximately as the first power of $\delta\rho$.

Figure 5 is a photograph of the reactor at the instant of a pulsing operation. The core region is flashing in light blue because of the Čherenkov effect.

Figure 6 shows the measured power shape with the reactivity insertion of 3.41% $\Delta k/k$ ($4.67) that is the maximum

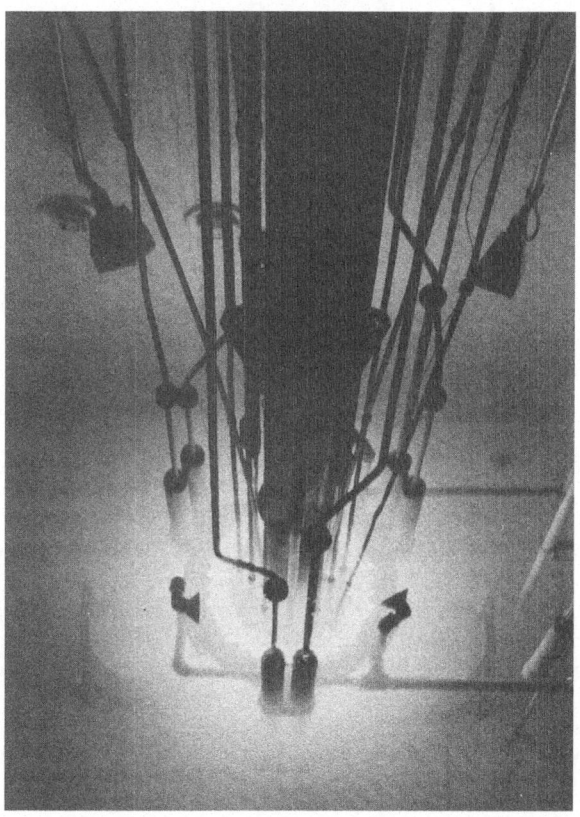

Figure 5

Photograph of the Reactor Core
At the Instant of a Pulsing Operation

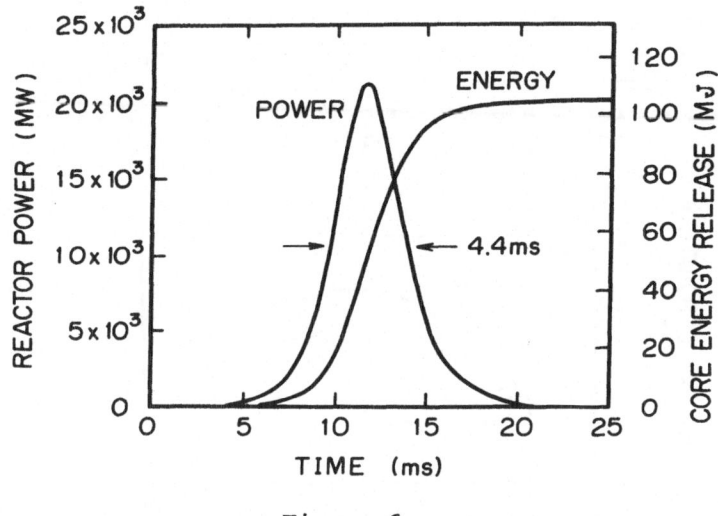

Figure 6

Reactor Power and Core Energy Release in the Pulsing Operation
With $4.67 Reactivity Insertion

value attained so far. The power shape is quite sharp,
with the pulse width at half maximum power of 4.4 ms.

In this pulse, the peak reactor power is 21,000 MW,
with a prompt energy release of 105 MJ and a minimum reactor
period of 1.13 ms. The minimum reactor period of the NSRR
is much shorter than those in hypothetical accidents in LWRs,
such as control rod drop in BWRs or control rod ejection in
PWRs, which are believed to be not less than 3 ms, and is
rather close to the periods in control rod withdrawal accidents in FBRs, which are estimated to be about 0.5 ms.

The fast neutron fluence in the experiment tube given
by the maximum pulse is estimated to be about 5×10^{18}
neutrons/m^2. The fast neutrons are thermalized by the moderator in a capsule or a loop to be utilized for nuclear
fissions in test fuel rods; hence, the heat deposition in

test fuel rods is dependent on the amount of moderating
material as well as the enrichment of test fuel. Figure 7
shows the relation between calculated heat deposition in a
single LWR fuel rod and the radial thickness of water surrounding the fuel rod in a capsule (8).

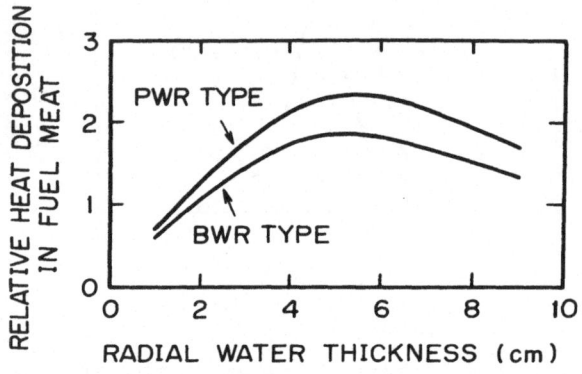

Figure 7

Relative Heat Deposition in an LWR Fuel Rod
As a Function of Radial Thickness of Water in a Capsule

     Heat deposition in a fuel rod reaches the maximum
value with the water thickness of about 60 mm. Based on
this result, the inner diameter of standard water capsules
has been decided to be 120 mm. Thermal neutron flux level
measured in this capsule is much higher than that in the
driver core as shown in Figure 8, and the average thermal
neutron flux in a 2.6% enriched BWR fuel rod contained in
a capsule is approximately 2.5 times higher than that in
the core region.

     This high flux in the capsule makes it easy to melt
test fuel without causing any damage to the driver core
fuel. Figure 9 shows adiabatic heat deposition in a single
fuel rod contained in a water capsule as a function of reactivity insertion for pulsing operation. With the maximum
pulse, a 2.6% enriched BWR type fuel rod gains about
1 MJ/kg UO$_2$, which corresponds to the maximum fuel enthalpy
assumed in a control rod drop accident, and a 5% enriched

PWR type fuel rod, for example, gains about 1.6 MJ/kg UO$_2$
which exceeds the melting enthalpy of UO$_2$ pellets.

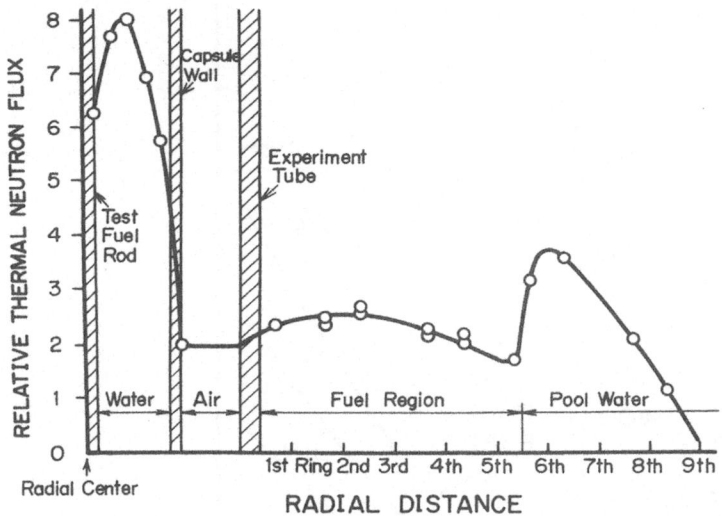

Figure 8

Radial Thermal Neutron Flux Distribution in the Driver Core
and the Standard Water Capsule Containing a 2.6% Enriched
UO$_2$ Fuel Rod

A fuel failure experiment is carried out not only using
a single test fuel rod but also using clustered test fuel
rods. Heat deposition in clustered rods, however, is considerably reduced in comparison with that in a single rod
because of thermal neutron flux depression in clustered
fuel region. Figure 10 shows the calculated thermal neutron flux distribution in a water capsule with five clustered 2.6% enriched BWR fuel rods, and with a single fuel
rod of the same type. The heat deposition in the clustered
rods shows a decrease of about 35% in the central rod and
about 25% in each of the surrounding rods as compared with
that in a single rod.

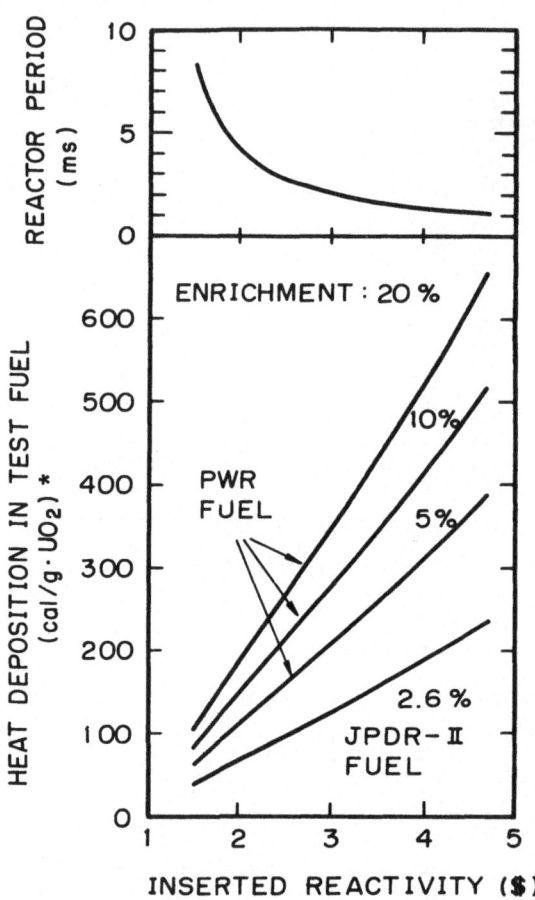

Figure 9

Reactor Period and Heat Deposition in a Single
Test Fuel Rod Contained in a Standard Water Capsule
as a Function of Reactivity Insertion
for Pulsing Operation
(* one calorie (cal) = 4.187 J)

In the experiments on FBR fuel behavior, capsules are filled with sodium. In this case, it is necessary to introduce some neutron-moderating material into the capsule, since sodium does not work as a moderator. Nuclear calculations indicate that more than 50% enrichment in ^{235}U is required to make an FBR type test fuel rod melt in a sodium capsule without a moderator. On the other hand, it was found that an approximate 20% enriched fuel rod can be melted with the maximum pulse if a 30 to 40 mm thick zirconium hydride moderator layer surrounds a capsule.

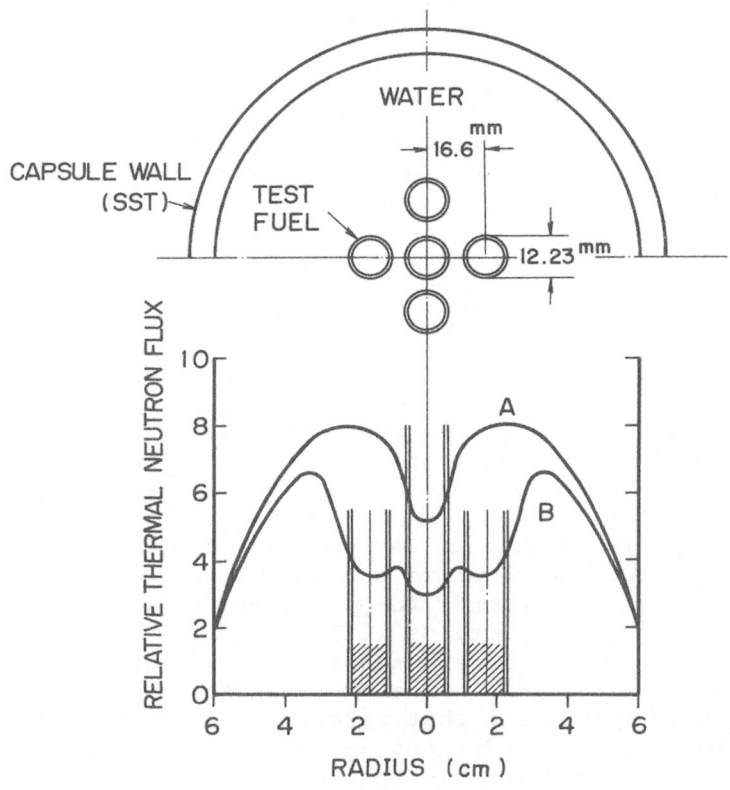

Figure 10

Calculated Thermal Neutron Flux Distribution in a Standard Water Capsule with a Single 2.6% Enriched UO2 Fuel Rod (A), and With Five Clustered Fuel Rods of the Same Type (B)

C.   Experimental Capsules and Loops

     The fuel failure experiments should be repeated so
many times the results can be reliable from a statistical
point of view. In addition, it is required to carry out
wide-range parametric studies on the factors pertaining to
fuel design that may affect fuel failure mechanisms such
as cladding material, heat treatment of cladding tubes,
fuel dimension, fuel internal pressure and so forth. In
order to carry out these numerous experiments efficiently,
standard capsules are mainly used. The standard capsule is
120 mm in diameter by 1.2 m in length. This dimension
allows easy insertion and removal from the offset loading
tube as discussed in Section IIA, and it is possible to
carry out two or three experiments in a day.

     Figure 11 shows the details of the standard capsule.
The capsule is made of stainless steel. A seal for instrumentation wires and a purge valve to release gaseous fission
products after the transient test are provided at the top of
the capsule. A sensor to measure capsule internal pressure
can be installed at the bottom of the capsule.

     The central part of the capsule body is 7 mm thick,
and the upper part is 28 mm thick. These thicknesses are
designed to withstand the pressure pulse of 4.3 MPa* at the
central part and the waterhammer of 24 MPa at the upper
part, respectively (11). The lower part of the capsule is
designed to fit the capsule gripping device. The total
weight of the standard capsule is approximately 100 kg.

     The standard capsule is designed for the experiments
with water at room temperature and ambient pressure. This
system condition corresponds to cold-startup conditions in
BWRs; hence, system conditions with high pressure and high
temperature, which are more general in power reactor operation, are also required in the experiments. In fact, safety
analyses on LWRs indicate that reactivity-initiated accidents in hot-standby conditions are much more severe in
most cases than those in cold startup conditions; therefore,
high-pressure capsules are being developed in the NSRR.

_____

*   1 pascal (Pa) = 1 N/m^2

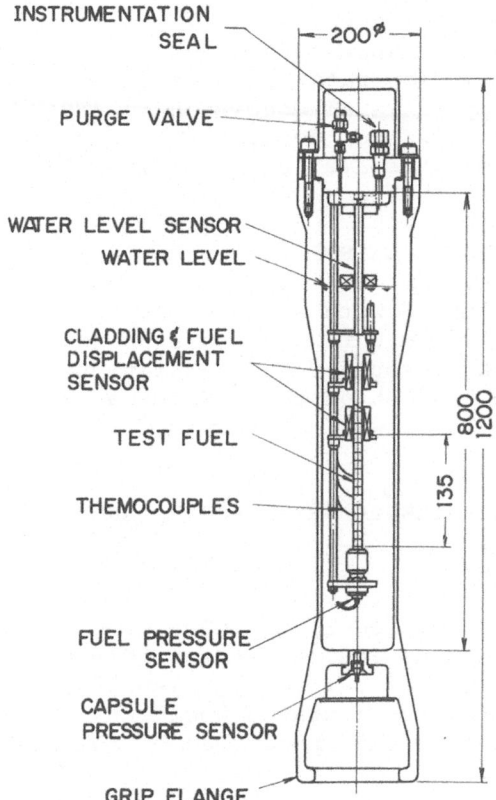

Figure 11

Standard Water Capsule
(unit = mm; φ = outside diameter)

    These capsules are designed to simulate system condi-
tions in BWRs and PWRs with respect to pressure and temperature, and they can be raised up to 15 MPa and its saturation
temperature of 350°C. Figure 12 shows the concept of the
high-pressure capsule design.

In addition, loops are being developed in order to simulate flow conditions as well as pressure and temperature of
coolant. The maximum flow rate in loops can be 5 m/s, which
corresponds to the operational condition in PWRs. Figure 13
shows the concept of the loop design. The high-pressure
capsules and loops will be available for in-pile experiments
from 1977 on.

For the experiments on FBR fuel behavior, capsules and
loops that contain sodium are being developed. A capsule or
a loop will be surrounded with a zirconium hydride moderator
layer approximately 40 mm thick to increase heat deposition
in test fuel rods, as discussed in the previous section.
Figure 14 shows the concept of the sodium capsule design.
Sodium capsules and loops will be served for in-pile experiments from 1980 on.

III.   EXPERIMENTAL RESEARCH PROGRAM

A.   Aims of the Research

As described in Section I, the NSRR program aims at
clarifying the thresholds and mechanisms of fuel failure
and the behavior of mechanical energy resulting from fuel
failure during reactivity-initiated accidents. In this
section, we will discuss the objectives of the research a
little more in detail.

In RIAs, reactor fuel is rapidly heated and can suffer
failure from various causes such as melting of cladding,
melting of fuel meat, mechanical interaction between fuel
meat and cladding, increase in pressure inside a fuel rod,
and so forth. In the case of fuel failure due to melting
of cladding, for example, fuel meat is promptly heated with
the power burst, but cladding temperature rises very little
during this period because of low gap conductance.* Then
the gap conductance increases rapidly as the fuel meat ex-
pands and gap gas temperature rises. This results in rapid
heat transfer from fuel meat to cladding, and cladding
temperature reaches its melting point.

---

*   A reactor period in a credible RIA in LWR that may cause
     fuel failure is believed to be about 3 to 10 ms; in this
     case, all of the burst energy is released within about
     30 to 100 ms.

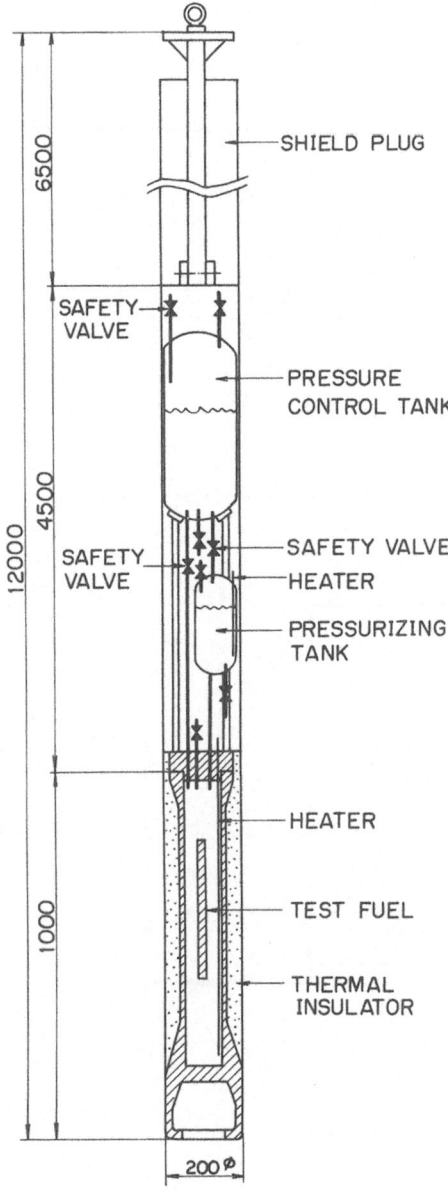

Figure 12

Concept of the High-Pressure Capsule Design
(unit = mm; φ = outside diameter)

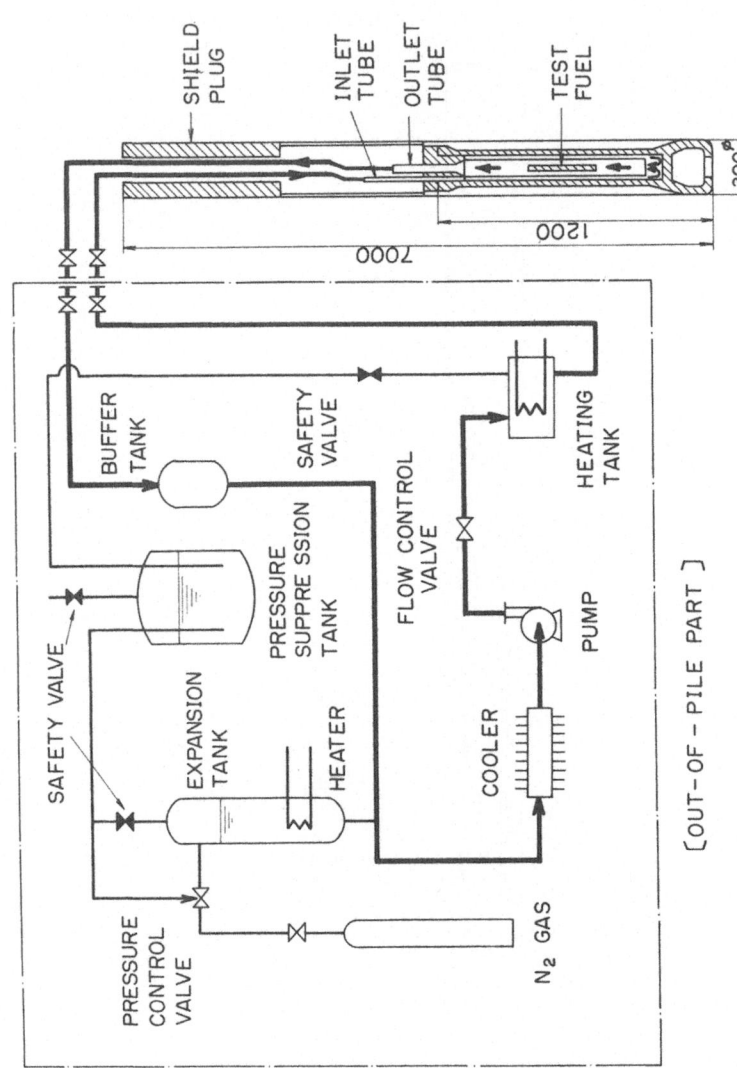

Figure 13

Concept of the Water Loop Design
(unit = mm; φ = outside diameter)

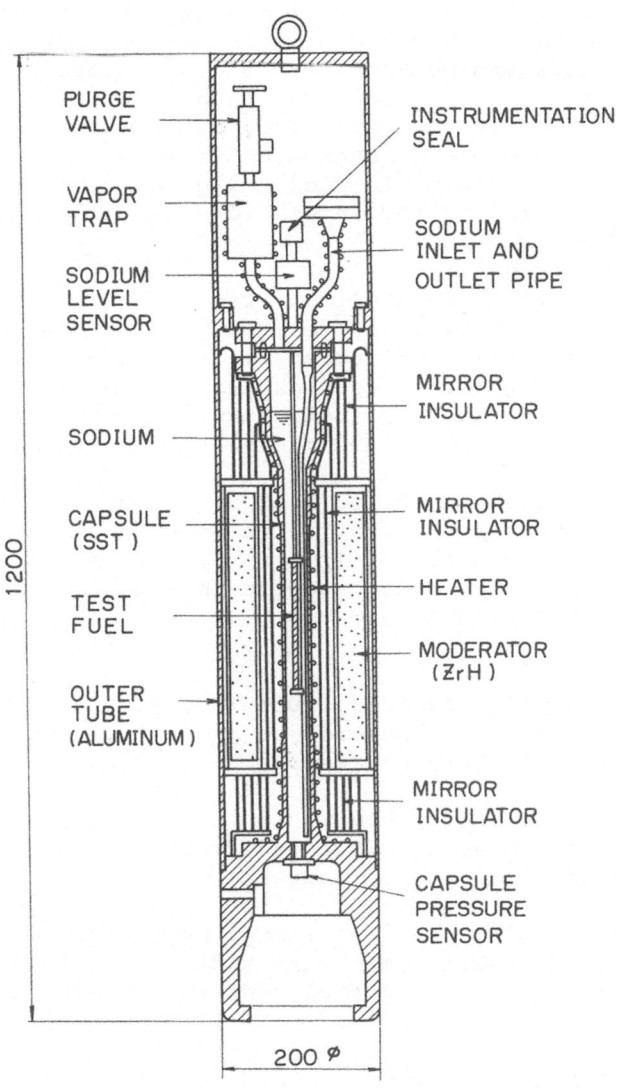

Figure 14

Concept of the Sodium Capsule Design
(unit = mm; φ = outside diameter)

The investigation of fuel failure mechanism of this
kind requires detailed study of transient heat transfer
problems such as gap conductance in a fuel rod and cooling
conditions of surrounding coolant; thus, the parametric
studies in which gap conductance is changed by changing
radial gap width or gap gas pressure, or in which coolant
condition is changed are of much significance. In this way,
many parameters pertaining to fuel design will be investigated so that mechanisms and thresholds of fuel failure may
become clarified.

For these parametric studies, standard capsules containing single test fuel rods are mainly used; however, since
fuel behavior is fairly influenced by cooling conditions,
experiments with high-pressure capsules and loops are required to study the effect of temperature, pressure and
flow rate of coolant on failure mechanisms and thresholds.

Experiments with clustered test fuel rods are also performed to simulate a part of a fuel element assembly in
commercial nuclear reactors. It is expected that neighboring fuel rods in the cluster can give thermal influence to
each other and this might lead to the change in failure
threshold.

Experiments with clustered fuel rods may be more significant in studying the behavior of mechanical energy resulting from fuel failure because:

1. mechanical energy generation can induce failure
    propagation among clustered fuel rods, resulting
    in mechanical energy increase

2. behavior of mechanical energy may suffer a
    change depending upon surrounding structures
    such as neighboring fuel rods and a channel
    box

In addition, system pressure and temperature are believed to have a strong influence on mechanical energy generation because the boiling condition is directly controlled
by these system conditions.

In the investigation concerning fuel failure problems
discussed above, a few research items of interest from the

point of view of safety are being planned. One of them is
the study on fuel behavior during overpower transients with
rather small reactivity insertions not enough to cause fuel
failure. Since current nuclear power reactors are designed
based on severe safety criteria, it is hard to believe that
fuel failure would occur even if an abnormal reactivity insertion takes place; hence, a practical problem may be
whether fuel rods can be used continuously after such an
abnormal condition. From this point of view, studies on
bending, deformation, oxidation and reduction in mechanical
strength of cladding will be carried out.

Studies on coolant behavior during an overpower transient are also of much importance, since the boiling condition with very rapid heating of cladding still has many
questions to be solved. Another item to be studied is the
melt-through phenomenon resulting from molten fuel debris,
although the experiments on this subject will be on a small
scale.

Along with the experiments, computer codes to evaluate
thermal and mechanical transient behavior of a fuel rod before failure and to evaluate pressure pulse resulting from
fuel-coolant interaction after failure are being developed.
These codes are used for deciding detailed experimental conditions, and for better interpretation of experimental results. The analytical models will be improved, based on the
experimental data so they may predict fuel behavior and consequences of an RIA more precisely.

Research items in the NSRR program discussed above are
summarized as follows:

1. Study of the following items with respect to
    fuel failure mechanisms and their thresholds:

    a. fuel failure mechanisms and their
        affecting factors

    b. fuel failure thresholds with various
        failure mechanisms

    c. influence of neighboring fuel rods on
        fuel failure thresholds

d.   influence of temperature, pressure and
     flow rate of surrounding coolant on fuel
     failure mechanisms and their thresholds

e.   effect of fuel burnup on fuel failure
     mechanisms and their thresholds

2.   Study of the following items with respect to
     mechanical energy generation resulting from
     fuel failure:

     a.   mechanisms of mechanical energy generation

     b.   relation between fuel failure thresholds
          and the intensity of mechanical energy

     c.   influence of mechanical energy on
          neighboring fuel rods and other structures

     d.   influence of clustered fuel rods and other
          structures on the behavior of mechanical
          energy

     e.   influence of temperature, pressure and flow
          rate of coolant on mechanical energy gener-
          ation

3.   Study of other items related to RIAs:

     a.   fuel behavior during rather small abnormal
          power transients

     b.   coolant behavior during overpower transients

     c.   possibility of a melt-through event

B.   Test Fuel Rods

     The final objective of the research in the NSRR program
should be the behavior of commercial reactor fuels under
reactivity-initiated accident conditions; however, there
are many differences in commercial reactor fuel design, depending on reactor types such as rod diameter, cladding material, fuel internal pressure, fuel enrichment and so forth.
It is believed that these differences may work as influen-

tial factors on fuel failure mechanisms and failure thresholds. Therefore, direct application of commercial reactor
fuels to the experiments would not bring basic information
on general fuel failure mechanisms, although it may bring
practical results on specific reactor fuels. For this
reason, wide-range parametric studies concerning the factors
related to fuel design are required so as to have fundamental knowledge of fuel failure mechanisms.

In order to accomplish this end effectively, the fuel
rods with a particular specification are required as the
standard in evaluating the effect of change in design parameters; thus, what are called standard test fuel rods have
been decided, having the dimensions of typical PWR fuel
rods but not pressurized; they are clad with Zircaloy-4
tubes as described in Table I. The effect of design parameters will be examined in the experiments using what are
called parameter test fuel rods that vary from the standard
test fuel specification. Table II summarizes the design
parameters to be examined in the experiments.

In addition to these parameters, it is believed that
fuel burnup has a significant influence on fuel failure
thresholds through embrittlement of cladding, increase in
internal pressure, and transformation of fuel meat. Experiments with irradiated fuel rods, however, require a well-
equipped hot cell; therefore, experiments with irradiated
fuel rods are scheduled for after the completion of the experiments with unirradiated fuel rods. In addition, since
it is quite difficult to affix instrumentation to irradiated fuel rods, information from the experiments with irradiated fuel rods will be very limited, and must depend much
upon visual inspection by well-experienced eyes through numbers of unirradiated fuel tests. This is another reason
why irradiated fuel experiments are scheduled for after
unirradiated fuel experiments. However, during the time
before irradiated fuel experiments, burnup simulation tests
are scheduled as a part of parameter tests. In these tests,
fuel rods with unirradiated UO_2 pellets clad with irradiated
tubes, together with appropriate gas pressure and gas composition will be used to simulate burnup effects.

Another significant item is the experiments with waterlogged fuel rods. Since pin holes or small cracks can take
place in the cladding tube with a certain probability during

TABLE I

CHARACTERISTICS OF THE STANDARD TEST FUEL ROD, WIDE-GAPPED
FUEL ROD, AND JPDR-II TYPE TEST FUEL ROD

	Standard Fuel	Wide-gapped Fuel	JPDR-II Fuel
UO_2 Pellet			
Diameter, mm	9.29	9.09	10.66
Length, mm	10	10	15
Density (theoretical)	95%	95%	95%
Enrichment in ^{235}U	10%	10%	2.6%
Shape	Chamfered	Chamfered	Flat
Cladding			
Material	Zircaloy-4	Zircaloy-4	Zircaloy-2
Outer Diameter, mm	10.72	10.72	12.23
Wall Thickness, mm	0.62	0.62	0.70
Fuel Element			
Overall Length, mm	265	265	265
Active Length, mm	135	135	126
Weight of Pellets, g	96	91	117
Radial Gap Width, mm	0.095	0.195	0.085
Gap Gas Composition (Atmospheric)	Helium	Helium	Helium

reactor operation, some fuel rods with these defects can
become waterlogged while the reactor is shut down. In general, waterlogged fuel rods can suffer failure with rather
low threshold enthalpy because of an increase in internal
pressure resulting from vaporization of internal water;

TABLE II

MAJOR EXPERIMENTAL PARAMETERS FOR THE RIA TESTS
IN THE FIRST PHASE PROGRAM

Fuel Rod Specification	Rod dimension Pellet density Pellet shape Cladding material Heat treatment of cladding Radial gap width Gap gas pressure Gap gas composition Extent of irradiation of cladding Water volume in a rod Dimension and position of a defect in cladding Number of rods in a cluster

Coolant Condition	Water Capsule and Loop:
	
	    Temperature: Room temperature to 350 °C
	
	    Pressure: Atmospheric, to 15 MPa
	
	    Flow velocity: 0 - 5 m/s
	Sodium Capsule and Loop:
	
	    Temperature: 300 °C to 600 °C
	
	    Pressure: Atmospheric
	
	    Flow velocity: 0 - 5 m/s

Heat Generation Condition	Heat generation condition is controlled by the amount of the reactivity insertion and the enrichment of the pellets.

hence, the effects of water volume, and of dimensions and
positions of defects in waterlogged fuel rods on fuel failure mechanisms and failure thresholds will be investigated.

C.   Test Procedures and Program Schedule

      The first-phase program of the NSRR experiments with
unirradiated fuel rods is scheduled to be continued for
seven years, from 1975 to 1981, as shown in Table III.   The
program can be divided into two categories:

      1.   the first five years of the experiments on
           LWR fuel behavior

      2.   the last two to three years for FBR fuel
           behavior

      In the early stage of the program, effects of fuel de-
sign parameters such as fuel dimension, pellet shape, pellet
enrichment, cladding material, heat treatment of cladding,
gap gas pressure, extent of irradiation of cladding, volume
of water in waterlogged fuel rods and so forth will be
studied to obtain fundamental knowledge of fuel failure
mechanisms.   In these tests, standard capsules are mainly
used.   The system pressure and temperature of a standard
capsule correspond to a cold startup condition in BWRs.

TABLE III

TIME SCHEDULE OF THE RIA TESTS IN THE FIRST PHASE PROGRAM

	1975	1976	1977	1978	1979	1980	1981
Tests for LWR Fuel	Tests with Standard Capsules							
	Tests with High-Pressure Capsules							
	Tests with Water Loops							
Tests for FBR Fuel					Tests with Sodium Capsules		
						Tests with Sodium Loops	

Simulation of hot standby and operating power conditions will be made in the experiments with high-pressure capsules and loops, which will be started in 1977 and 1978, respectively. In these tests, the effects of pressure, temperature, subcooling and flow rate of coolant on fuel failure mechanisms and failure thresholds will be studied. Specifically, the effects of these system conditions on mechanical energy generation resulting from fuel failure will be intensively studied.

The experiments on FBR fuel behavior will be started in 1979. For these tests, sodium capsules will be used from 1979, and sodium loops from 1980. A total of about 400 tests with about 900 fuel rods will be carried out in the first-phase program.

IV. SOME RESULTS OF THE FUEL FAILURE EXPERIMENTS

The NSRR achieved its first criticality on June 30,1975, and following the three months of reactor performance tests, the experiments were begun in October on fuel behavior under RIA conditions. During the period from October, 1975 through June, 1976, a total of 37 tests were carried out using standard capsules containing single LWR fuel rods and water at room temperature and atmospheric pressure. During the transient tests, the responses of fuel rods and surrounding environments such as cladding surface temperature, water temperature, fuel internal pressure, water column motion, capsule internal pressure, and capsule strain were measured using the instrumentation fitted into the test fuel rods and capsules. These data were recorded in the data acquisition and processing system and in a visicorder. In addition, in one to two weeks after the transient tests, the capsules were disassembled, and test fuel rods were subjected to visual and dimensional inspection. The experiments performed during the aforementioned period consisted of the following three test series: (1) scoping tests; (2) wide-gapped fuel tests; and (3) waterlogged fuel tests. In this chapter, the major results of these tests are discussed so that the intention of the experiments may be better understood, although some of the test results still require more precise examination.

The results of these test series (12,13) are summarized in Table IV.

TABLE IV

RESULTS OF THE NSRR EXPERIMENTS (October 1975 ∿ June 1976)

Test Series	Test No.	Date	Type of Test Fuel	Reactor Period (msec)	Heat Deposition in Test Fuel (cal/g.UO_2)	Maximum Cladding Surface Temperature (°C)	Fuel Failure	Water Column Velocity (m/sec)	Capsule Internal Pressure (kg/cm^2)	Nuclear to Mechanical Energy Conversion Ratio (%)	Post-Test Observation	Remarks
	111-1	10.2.75	JPDR-II(2.6%E)	9.14	44	90	No failure	0	-	0	No visible change.	
	111-2	10.15.75	"	2.29	117	140	"	0	-	0	"	
	111-9	5.12.76	"	1.54	160	-	"	-	-	-	Fully oxidized, slight bowing.	
	200-3	12.3.75	Std. (10%E)	3.33	177	1200	"	-	-	-	Partially oxidized.	
	111-3	10.20.75	"	3.33	183	∿1200	"	0	-	0	Fully oxidized.	
	209-1-1 ⌇ 209-1-5	5.13.76	"	3.49	178 ⌇ 184	-	"	-	-	-	Fully oxidized, rather large bowing.	Five time irradiation.
Scoping Tests	111-10	5.12.76	JPDR-II(2.6%E)	1.19	211	1330	"	-	-	-	Fully oxidized.	
	200-4	2.20.76	Std.(10%E)	2.28	234	1680	"	-	-	-	Break-away in cladding surface.	
	200-1-1	2.26.76	"	27.0	39	147	"	-	-	-	"	Twice irradiation.
	200-1-2		"	2.28	242	1700	"	-	-	-	"	
	111-4	10.24.75	"	2.41	246	1690	"	0	-	0	"	
	201-1	5.14.76	"	2.32	246 + 39	1770	"	-	-	-	"	Irradiation with long run-out power.
	200-5	3.11.76	"	2.03	267	>1780	Failure	-	-	-	Penetrating circumferential crack in cladding.	
	200-5b	5.19.76	"	2.03	269	>1700	"	-	-	-	"	
	111-5	10.31.76	"	1.93	272	1720	"	0	-	0	"	
	200-6	3.5.76	"	1.93	273	>1780	"	0	-	0	"	
	200-2-1	2.19.76	"	5.79	117	179	"	-	-	-	Broken into 2 pieces while disassembling the capsule.	Twice irradiation.
	200-2-2		"	1.98	273	>1700	"	-	0	-	"	
	200-6b	5.20.76	"	1.95	278	>1800	"	0	0	0	"	
	200-7	3.12.76	"	1.77	298	>1800	"	0	0	0	Broken into 3 pieces while disassembling the capsule.	

Test Series	Test No.	Date	Type of Test Fuel	Reactor Period (m/sec)	Heat Deposition in Test Fuel (cal/g.UO_2)	Maximum Cladding Surface Temperature (°C)	Fuel Failure	Water Column Velocity (m/sec)	Capsule Internal Pressure (kg/cm²)	Nuclear to Mechanical Energy Conversion Ratio (%)	Post-Test Observation	Remarks
Scoping Tests	111-6	11.7.75	Std. (10%E)	1.61	335	>1720	Failure	0	-	0	Broken into 5 pieces during the transient test.	
	111-7	2.10.76	"	1.38	379	>1700	"	1.8	~10	0.02	Broken into fine particles.	
	111-8	6.8.76	"	1.32	437	-	"	0.7	-	6x10⁻³	"	
	232-1	12.11.75	Std. (10%E)	3.36	183	136	No failure	-	-	-	Partially oxidized.	
	232-2	3.4.76	"	2.42	238	1680	"	-	-	-	Break-away in cladding surface.	
Wide-gapped Fuel Tests	232-3	3.11.76	"	2.15	263	1745	"	-	-	-	Fully oxidized.	
	232-4	2.27.76	"	1.93	276	1770	Failure	-	0	0	Penetrating circumferential crack in cladding.	
	232-6	3.23.76	"	1.77	296	>1720	"	0	0	0	Broken into 2 pieces while disassembling the capsule.	
	232-5	3.18.76	"	1.63	329	>1720	"	0	0	0	Broken into 7 pieces during the transient test.	
	402-1*	1.22.76	JPDR-II(2.6%E)	5.41	48	240	No failure	0	8	0	No visible change.	
	402-2*	1.22.76	"	2.20	105	540	"	0	15	0	Partially oxidized.	
	402-3*	2.5.76	"	1.51	156	560	"	0	-	0	Partially oxidized, slight bowing.	
Waterlogged Fuel Tests	401-1**	12.25.75	"	5.08	53	180	"	0	5	0	No visible change.	
	401-2**	12.26.75	"	2.25	108	270	"	0	10	0	"	
	401-3**	1.30.76	"	1.50	156 ***(112)	170	Failure	2.9	55	0.02	Ruptured with longitudinal breach.	
	401-3b**	5.28.76	"	1.53	152 ***(144)	310	"	2.7	56	0.02	"	
	411-3**	6.4.76	"	1.53	154 ***(124)	215	"	5.1	65	0.09	"	0.4mmφ hole at middle part.
	421-3**	6.2.76	"	1.53	154	325	No failure	0	3	0	Slight bowing.	0.4mmφ hole at upper part.
Bundled Fuel Tests	311-1	3.25.76	JPDR-II(2.6%E)	5.21	36 ~ 44	120	No failure	-	-	-	No visible change.	Five pin bundle

* Radial gap of fuel rods was filled with water. (Tests 402-1~402-3)
** All space of fuel rods including upper plenum portion was filled with water. (Tests 401-1~421-3)
*** Numbers in parentheses are heat deposition in test fuel at time of fuel failure.

A.   Scoping Tests

The purpose of this test series is to obtain the basic
information concerning the response of the 10% enriched
standard test fuel rods to rapid, overpower transients with
the reactor periods of about 30 ms to 1.3 ms. The heat deposition in a fuel rod given by a single pulse was about
0.17 MJ/kg UO_2 to 1.8 MJ/kg UO_2.* For the tests at rather
low heat depositions, 2.6% enriched BWR type fuel rods
(JPDR-II fuels) were also used (detailed specifications of
the test fuel rods are described in Table I).

These experimental conditions cover the range of heat
depositions anticipated in RIAs in LWRs. For example, in
an accident with continuous withdrawal of a control rod in
an LWR, the maximum heat deposition in reactor fuel is believed to be not more than 0.7 MJ/kg UO_2, and even in a
hypothetical design-basis accident such as control rod
ejection in a PWR or control rod drop in a BWR, the heat
deposition in the hot channel fuel would be about 1 MJ/kg
UO_2 at most.

General Fuel Behavior. Figure 15 is a post-test photograph of the fuel rods showing typical fuel behavior with
various heat depositions. As shown in this figure, no visible change took place in the fuel rods following the tests
at heat depositions not more than 0.5 MJ/kg UO_2. At 0.67
MJ/kg UO_2 (Test 111-9), the cladding surface of a JPDR-II
fuel rod was uniformly oxidized and discolored in black over
the active region. The cladding surface of a standard test
fuel rod was partially oxidized at 0.74 MJ/kg UO_2 (Test
200-3). At the heat depositions over 1 MJ/kg UO_2, oxidation
of cladding was such that a very thin layer of part of the
cladding surface dropped away, as shown in Figure 17. In
this heat deposition range, the fuel rods also sustained
rather large bending.

---

* In safety evaluation on RIAs, the heat deposition in fuel
  pellets often is expressed in terms of 1 cal/g $\simeq$ 4.2 kJ/kg
  of UO_2. Since the specific heat of UO_2 is approximately
  0.3 kJ/kgK, the adiabatic temperature of UO_2 in kelvin is
  approximately three times the value of the heat deposition
  in terms of KJ/kg UO_2.

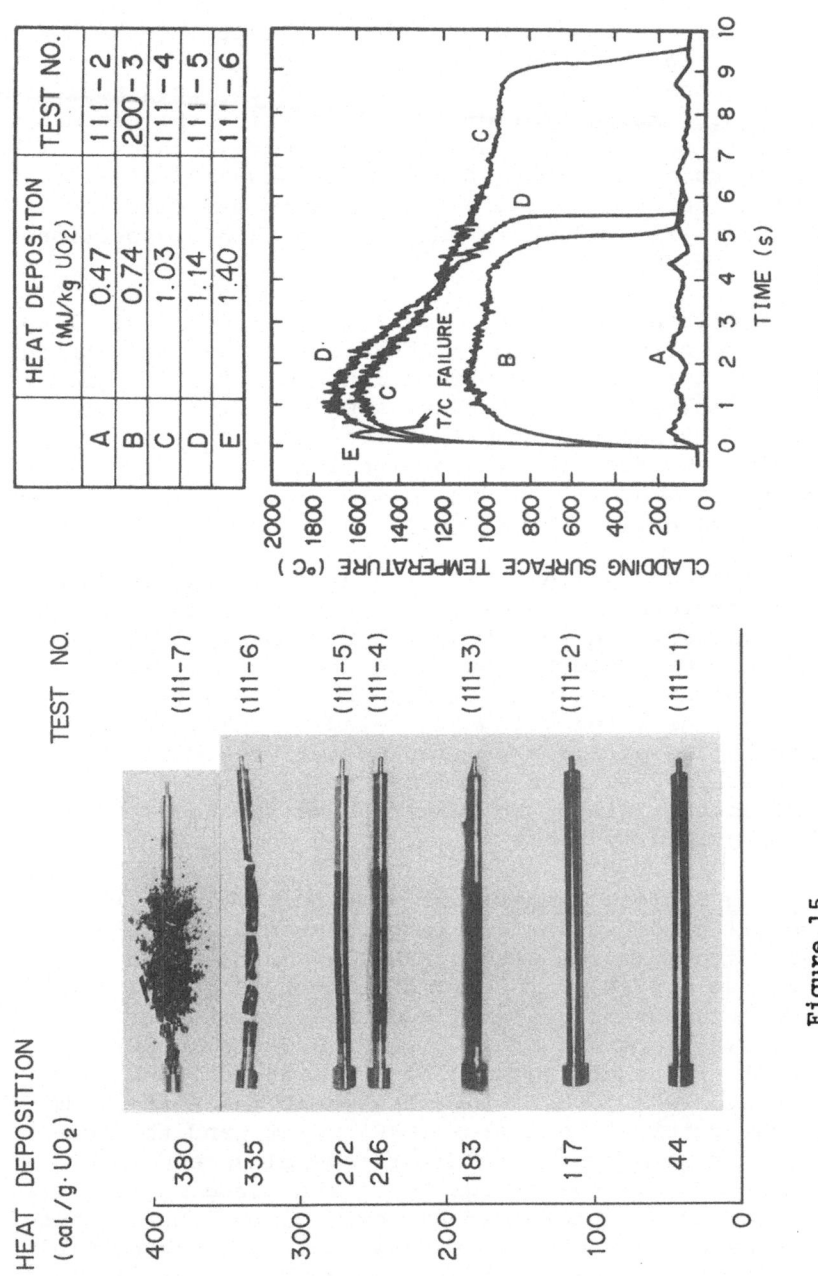

	HEAT DEPOSITON (MJ/kg UO$_2$)	TEST NO.
A	0.47	111 – 2
B	0.74	200 – 3
C	1.03	111 – 4
D	1.14	111 – 5
E	1.40	111 – 6

Figure 16

Cladding Surface Temperatures of Standard
Test Fuel Rods During the Scoping Tests

Figure 15

Test Fuel Rods Following the Scoping Tests
(* 1 cal/g UO$_2$ ≈ 4.2 kJ/kg)

The failure threshold of standard test fuel rods was
found to be in the heat deposition range of 1.03 to 1.12
MJ/kg UO_2; at the higher end (Test 200-5), the fuel rod
failed with circumferential cracks penetrating the cladding
tube. In this test as well as in other tests, cracks in
the cladding tube were located at pellet stack boundaries.
It was found that the pellet outer surfaces and cladding
inner surfaces were melted in the tests at heat depositions
over 1.12 MJ/kg UO_2.

At somewhat higher heat depositions, from 1.14 to 1.25
MJ/kg UO_2, the cladding tube had lost its mechanical strength
so much that fuel rods were broken into two or three pieces
while the capsules were being disassembled after the transient tests. During the transient test at 1.4 MJ/kg UO_2
(Test 111-6), the fuel rod was broken into five large pieces
with melting of most of the UO_2 pellets and cladding; however, there was no mechanical energy generation in this test.

At 1.6 MJ/kg UO_2 (Test 111-7), the fuel rod was broken
into small particles. The particle sizes were about 0.1 to
10 mm in diameter. In this test, approximately 1 MPa of
pressure pulse was detected at the bottom of the capsule.
Jumping of the water column was also detected with a float-
type sensor. The water column traveled by about 40 mm, with
the maximum speed of 1.8 m/s. The nuclear-to-mechanical
energy conversion ratio is estimated to be about 0.02%, with
an assumption that all of the water above the fuel rod
(about 26ℓ) jumped by about 40 mm.

Cladding Surface Temperature Behavior. Figure 16
shows typical cladding surface temperatures during the
transient tests measured with Pt/Pt-13% Rh thermocouples.
In the test at 0.47 MJ/kg UO_2 (Test 111-2), the fuel rod
was cooled by nucleate boiling, and the cladding surface
temperature was kept around 140°C. At 0.74 MJ/kg UO_2
(Test 200-3), the cladding surface temperature promptly
reached about 1200°C, which indicates that the heat flux
exceeded the critical heat flux (CHF) point, and the temperature was kept around 1000°C for several seconds. It
should be noted that the integrity of the cladding tube has
not been lost as described before, even though the condition
exceeding the CHF lasted for several seconds (only partial
oxidation on the cladding surface occurred in this test).
This fact may provide useful information in evaluating the
fuel behavior during other accidents such as power-cooling

mismatch accidents in which departure from nucleate boiling
is believed to be a prime factor in causing fuel failure.

In the tests with heat deposition near the fuel failure
threshold, cladding surface temperatures reached the value
close to the maximum measurable temperature of Pt/Pt-Rh
thermocouples (~1800°C). In case the heat flux exceeds
the CHF point, the cladding surface temperature reaches the
maximum value and keeps the value over 1000°C for several
seconds with film boiling. Then the temperature rapidly
decreases down to around 150°C (nucleate boiling region).

Based on the detailed examination of experimental data
obtained so far, it has been confirmed that the process of
temperature drop takes two phases. The first phase is a
rather slow temperature drop from about 900°C down to 500 -
600°C. The second phase is a rapid temperature drop down
to the nucleate boiling region following the first phase as
a result of the rewetting of the cladding surface. The cooling in the first phase is interpreted as the result of the
disturbance given to the vapor film covering the cladding
tube by the surrounding cold water. The initiation temperature of the first phase confirmed so far is 920°C ± 100°C.
Existence of a cooling phase preceding the quenching is an
interesting finding in these experiments.

Fuel Failure Mechanisms. As described in the previous
section, standard test fuel rods failed with circumferential
cracks in the cladding tube at the heat depositions over
1.12 MJ/kg UO_2. Very thin layers of the cladding inner surfaces were melted. At 1.4 MJ/kg UO_2 (Test 111-6), the fuel
rod was borken into five large pieces. Most of the pellets
and cladding were melted as shown in Figure 18.

Section A in this figure is the broken surface at the
lower part of the fuel rod. The central black area in this
section is the fuel meat that was not melted, and the surrounding gray area is the melted fuel meat. This difference
in melting condition primarily results from the radial power
distribution in the fuel rod. Section C in the figure is
the broken surface at the middle part of the fuel rod, where
all of the fuel meat region was melted. In this section, the
outer deformed layer, partly thin and partly thick, is the
cladding material. It seems that a very thin oxide film was
formed on the cladding surface and worked as a kind of cru-
cible in which Zircaloy was completely melted.

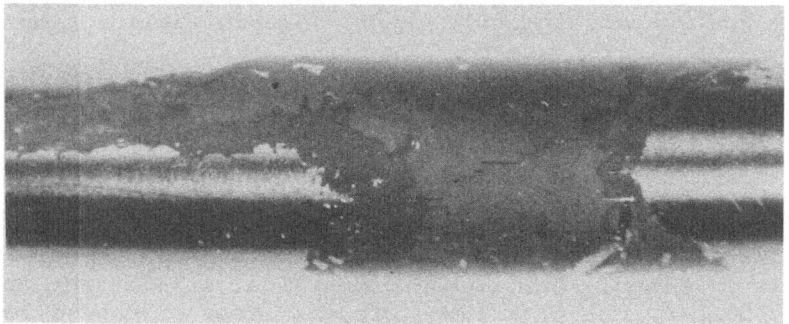

Figure 17

Cladding Surface of the Standard Test Fuel Rod
Following Test 111-4 (1.03 MJ/kg UO_2)

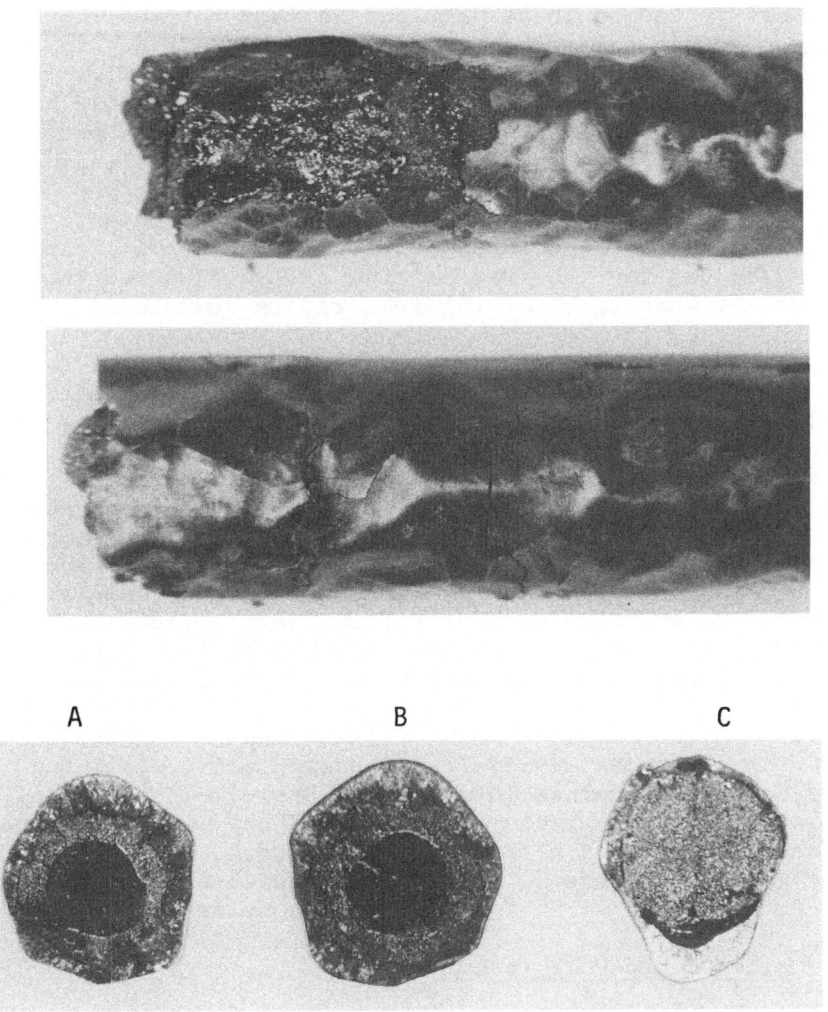

Figure 18

Fragments and Broken Surfaces of the Standard Test Fuel Rod
Following Test 111-6 (1.4 MJ/kg UO_2)

Section B is the broken surface that was once faced with Section A. An interesting fact related with broken surfaces including these sections is that neighboring surfaces resemble each other in appearance. This may indicate that the fuel rod was fractured during or after the quenching; otherwise, the neighboring surfaces would have been changed independently as a result of the contact with cold water. This assumption can give a reasonable explanation to another fact that mechanical energy was not generated in this test. As a conclusion, it can be said that fuel failure in the heat deposition range from 1.12 to 1.4 MJ/kg UO_2 results from melting of the cladding tube and occurs when or after the fuel rod is quenched.

In the test at 1.6 MJ/kg UO_2 (Test 111-7), the fuel rod was broken into small particles as shown in Figure 19. From the transient data on water level motion, it was estimated that the fuel failure occurred within about 0.1 s after the peak power time. Among the fuel particles, there are several pieces of cladding whose outer surfaces were not melted, which is completely different from the appearance of the cladding tube in the aforementioned test at 1.4 MJ/kg UO_2 (Test 111-6). The cladding surface temperatures measured at the time of fuel failure were approximately 1700°C. This may indicate the possibility that increase in internal pressure contributed to the fuel failure in this test.

The experimental results discussed above still require more detailed examination to obtain decisive conclusions. In addition, these tests are concerned only with cold start-up conditions. The effects of temperature, pressure and flow rate of coolant on fuel behavior will be studied in the future tests with high-pressure capsules and loops.

B.   Wide-Gapped Fuel Tests

The wide-gapped fuel tests are one of the parameter tests in which the effect of gap conductance of a fuel rod on failure mechanisms is studied. Since a radial gas gap in a fuel rod works as a thermal resistance, it is expected that a wide gas gap would reduce heat transfer from fuel pellets to cladding, and thus would change the threshold of fuel failure resulting from melting of cladding. The fuel rods used in this test series are a variation of the

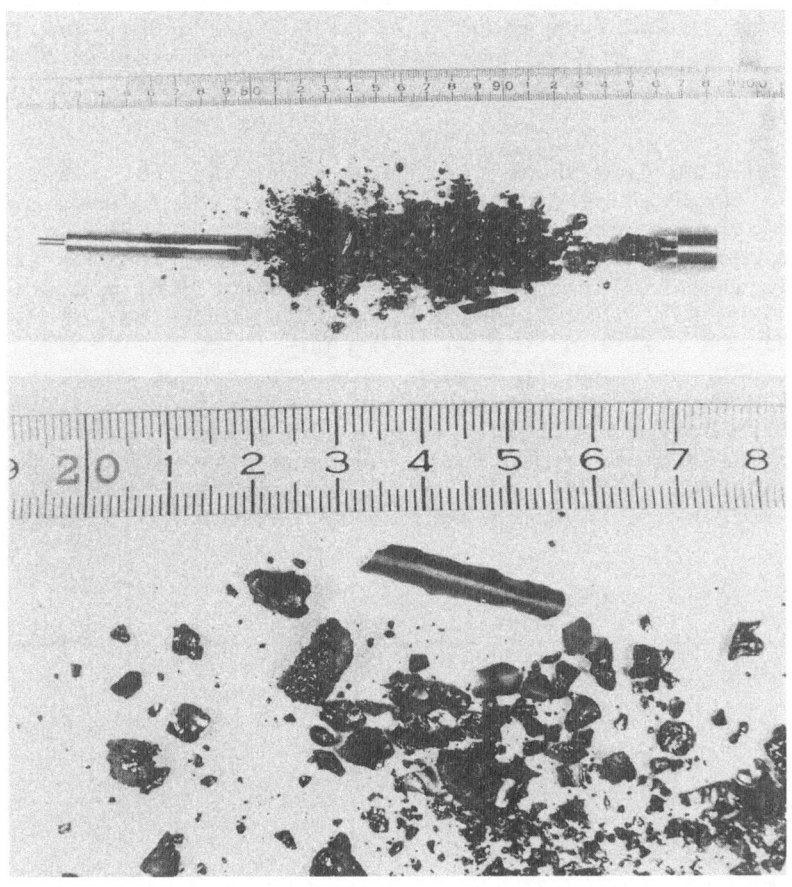

Figure 19

The Standard Test Fuel Rod and Its Fragments
Following Test 111-7 (1.6 MJ/kg UO_2)

standard test fuel rods, and have a radial gas gap as wide
as 0.195 mm which is twice the width of the radial gap of
the standard test fuel rods.

The heat deposition in a fuel rod given by a single
pulse was varied from about 0.75 to 1.2 MJ/kg UO_2, and the
effect of gas gap width on the transient behavior of cladding surface temperature and on the failure threshold was
studied.

The data from these tests indicate that the cladding
surface temperature of a wide-gapped fuel rod reduced much
in comparison with that of a standard test fuel rod at the
heat deposition of about 0.75 MJ/kg UO_2. The maximum cladding surface temperature of a standard test fuel rod at
0.74 MJ/kg UO_2 (Test 200-3) was 1200°C, while that of a
wide-gapped fuel rod at 0.77 MJ/kg UO_2 (Test 232-1) was
only 136°C, as shown in Figure 20. But at heat depositions
over 1 MJ/kg UO_2, the effect of the difference in initial
gap width became negligibly small, and cladding surface temperatures of wide-gapped fuel rods were very close to those
of standard test fuel rods, as illustrated in Figure 21. A
major difference in temperature rise between a wide-gapped

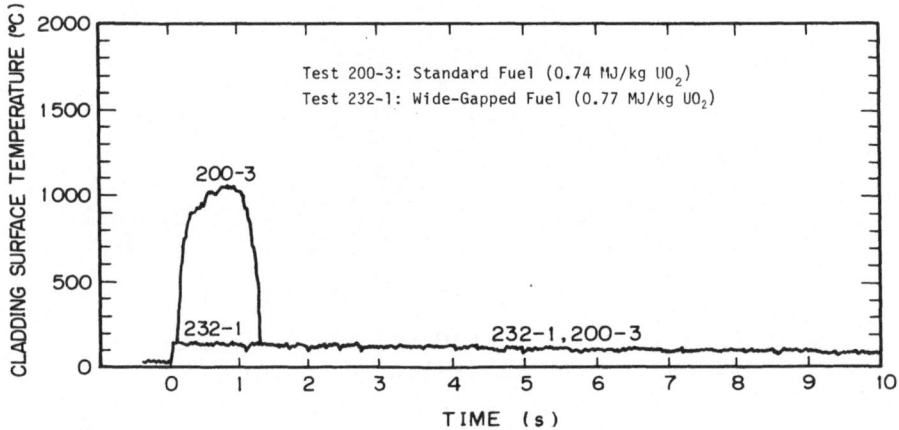

Figure 20

Cladding Surface Temperature of the Standard Test Fuel Rod
and the Wide-gapped Fuel Rod Subjected to ~0.75 MJ/kg UO_2

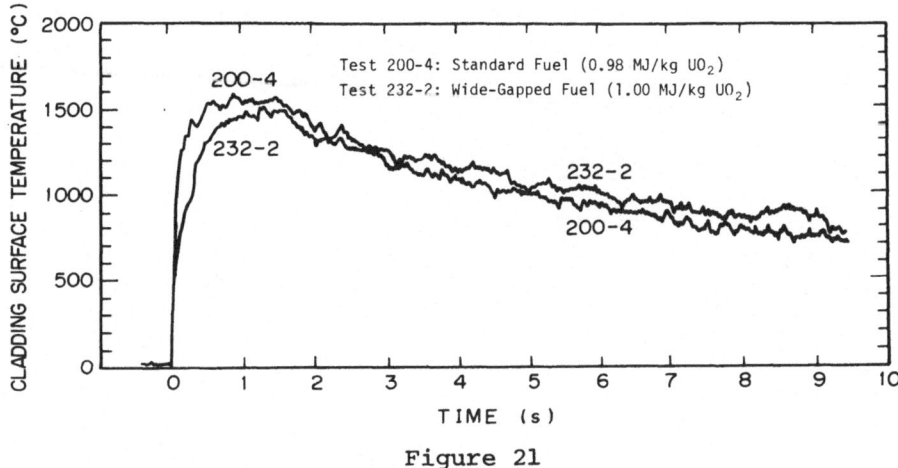

Figure 21

Cladding Surface Temperature of the Standard Test Fuel Rod
and the Wide-Gapped Fuel Rod Subjected to ~1 MJ/kg UO_2

fuel rod and a standard fuel rod is that the cladding surface temperature of the wide-gapped fuel has a slower rise
rate due to a larger thermal resistance of the radial gap.
A detailed calculation indicates that in a wide-gapped fuel
rod, fuel pellets come into contact with the cladding in
the heat deposition range over 0.96 MJ/kg UO_2.

The failure threshold and failure mechanism of wide-
gapped fuel rods observed so far are quite similar to those
of standard test fuel rods. The failure threshold is in the
heat deposition range of 1.1 to 1.2 MJ/kg UO_2; at 1.2 MJ/kg
UO_2, a wide-gapped fuel rod failed with a circumferential
crack penetrating the cladding (Test 232-4). A fuel rod
was broken into two pieces while disassembling the capsule
after the transient test at 1.24 MJ/kg UO_2 (Test 232-6).
The cladding tube was quite flattened around the broken area.
During the transient test at 1.38 MJ/kg UO_2 (Test 232-5), a
fuel rod was broken into seven pieces with melting of most
of the fuel meat and cladding. There was no mechanical
energy generation in this test.

In conclusion, it can be said that the initial gap
width has little influence on fuel failure mechanisms and
thresholds as far as these tests are concerned, although

the number of tests performed so far is not sufficient. In
the future, tests with the fuel rods with a very narrow radial gap will be carried out to confirm the effect of gap
width.

C.   Waterlogged Fuel Tests

There is a certain probability that fuel rods in a reactor core can have pinholes or hairline cracks around welded portions and water can penetrate into the rods through
these defects while the reactor is shut down. These rods
are called "waterlogged fuel rods". During a normal reactor
startup, power increase is so slow that the vapor formed in
these rods may be released into the coolant and there may
be no fuel failure. On the other hand, in the case of an
RIA, vapor formation in the waterlogged fuel rods is so
rapid that these rods can rupture with rather small heat
deposition due to the sudden increase in internal pressure.
In addition, it is anticipated that relatively large pressure pulses can be generated following the rupture of waterlogged fuel rods because the fragmented fuel meat expelled
from the cladding by the internal pressure rapidly heats
the surrounding water.

So far, nine tests with waterlogged JPDR-II fuel rods
were carried out in the NSRR. In these tests, two kinds of
parametric studies were made. One was to study the influence of water volume in fuel rods on failure threshold, and
the other was to study the effect of pinhole positions in
the cladding tube on fuel failure behavior.

For the former tests, two types of test fuel rods were
used. In one type, all the space inside the rod, including
the upper plenum portion as well as the radial gap was
filled with water; in the other type, the radial gap only
was filled with water. The water was injected into the rods
through holes at the rod bottoms where fuel pressure sensors
were fitted after injecting the water; thus, these rods did
not have defects in the cladding tube before the tests. The
heat deposition given in the test was approximately 0.2 to
0.65 MJ/kg UO_2.

At the maximum heat deposition, a partially waterlogged
fuel rod did not suffer failure but underwent slight swelling and partial discoloration in the cladding (Test 402-3).

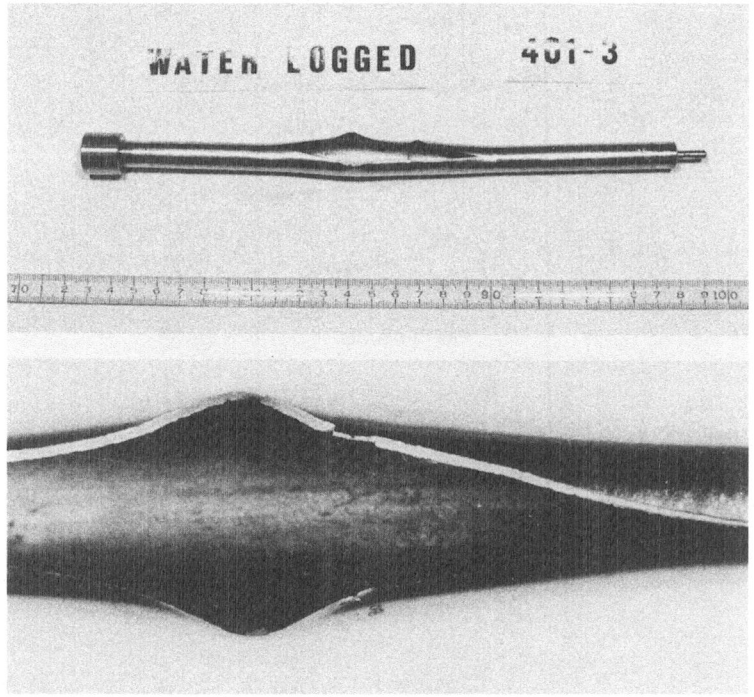

Figure 22

Cladding Tube of the Fully Waterlogged JPDR-II Fuel Rod
Following Test 401-3 (heat deposition at the time of fuel
failure = 0.47 MJ/kg UO_2)

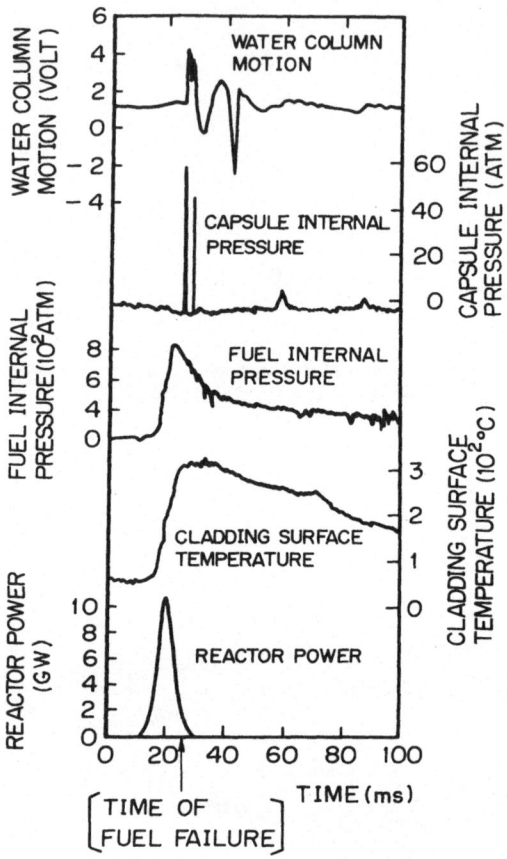

Figure 23

Transient Data During the Waterlogged Fuel Test
At Heat Deposition of 0.64 MJ/kg UO_2 (Test 401-3B)

In the tests at the same heat depositions, fully waterlogged
fuel rods failed, with large longitudinal breaches (about
85 mm long) as shown in Figure 22, and almost all of the
fuel meat was expelled into the water (Tests 401-3, 401-3B).
The fuel meat was fractured into fine particles, and pressure
pulses of about 5 MPa, and water column jumping with the
maximum speed of about 3 m/s were observed. The maximum

fuel internal pressure measured in Test 401-3B was approximately 130 MPa, as shown in Figure 23. The heat deposition at the time of fuel failure was 0.47 MJ/kg UO_2 in
Test 401-3 and 0.6 MJ/kg UO_2 in Test 401-3B, respectively;
thus, the cladding tube did not sustain discoloration. The
difference in the fuel failure thresholds in these two tests
is rather large, but its reason is not yet well understood.
The only difference in these tests is that the cladding of
the fuel rod employed in Test 401-3B was an autoclaved
Zircaloy-2 tube, but the cladding for Test 401-3 was not
autoclaved. The nuclear-to-mechanical energy conversion
ratio in these tests is estimated at approximately 0.02%.

These results, in general, are rather similar to the
results of the SPERT-CDC experiments on fully waterlogged
fuel rods clad with either cold-worked Zircaloy-2 tubes or
annealed Zircaloy-2 tubes (14).

For the other tests, in order to study the effect of
pinhole positions in the cladding tube on fuel failure behavior, two types of fuel rods with different pinhole positions were made. One was a rod with a 0.4 mm diameter
hole drilled near the top of the upper plenum portion, and
the other was a rod with the same diameter hole drilled at
the axial center of the pellet stack region. The holes
were patched with vinyl tape after filling the rods with
water. The cladding of these rods was autoclaved Zircaloy-
2 tubes. Both fuel rods were subjected to the tests at
heat depositions of 0.6 MJ/kg UO_2.

The fuel rod with the pinhole at the top portion did
not suffer failure but sustained slight bowing and swelling
(Test 421-3), while the fuel rod with the pinhole at the
middle portion ruptured with a 60 mm long breach (Test
411-3). A pressure pulse of 6 MPa and water-column jumping
with the maximum speed of about 5 m/s were observed. The
heat deposition at the time of fuel failure was 0.5 MJ/kg
UO_2. The nuclear-to-mechanical energy conversion ratio is
estimated to be approximately 0.09%. An interesting fact
in this test is that the breach in the cladding tube was
located at the opposite side of the pinhole position.

In future experiments on waterlogged fuel rods, the
following items will be investigated in detail:

1. effect of water volume in fuel rods on
    failure threshold

2. effect of positions and dimensions of
    defects in cladding on failure behavior

3. effect of the additional heating of fuel
    meat after failure on the behavior of
    mechanical energy

4. effect of the mechanical energy on the
    surrounding fuel rods as a result of the
    failure of a waterlogged fuel rod

5. behavior of the mechanical energy subsequent
    to fuel rupture in a capsule without free
    surface

V.   FUTURE PLANS

The first phase experiments of the NSRR program related
to reactivity-initiated accidents (RIAs) are being carried
out using unirradiated fuel rods until 1981. According to
an original plan, approximately five years of the experiments related to power-cooling mismatch (PCM) accidents
including loss of coolant accidents (LOCAs) using unirradiated fuel rods were scheduled as the second-phase experiments following the first phase. The second-phase experiments require the modification of the driver core so that
the steady state power level may be increased to more than
6 MW from the present maximum steady power of 300 kW. The
experiments related to RIAs and PCM accidents using irradiated fuel rods were scheduled following the second-phase
program. However, the Japan Atomic Energy Research Institute
and the United States Nuclear Regulatory Commission recently
have made (March 1976) an agreement on technical collaboration between the NSRR program and the Power Burst Facility
(PBF) program. The PBF program of the United States mainly
concentrates on the in-pile experiments related to fuel behavior during PCM accidents and LOCAs (15). It is expected
that active exchange of the technical information as well
as the experimental data between the two programs will be
realized. In addition, the possibility of other technical
collaboration with similar experimental programs of European
countries is being examined. Therefore, the necessity of

the PCM experiments in the NSRR is being reexamined from various points of view. There is a high probability that execution of the RIA experiments using irradiated fuel rods, which was once scheduled as the third-phase program, will be expedited as the second-phase program.

We hope that the NSRR program, together with other research programs, may provide useful information to better resolve fuel failure problems during abnormal reactor operations and contribute to securing further nuclear safety.

REFERENCES

1. Dietrich, J. R., "Experimental Investigation of the Self-limitation of Power Driving Reactivity Transients in a Subcooled, Water-moderated Reactor," USAEC Report ANL-5323, Argonne National Laboratory, 1954.

2. Spano, A. H., Barry, J. E., Stephan, L. A. and Young, J. C., "Self-limiting Power Excursion Tests of a Water-moderated Low-enrichment UO_2 Core in SPERT-1," USAEC Report IDO-16751, Phillips Petroleum Co., 1962.

3. "IDO Report on the Nuclear Incident at the SL-1 Reactor, January 3, 1961, at the National Reactor Testing Station," USAEC Report IDO-19302, Idaho Operations Office, 1962.

4. Miller, R. W., Sola, A. and McCardell, R. K., "Report of the SPERT-1 Destructive Test Program on an Aluminum, Plate-Type, Water-moderated Reactor," USAEC Report IDO-16883, Phillips Petroleum Co., 1964.

5. Neal, W. J. (Editor), "Safety Analysis Report SNAPTRAN 2/10A-1 and -2 Safety Tests," USAEC Report IDO-17076, Phillips Petroleum Co., 1965.

6. Miller, R. W., McCardell, R. K. and Lagier, T. F., "Addendum to the SPERT-IV Hazards Summary Report - Capsule Driver Core," USAEC Report IDO-17002, Phillips Petroleum Co., 1965.

7. Hasenkamp, A. (Editor), "Final Safety Analysis: Annular Core Pulse Reactor," SC-RR-66-2609, Sandia Corporation, 1966.

8. Ise, T., Inabe, T. and Nakahara, Y., "Reactor Physics
    Characteristics of the NSRR," J. Atomic Energy Society
    Japan (in Japanese), 17 (6), PP 314-321, 1975.

9. "NSRR Startup Test Report," JAERI-M Report 6791 (in
    Japanese), NSRR Operation Section and Reactivity Acci-
    dent Laboratory, Japan Atomic Energy Research Institute,
    1976.

10. Saito, S., Inabe, T., Fujishiro, T., Ohnishi, N. and
    Hoshi, T., "Measurement and Evaluation on Pulsing
    Characteristics and Experimental Capability of NSRR,"
    Nuclear Science Technology, 14, 3, 226-238, 1977.

11. Fujishiro, T., Iwata, K., Kikuchi, T., Yoshihara, F.
    and Hoshi, T., "Explosive Proof Test of NSRR Experi-
    mental Capsule," JAERI-M Report 5861 (in Japanese),
    Japan Atomic Energy Research Institute, 1974.

12. Ishikawa, M. and Tomii, K. (Editors), "Quarterly Pro-
    gress Report on the NSRR Experiments (1), Combined,
    October 1975 - March 1976," JAERI-M Report 6635 (in
    Japanese), Japan Atomic Energy Research Institute,1976.

13. Ishikawa, M. and Tomii, K. (Editors), "Quarterly Pro-
    gress Report on the NSRR Experiments (2), April - June
    1976," JAERI-M Report 6790 (in Japanese), Japan Atomic
    Energy Research Institute, 1976.

14. Stephan, L. A., "The Effects of Cladding Material and
    Heat Treatment on the Response of Waterlogged UO_2 Fuel
    Rods to Power Bursts," USAEC Report IN-ITR-111, Idaho
    Nuclear Corporation, 1970.

15. "Final Safety Analysis Report for the Power Burst
    Facility," ANCR-1011, Aerojet Nuclear Co., 1971.

PRACTICAL USAGE OF PLUTONIUM IN POWER REACTOR SYSTEMS

Karl H. Puechl
Nuclear Consultant, Atlanta, Georgia
(Currently, Director of Advanced Development
Nuclear Power Systems, Combustion Engineering, Inc.,
Windsor, Connecticut)

I. POPULAR NOTIONS ABOUT PLUTONIUM

Over the past two years, plutonium has been discussed in the popular literature almost as often as in technical publications. Perhaps the mythological derivative of its name has a special kind of appeal to society. Does plutonium, like Pluto the god of the underworld, have the potential to destroy civilization? The popular concept is that it does. It is "one of the most hazardous materials known to man." Anyone with "a smattering of technical knowledge can make a plutonium bomb." Before delving into the technical aspects of plutonium utilization in power reactors, let us briefly examine the validity of these popular notions. After all, it is these popular conceptions that are impeding the utilization of plutonium as a reactor fuel; the technical problems have been solved.

A. Is Plutonium One of the Most Hazardous Materials Known to Man?

From Day 1 of the Atomic Age, scientists have been concerned with hazards and safety. This concern has given the nuclear power industry a better safety record, by far, than any other comparable human endeavor. In spite of this admirable record, the public is not convinced that the atom can be really tamed.

A major problem is one of semantics. When a nuclear investigator speaks of a "maximum credible accident", these

words have particular meaning for him. To the general public, the words connote nothing more than their meaning in
normal, every-day usage. Misconstruction is always possible
when a scientist uses nonscientific terms to describe something. The classic example, although it is not germane to
the present discussion, is Einstein's often-quoted remark
about quantum theory, "God does not play dice with the world".
To the layman, this suggests that Einstein believed in an
anthropomorphic God or in determinism, or in both; Einstein
had neither connotation in mind. The same semantic problem
exists relative to the definition of radiological hazard.
To allow quantitative discussion and evaluation, scientific
investigators defined a "maximum permissible body burden".
Literal interpretation of these words implies that the term
defines the boundary between a nonhazardous and a hazardous
quantity of nuclear material; consequently, the public has
equated "maximum permissible body burden" with the more common chemical concept, "lethal dose". This equation is false;
can be readily shown to be false; yet it is generally accepted as being true in popularized assessments of the radioactivity hazard (1).

The concept, "maximum permissible body burden" (MPBB),
was introduced to allow quantitative control of occupational
exposures. If sufficient radioactivity could settle in a
worker's body over the duration of his occupational lifetime,
his health could become impaired. An MPBB is that amount
which, if settled in the body, would make working with radioactive materials no more dangerous than other occupations;
in short, it is the amount that can be contained in the body
without causing known unacceptable effect. The concept of
MPBB allows definition of safe working levels of radioactivity (2,3), in sievert (Sv) ($\equiv$ 100 rem), i.e., J/kg.

Generally, for inhaled or ingested radioisotopes, the
MPBB is defined as that concentration that gives a dose of
0.15 Sv/year to the critical organ. This certainly is not
an acute dangerous level of exposure. It is generally accepted that a dose of 4 Sv delivered to the total human body
(not just to an individual organ) in an hour or less will
result in a 50% probability of a fatality. Four sieverts
delivered over a one-week time span will produce a negligible
number of fatalities (4). Even the most severe critics of
current radiological exposure practices admit that a continuous dose rate of less than 10 Sv/year to lung tissue would
probably not cause cancer (5).

Based on these consensus data, a "lethal dose" to a vital organ appears to be somewhat more than 10 Sv/week. Even then, an acute uptake of plutonium, consistent with the concept of "lethal dose" would have to be somewhat greater, since there is time for ameliorating action. For example, chelating agents such as calcium ethylenediamine-tetraacetic acid (EDTA) and trisodium calcium diethylene-triaminepentaacetic acid (DTPA), taken orally or intraven-ously, could be administered to enhance the elimination of plutonium from the body (6). New chelating agents (Puchel elē) are being developed (54).

It is to be noted that this "lethal dose" radiation level of 10 Sv/week is 3500 times greater than the 0.15 Sv/year level that persists when the human body contains an MPBB. In short, the popular conception about plutonium radiological toxicity makes plutonium seem to be 3500 times more toxic than it really is. In actuality, the discrep-ancy is even greater, since "lethal dose" implies an uptake quantity, while the above conversion was done on the basis of material already settled in the body. Also, when plu-tonium is used in power reactors, the plutonium rarely is in pure form; in proposed practice, it is always used in conjunction with uranium (at most, 5% plutonium content for thermal reactor use) and the mixture of uranium and plutonium can be made intimate on a microscopic, almost atomistic scale. When these factors also are taken into account, one calculates that the "lethal dose" for ingestion is about 400 grams of plutonium-bearing fuel material and for inhal-ation, about 0.1 gram. Eating a pound of plutonium-bearing fuel material can kill you. Inhaling 0.1 gram can do the same. When one realizes that an unburned cigarette weighs only about one gram, it becomes pictorially clear that plu-tonium, in the form in which it will be used as a power re-actor fuel, is not a particularly toxic material, certainly not one of the most hazardous materials known to man.

This is not to imply that plutonium is radiologically innocuous. All radiation is harmful, and human exposure should be kept to an absolute minimum. Plutonium should be handled with due respect, even though the popular concept grossly exaggerates its toxicity.

B.   Is it Easy to Make a Plutonium Bomb?

To make a simple nuclear device requires sufficient fissile material to go promptly critical, and a surrounding

chemical explosive to implode the critical mass and to con-
fine it for an adequate period of time. The fissile material in A-bombs is in metallic form and there is strict control on impurity content, especially on impurities that can
result in neutron generation prior to going critical. The
neutron background consideration places strict limits on
nonfissile isotope content, especially on isotopes that
undergo decay by spontaneous fission, and on light element
content, since neutrons then can be generated by alpha-
neutron reactions. Generally speaking, the higher the background neutron level, the lower the potential fission or
explosive yield.

The common fuel material contemplated for nuclear power
generation is the mixed (uranium/plutonium) oxide with the
plutonium usually containing about 20% or more nonfissile
isotopes (primarily ^{240}Pu). It is generally believed that
a terrorist organization would be capable of chemically
separating the uranium from the plutonium and that the separated plutonium in oxide form would then be used as the
nuclear explosive. The potential fission yield from a crude
device made with adequate nuclear material of this type and
adequate imploding chemical explosive is difficult to ascertain. Given a specific design configuration, it is doubtful
that even experts in weapons technology would agree; estimates made without elaborate computer simulation probably
would range from zero to a few hundred tons TNT equivalent.
Since 10^{18} fissions can result from accidental criticality
(7) there certainly is some probability that more than 10^{20}
atoms can be made to fission in a crude device, thereby
yielding a release of close to one ton of TNT equivalent.
Discharge of such a device, even in a populous area, would
not be catastrophic. There would be local damage due to the
explosive force; there would be local overexposure due to
the release of neutrons and gamma rays; and there would be
dispersal of the plutonium, perhaps, resulting in local air
concentration greater than the maximum permissible. However,
as noted above and as will be detailed further, the probability of anyone receiving more than a "lethal dose" or
even more than an MPBB of plutonium is extremely small.
Relative to terrorist action, the greatest concern would be
fear of the unknown; one would always have the nagging
thought that perhaps the terrorists could have "lucked out"
in the construction of the device or that they were much
more sophisticated than can be reasonably expected.

Perhaps a more legitimate concern is that other nations might use the reactor-produced plutonium to manufacture nuclear weapons. This they could do in a clandestine manner with technical talent, in suitable facilities, and with adequate time; however, even in this case, unless they drastically altered reactor operations (and such alteration could be readily detected because of the large associated fuel requirements), they would be limited to making devices with nuclear material that contained more than 20% ^{240}Pu. From the military standpoint, such material probably would not be particularly devastating as an A-bomb; however, it could make a satisfactory initiator for an H-bomb. If the country, in addition to plutonium, had facilities for separating lithium isotopes and for producing and handling tritium, weapons with adequate military effectiveness probably could be produced. It is pointed out that considerable technology would be required for such an endeavor.

II. HISTORICAL PERSPECTIVE AND REVIEW OF PLUTONIUM PROPERTIES

In the early days of the development of commercial nuclear power, effort was directed at maximizing the production and utilization of plutonium. In the weapons program, plutonium in pure form was being routinely processed in large quantity, and while some personnel overexposure did occur (8), experience indicated that such processing could be carried out safely. Nuclear proliferation was a concern even in the early 1950's; however, weapons were being manufactured with "weapons material" (that which had less than 6% ^{240}Pu content) and "commercial" plutonium (that which had greater than 20% ^{240}Pu content) was not considered to be suitable for weapons use. Terrorism was then not foreseen and no one was particularly concerned about the consequences of someone's making a "crude device".

Use of plutonium was always considered necessary for the commercialization of nuclear power, since the cost of "burning" only ^{235}U appeared to be prohibitive. Also, since ^{235}U comprises only 0.71% of the natural uranium weight, it appeared that the earth's uranium resources were too limited to sustain a nuclear industry based only on the consumption of ^{235}U. The manner in which plutonium could be used best was not clear.

If power was to be generated in reactors that used
natural or slightly enriched uranium as fuel, plutonium
would be produced as a byproduct, due to neutron absorption in ^{238}U, and some of the produced plutonium would
be fissioned *in situ*. Extending the reactivity life and
the irradiation damage limit of the fuel elements could
maximize *in situ* plutonium utilization; however, material
damage severely limited achievable burnup, and in addition,
the fissile material inventory required to counteract the
fission product poisoning was economically prohibitive.
Accordingly, in practical reactor systems, it was found
that *in situ* plutonium utilization could, at most, double
the fission energy obtainable from "burning" ^{235}U alone.

The plutonium produced in thermal reactors could,
following fuel reprocessing and plutonium recovery, be
used to replace ^{235}U in thermal reactors or the plutonium
could be used to fuel fast breeder reactors. The latter
alternative had particular attraction since all the ^{238}U
could be eventually converted and "burned" in these systems
wherein the conversion rate was expected to be greater than
the fission rate, or the plutonium destruction rate. The
problem was that fast breeder reactors had to utilize a
heat transfer technology other than water and such could
not be developed and commercially instituted within a short
period of time. Consequently, major plutonium utilization
development effort during the past 20 years has been directed at reinsertion of the plutonium into the thermal reactors wherein it had been produced (plutonium recycle).
Major emphasis of this presentation is directed at such
plutonium utilization.

A.   Evolution of Plutonium Utilization

In the United States, serious effort directed at plutonium
utilization in thermal reactors was initiated at the govern-
ment-owned Hanford Laboratories in 1956. Such effort consisted of core physics and economics evaluation, criticality studies, material fabrication investigation and material
damage assessment (9,10).

The physics studies, along with supportive measurement,
soon gave indication that plutonium would be a satisfactory
fuel material in thermal reactors. The fission cross-
sections of the fissile isotopes, ^{239}Pu and ^{241}Pu, were

approximately twice as high as that of ^{235}U; these cross-sections, in conjunction with a relatively high neutron release rate (2.9 neutrons released per fission, compared to 2.5 for ^{235}U) were adequate to counterbalance the relatively high nonfission cross-sections of the fissile isotopes as well as the conversion cross-sections of the nonfissile isotopes (primarily ^{240}Pu). The higher cross-sections would make control materials relatively less effective, but such could be accommodated by appropriate fuel/core design modification. Also, since plutonium fissioning had an associated smaller delayed neutron fraction than uranium fissioning, a somewhat shorter reaction time could be anticipated; however, assessments indicated that the faster response would not result in kinetic instabilities.

Fuel fabrication studies were directed primarily at the development of alternate fabrication techniques, so the fabrication cost would not be prohibitive when operations were carried out in remote or semi-remote fashion as was dictated by the plutonium toxicity. Since the pelletizing process had been satisfactorily developed for uranium fuel, this process was not given particular attention; rather, other schematically simpler techniques were given primary emphasis. These were vibratory compaction of powders (Vipac fuel) and later, hot pressing.

Material irradiation evaluations were undertaken primarily in the Plutonium Recycle Test Reactor (PRTR), a D_2O moderated and cooled reactor designed specifically for this purpose. This reactor was operated from 1961 through 1964, and irradiation of a variety of materials gave indication that plutonium-bearing materials behavior was not significantly different from uranium fuel performance.

In 1959 and thereafter, this effort in government laboratories was supplemented by AEC-supported commercial activities. Work at U. S. commercial installations initially diverged from the National Laboratory effort in both the core physics and fuel fabrication areas. Core physics studies were directed at making fruitful use of the inherently high separative work content of the plutonium (11,12). Fabrication studies and irradiation performance evaluations were directed primarily at using the pellet-in-tube process that had been proven so successful with uranium fuel (13).

Work in government laboratories and in industrial facilities was also begun in Europe at about the same time
(14,15). For recycle in thermal reactors, the fabrication
effort was concentrated on the vibratory compaction process;
subsequently, this was redirected toward pelletizing. The
work at the government laboratories was directed more at
the longer-range fast breeder development effort, leaving
recycle fuel development primarily to the commercial parti-
cipants.

As time went on, the trend of plutonium fuel development tended to copy the already-developed uranium fuel
bases. Industry became conservative, not wanting to deviate
from established and proven processes. Accordingly, for
recycle, industry chose to defer the question concerning
how plutonium can best be recycled in favor of the question
of how plutonium can be recycled with minimum perturbation of
uranium fuel cycles, uranium fabrication techniques, etc.
Specifically, physics effort was directed at using plutonium in place of some of the ^{235}U in the reactor core without
changing the length of the reactor operating cycle and without changing fuel rod spacing or control rod configuration.
Fabrication effort was directed at the blending and homogen-
ization of UO_2 and PuO_2 powders with subsequent pelletization
and sintering under conditions nearly identical to those used
for the fabrication of uranium fuel.

B.   Radiological and Nuclear Properties of Plutonium

Since it is impossible to discuss plutonium utilization
without consideration of its radiological and nuclear properties, pertinent characteristics are presented below.

The radiological hazard is due primarily to alpha emission. Radiological properties and related occupational
exposure criteria for ^{239}Pu are presented in Table I. The
values are nearly identical for the other alpha emitting
isotopes except for slight modification due to differences
in half-life. The beta-decaying ^{241}Pu isotope with relatively short half-life (13 years) has significantly less-
stringent exposure criteria.

In both the thermal and fast breeder reactor fuel cycles
one is primarily concerned with insoluble plutonium, specifically plutonium in oxide form. In aqueous reprocessing,

TABLE I

RADIOLOGICAL PROPERTIES OF ^{239}Pu

	Soluble	Insoluble
Maximum Permissible Body Burden, MBq/m^3		
If Inhaled	1500	600
If Ingested	1500	-
Critical Organ:		
If Inhaled	Bone	Lung
If Ingested	Bone	GI Tract
Fraction Reaching Critical Organ:		
By Inhalation	0.18	0.12
By Ingestion	0.0001	-
Maximum Permissible Concentrations in Air and Water, 40 hr/wk Exposure, MBq/m^3:		
(MPC)$_a$	0.075 x 10^{-6}	1.5 x 10^{-6}
(MPC)$_w$	3.7	3.0

n.b.   1 becquerel (Bq) is one disintegration/s

37 x 10^9 Bq is one curie (Ci).

the plutonium is recovered as nitrate solution, but this can
be almost immediately converted to PuO_2 or blended with uranium solution and coprecipitated to the mixed-crystal,
$(U,Pu)O_2$. The previous introductory discussion relative to
the plutonium hazard was based upon taking credit for the
insolubility of the anticipated material of commerce. Relative to the radiological hazard of plutonium, it is pointed
out that there is no history of anyone's ever receiving more
than a maximum permissible lung burden (600 MBq/m^3) when
working with insoluble compounds of plutonium. A cloud of
plutonium oxide powder has quite frequently been released
from containment structures (glove boxes) but the resultant
inhalations, although at greater than maximum permissible
air concentrations, have never resulted in an integrated uptake that approached the MPBB. This is considerably different from the history of work with soluble compounds where exposure has resulted in body burdens a few times higher than
the MPBB levels (8). However, even in such instances, the
exposed individuals have exhibited no physiological consequences.

Recently, there has been much discussion relative to
the interpretation of the radiological hazard. For insoluble
compounds, the values presented in Table I are based on
giving to all lung tissue an average dose of no more than
0.15 Sv/yr exposure; however, in the vicinity of a plutonium-
bearing particle, exposure to immediately-surrounding tissue
can be much greater. The question is whether high exposure
to a very small amount of tissue is more likely to produce
cancer than lower exposure of a larger amount of tissue.
The pros and cos of this debate are presented in referenced
documents (5,16,17). Governmental authorities generally have
concluded that there is no evidence to indicate that more
stringent criteria are warranted.

Nuclear reaction (fission and absorption) cross-sections
for the major plutonium isotopes and for [235]U are presented
in Figures 1 and 2, covering the thermal energy range.
These data were taken from the ENDF/B-II compilation (18).
Of particular interest are the significantly higher values
for the fissile plutonium isotopes (compared with [235]U) and
the existence of low-lying resonances in these isotopes as
well as in [240]Pu. Utilization of plutonium by simply replacing [235]U in thermal reactors, therefore, results in hardening of the neutron spectrum and substantial resonance

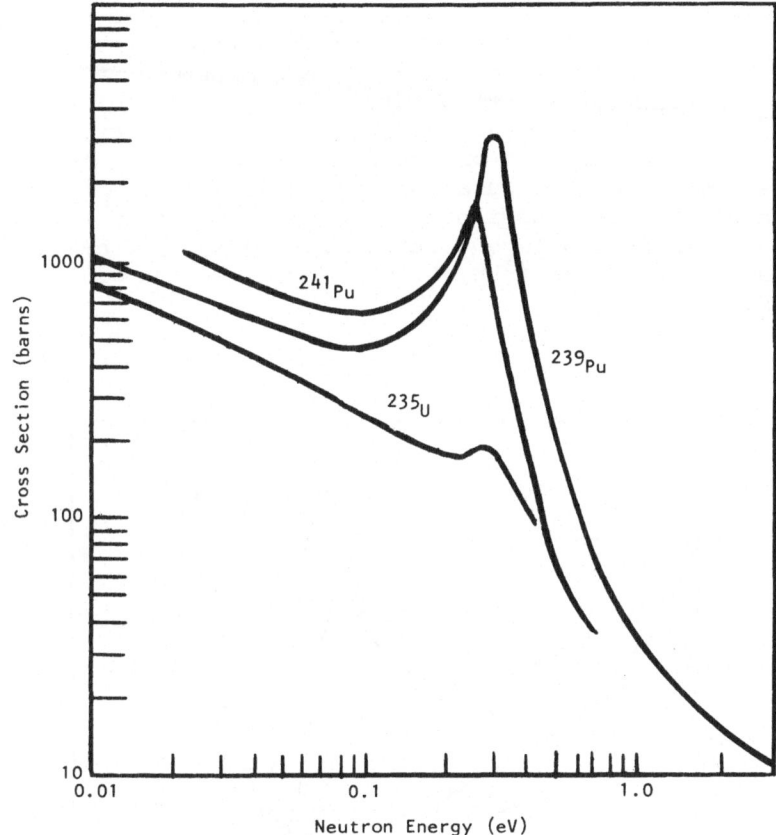

Figure 1

Fission Cross-sections of Plutonium and Uranium

absorption. Resonance absorption in ^{240}Pu is particularly
pronounced and such absorption, because of its high peak
and its low energy (1.055 eV)* location, can vary considerably with burnup due to self-shielding and neutron temperature
variations. Optimization of plutonium utilization in thermal reactors is largely directed at using this resonance

---

*   1 eV = 0.1602 aJ.

absorption and its variation with burnup to maximum advantage (11,12). The presence of these low-lying capture resonances also gives lattices that contain plutonium a greater degree of stability; all temperature coefficients of reactivity (moderator, void and Doppler) are more negative than for lattices that contain only uranium fuel (19). However, kinetics comparison with uranium lattices is made more complex since the plutonium isotopes, when fissioning, release a smaller number of delayed neutrons. The delayed neutron fraction is 0.0067 for ^{235}U, 0.0022 for ^{239}Pu and 0.0054 for ^{241}Pu.

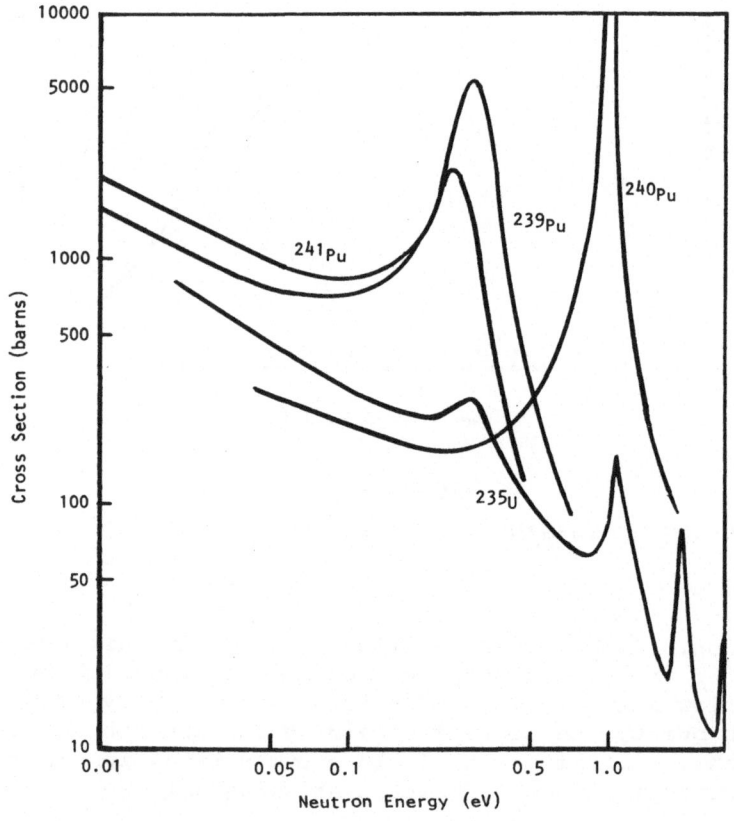

Figure 2

Absorption Cross-sections of Plutonium and Uranium

TABLE II

MINIMUM CRITICAL MASSES (kg)

	^{235}U	^{239}Pu
Solution	0.82	0.51
Metal	22.8	
alpha phase (ρ = 19.6 Mg/m^3)		5.6
delta phase (ρ = 15.8 Mg/m^3)		7.6
Oxide		
sintered ($\rho \sim$ 10 Mg/m^3)	$\sim$ 35.0	$\sim$12.7
green pressed ($\rho \sim$5 Mg/m^3)	>100	$\sim$ 32.0

(ρ = density)

Plutonium cross-sections in the high energy range, as anticipated, are smoothly varying. Average group fission cross-sections over the energy range 0.825 to 1.35 MeV are 1.81, 1.39, 1.83 and 1.57 barns* for ^{239}Pu, ^{240}Pu, ^{241}Pu and ^{242}Pu, respectively. Similarly, capture cross-sections are 0.104, 0.226, 0.077 and 0.226 barns, respectively (20). Of interest is the fact that all isotopes are highly fissionable; hence, for fast systems such as nuclear devices and fast breeder reactors, the nonfissile isotopes are not significant poisons.

For criticality control and to assess feasibility of device construction, minimum critical masses are of interest for both well-moderated and unmoderated systems. Pertinent values for ^{235}U and ^{239}Pu with thick hydrogenous reflectors or equivalent are presented in Table II (21,22). It is to be noted that material density has a strong effect on the minimum critical mass for unmoderated systems. The nonfissile isotope content has significant effect on the moderated critical mass (23,24).

---

* 1 barn = 10^{-28}m^2.

Nuclear analysis of plutonium systems is complicated not only because of the cross-section structure (Figures 1 and 2) but also because the isotopic composition of plutonium changes with fuel burnup in an intricate manner. Typical variation of the isotopic content of plutonium produced during the burnup of uranium in light water reactors is presented in Figure 3. It is seen that the primary nonfissile isotope, ^{240}Pu, tends to reach an equilibrium concentration at somewhat greater than 20%. The exact amount is dependent upon all core design variables that effect resonance absorption.

When plutonium is recycled in thermal reactors, i.e., when it is used to replace some of the ^{235}U, there is continuous variation in the isotopic composition. This variation is strongly dependent upon the concentration of plutonium (in uranium) that is utilized. When the plutonium concentration in uranium is relatively small, the infinite dilution resonance integral is applicable and there is little self-shielding; consequently, the effective ^{240}Pu cross-section is quite large, and rapid conversion to ^{241}Pu results in a relatively low ^{240}Pu equilibrium concentration. In fact, the ^{240}Pu concentration in the loaded plutonium may immediately begin to decrease in order to seek the lower equilibrium value. Conversely, in uranium with relatively high plutonium content, the effective ^{240}Pu cross-section is much lower; hence, the equilibrium concentration becomes relatively high.

Multiple recycling of plutonium in thermal reactors, therefore, can result in broad variability of the plutonium isoltopic composition. When the plutonium is blended with natural uranium to yield a mixed-oxide having approximately 3% fissile content, the variation of the plutonium isotopic composition by cycle is approximately as shown in Table III.

III. PLUTONIUM RECYCLE ALTERNATIVES

A. General Considerations

Plutonium is produced in all reactors that contain ^{238}U; consequently, plutonium is a byproduct of all commercial reactors that operate on the uranium fuel cycle. The following discussion is directed primarily at plutonium production and utilization in light water reactors (LWRs),

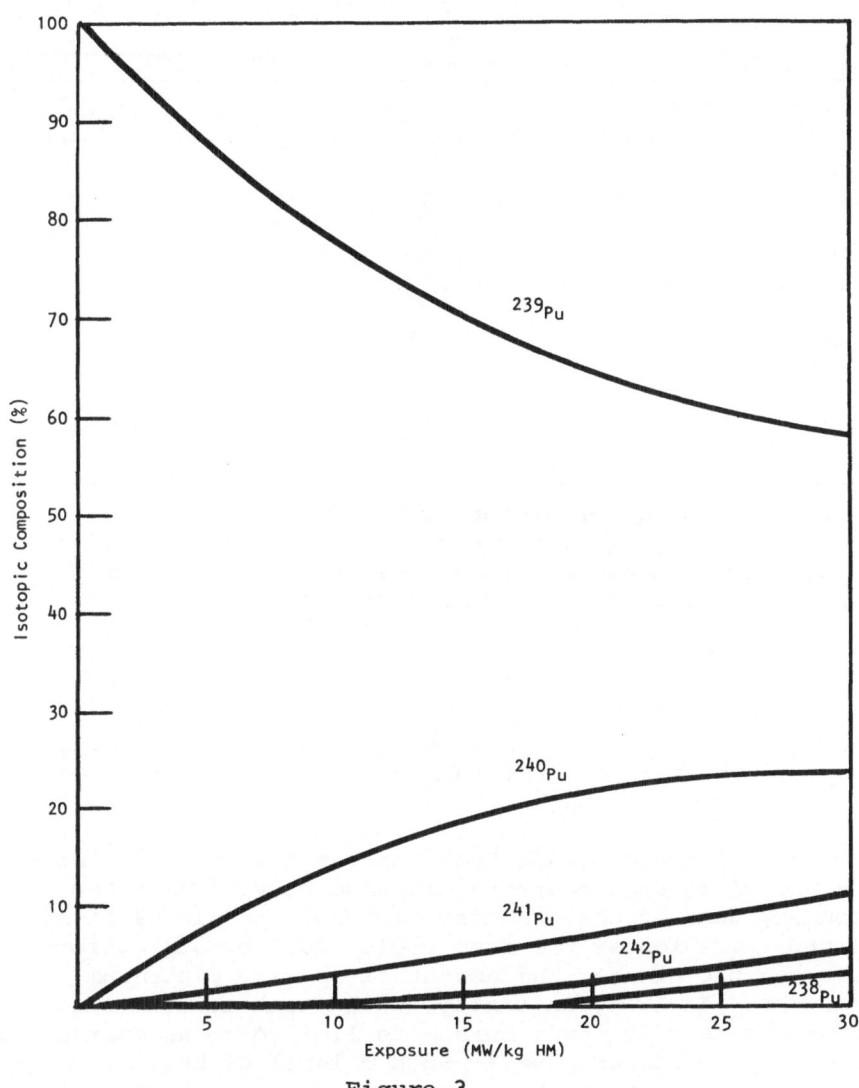

Figure 3

Plutonium Isotopic Composition
Versus Uranium Fuel Exposure in a Typical LWR

TABLE III

PLUTONIUM ISOTOPIC COMPOSITION AT REACTOR DISCHARGE[(a)]
(%)

Isotope	U Cycle	1st Pu Recycle	2nd Pu Recycle
238	2	3	5
239	58	40	31
240	24	29	32
241	11	17	17
242	5	11	15
Fissile	69	57	48

(a)   Based upon use of plutonium blended with natural
      uranium to yield a mixed-oxide fuel material having
      3% fissile content with all discharged plutonium
      being recycled in the same reactor.

although many of the discussed effects and peculiarities of
plutonium behavior are applicable to all thermal reactor
systems.

    Since plutonium production implies a presence of neu-
trons and since such neutrons can also cause fissioning of
plutonium, some of the produced plutonium is always fissioned
in situ . The longer the fuel cycle (fuel burnup) prior to
discharge, the greater the amount of *in situ* plutonium
fissioning. In LWRs, initially fueled with uranium enriched
to approximately 3%, fuel burnup is limited to an average of
about 30 MW-day/kg uranium.* Such a level of burnup implies
fissioning of about 3% of the contained heavy atoms. Since
the uranium enrichment is still about 0.8% at discharge,
fissioning of ^{235}U contributes about 22 MW-day/kg U heat

─────────────────────
*   1 MW-day/kg U = 1,000 MW-day/tonne U or metric ton.

generation; fissioning of plutonium contributes the other
8 MW-day/kg U. More than 25% of the heat generated in LWRs
results from the *in situ* fissioning of plutonium.

At least conceptually, the simplest method of plutonium utilization would be to increase the *in situ* fissioning; accordingly, this merits some discussion. Since, starting with uranium fuel, the plutonium concentration is
initially zero and increases with burnup (Figure 3), the
obvious way to increase *in situ* plutonium fissioning is to
increase the cycle length.

From the physics standpoint, this can be readily accom-
plished by increasing the initial uranium enrichment. Such
increased fissile material loading would result in higher
initial reactivity and this would have to be held down with
additional control rods or added solution poison. This waste
of neutrons would be economically prohibitive unless fertile
material rather than parasitic poisons were used for control
(as is contemplated for similar neutron conservation reasons
in the light water breeder reactor that operates on the
thorium/^{233}U fuel cycle) (*25*). However, even then, the added
fissile material inventory required to override the increased
fission-product poisoning would be economically prohibitive
under present accounting methods that attribute a sizable
carrying charge to in-core fuel inventory. It is pointed
out that this apparent economic limitation on fuel burnup
due to increased inventory carrying charge is not a technical
limitation but an accounting aberration. For example, if
fissile material were on hand (stored in a stockpile), the
cost of storage in a reactor would be no greater than the
cost of storage in a warehouse (or in obsolete weapons);
under such conditions, application of an inventory carrying
charge for in-reactor storage would be unwarranted, even
from a strict accounting standpoint. With revised accounting, one could afford higher in-core inventory, and thereby,
from the physics standpoint, realize longer operating cycles
and greater *in situ* plutonium utilization. Under such
accounting, one could rationally direct development, as has
been done in the U. S. naval reactors program, toward the
attainment of fuel life commensurate with the life of the
power plant (*26*).

The technical problem with this approach lies not in
the field of physics but rather in materials technology.
The LWR pellet-in-tube oxide fuel elements operate satis-

factorily to an average exposure of 30 MW-day/kg U, with
some elements experiencing maximum exposure of about 50
MW-day/kg U. Similar fuel elements, clad with stainless
steel rather than Zircaloy, are being developed for the
fast breeder reactor and for this application, the target
irradiation level is 100 MW-day/kg heavy metals (HM). If
one strove to go much beyond this limit, fuel materials
would have to contain much more structural material (perhaps
dispersion-type fuel) so as to maintain geometric integrity
even after substantially more than 10% of the contained
heavy atoms had fissioned.

Since current fuel cycle accounting practices indicate
poor economics, development effort toward increased
plutonium utilization has not been undertaken. Accordingly,
all effort has been directed at approximately annual discharge of about one-fourth the reactor fuel loading with
subsequent reprocessing and recovery of the residual uranium
and byproduct plutonium values. These recovered materials
would then be available for reintroduction into the same
reactors or into other reactors wherein utilization would
be of greater economic benefit.

Relative to the recovered plutonium, maximum benefit
would be derived by utilization of plutonium in fast reactor
systems wherein breeding is theoretically possible due to the
relatively high fission cross-sections and ν values; however,
to achieve economic breeding requires high power density and
an exotic coolant (liquid sodium being preferred). The time
frame for development and commercialization of this novel
technology is inconsistent with the time frame for plutonium
availability, which is as soon as light water reactor fuel
can be reprocessed. Accordingly, even though fast breeder
reactor development is being undertaken, for near-term plutonium utilization, development effort has been concentrated
on reintroduction of the plutonium into LWRs.

Technical Constraints. Since not all fuel is unloaded
from a reactor at any one time, fresh fuel containing plutonium must be compatible with the partially-burned fuel that
remains in the core. The major problem is local perturbation
of the power distribution; also of concern is the locally-
hardened neutron spectrum that influences neutron absorption
in surrounding fuel and control material. As indicated previously, there are other effects that have influence on reactor kinetics and time-varying reactivity behavior; however,

for reasonable design conditions, these effects are not con-
straining.

Generally, the tendency toward nonuniform power distribution can be ameliorated by limiting the amount of
plutonium contained in each fuel rod and/or by varying the
amount of fissile material (plutonium and/or ^{235}U) from rod-
to-rod within a fuel assembly. Another possible remedy, although given less emphasis, is to include varying amounts
of burnable poison within the fuel rods.

The negative effects of a hardened neutron spectrum
can be ameliorated by leaving a water hole in one or more
fuel rod locations within a fuel assembly and/or by surrounding the control elements with nonplutonium-bearing fuel
rods.

Detailed neutronic analysis is required to ascertain
the appropriate location of fuel assemblies that contain
plutonium as well as to determine the fissile element content and fuel rod (or water hole) distribution required to
yield acceptably flat heat generation and acceptable control rod worth.

Radiological Considerations. Sizable quantities of
plutonium were first used in the manufacture of nuclear
weapons. Accordingly, many of the plutonium handling and
fabrication practices, expected to be utilized for commercial application, were derived from practices initiated
in the weapons program.

Primarily, it was determined that one could not handle
pure plutonium metal or compounds out in the open; the
maximum permissible air concentrations (Table I) would be
exceeded within the breathing zone. Consequently, the philosophy was developed that plutonium should never be allowed
to "see" the breathing atmosphere. Implementation of this
philosophy required the use of air-tight structures to contain the plutonium and the related fabrication equipment.
Special precautions and procedures had to be devised to
allow removal of plutonium and/or contaminated equipment
from these structures without, at that time, releasing
plutonium to the breathing atmosphere.

Typically, glove box operations evolved. In this type
of facility, manipulation of the material and repair of

equipment located within the glove boxes can be effected
manually by working through attached gloves. During such
operation, the operator is exposed to external radiation
(neutrons and gamma rays) emitted by the plutonium or associated elements. With plutonium composed primarily of the
isotope ^{239}Pu (weapons material), this is not a major consideration, since the ^{239}Pu gamma rays have relatively low
energy and are satisfactorily attenuated by the glove box
walls, windows or the gloves themselves. Also, neutron generation is relatively low since ^{239}Pu does not undergo spontaneous fissioning to any great extent nor does it emit a
profusion of alpha particles (because of its relatively long
half-life) which can lead to neutron generation via the
alpha-neutron reaction in light elements (primarily in oxygen contained in the oxide fuel material).

Plutonium of the "commercial" type contains significant
amounts of other plutonium isotopes, however, as indicated
in Table III. Detailed assessment of shielding requirements
(28) based upon anlysis and supportive measurements (27)
indicate that mixed oxide containing 5% plutonium, after
multiple recycles, will not be capable of being handled in
glove boxes without supplementary neutron and gamma ray
shielding. For this mixed-oxide composition, attenuation
by more than a factor of 10 (in addition to that afforded
by the normal glove box structure, gloves and source-to-
operator distance) must be realized for each type radiation.

For gamma rays, such attenuation can be readily achieved
with thin sheets of steel or lead and with lead-impregnated
gloves; however, neutron attenuation requires greater thicknesses of water, plastic or concrete (about 6-12 inches).
These thick shielding requirements mitigate against glove
box operations. Accordingly, canyon-type facilities have
been proposed (27). In such a facility, minimal shielding
is provided on the containment structures; shielding is provided by neighboring walls. The process lines are highly
automated and equipment repair is effected by remotely re-
locating failed equipment to a specially-designed repair
cell. Gloves are provided on the containment structures
and surrounding corridors are provided for access; however,
glove operations are limited to emergency situations. During normal operation, access to the rooms (inside the shield
walls) that contain the process lines and surrounding en-
closures is restricted because of the high radiation levels.

Another consequence of the radiological hazard is concern for the public following potential upsets and release of plutonium to the general environment. This concern has resulted in the dictation of natural phenomena protection for plutonium plants. Specifically, plutonium fuel manufacturing facilities must be designed to withstand earthquakes, flooding and tornadoes without release of plutonium to the general environs.

It is quite obvious from these discussions that plutonium fuel fabrication is expected to be significantly more costly than comparable uranium fuel fabrication.

Economic Considerations. Obviously, plutonium recycle is economically desirable since ^{235}U requirements are reduced. Such reduction in ^{235}U requirements results in commensurate savings of virgin uranium and separative work. Savings that can accrue when byproduct plutonium is blended with natural uranium and reintroduced into the reactor from which it came are indicated in Table IV for large (1,000 MWe) LWRs (29). In developing the tabulated values, allowance has been made for reprocessing and fuel fabrication prior to the timing of plutonium availability; hence, savings are indicated as beginning 2.5 years after reactor startup. It is to be noted that eventually, enriched uranium requirements can be reduced by about 27%. From an economic standpoint, this potential saving must be balanced against the cost of reprocessing and plutonium recovery and the increased cost of fuel fabrication. In the voluminous generic environmental statement on the use of mixed oxide fuel (30), it is concluded that the potential savings far outweigh the added costs.

The economics question that is addressed herein is not the desirability of plutonium recycle per se but rather, the manner in which plutonium can be recycled or should be recycled. The economic aspects of such evaluation revolve around two major issues: the fuel fabrication cost and the wasteful destruction of separative work.

Based upon the radiological aspects described above, numerous investigators have estimated the probable cost of plutonium fuel fabrication (31-34). These estimates have changed with time because of changing environmental emphasis and changing regulatory requirements. Primarily, the conclusion has been that fabrication cost will be high;

TABLE IV

REDUCTION IN U FUEL REQUIREMENTS
AFFORDED BY Pu RECYCLE IN LARGE LWRs; SELF-GENERATION ONLY

Year From Startup	Fissile Pu Discharged kg	Equivalent Enriched U, kg	Fuel Loading Required kg	Reduction in Enrich-ed U require-ments, %
PWR				
0	0	0	93,000	0
1.5	160	0	31,000	0
2.5	200	5,000	31,000	16
3.5	200	6,200	31,000	20
4.5	220	6,200	31,000	20
5.5	270	7,000	31,000	33
6.5	270	8,400	31,000	27
7.5	270	8,400	31,000	27
BWR				
0	0	0	137,000	0
1.5	200	0	53,000	0
2.5	200	8,000	40,000	20
3.5	200	8,000	40,000	20
4.5	210	8,000	40,000	20
5.5	270	8,400	40,000	21
6.5	270	10,800	40,000	27
7.5	270	10,800	40,000	27

therefore, as much plutonium as possible should be concentrated within as little fuel material as possible. Since a specific amount of plutonium is produced by reactor operation, the average fuel fabrication cost for the industry will be lower if the amount of mixed-oxide fabrication is minimized. While this conclusion appears to be valid on the surface, recent broader considerations have questioned its validity (35). These will be discussed further.

The economic issue of enrichment degradation or separative work destruction has never been clearly presented; therefore, let us now attempt to remedy this situation. To do this, let us make the simplifying assumption that an atom of fissile plutonium (^{239}Pu and ^{241}Pu) is equivalent to an atom of ^{235}U, and that an atom of nonfissile plutonium is equivalent to the fertile atom, ^{238}U. Since plutonium discharged from LWR uranium cycle operations contains about 70% fissile atoms (Table III), under this assumption such plutonium can be considered equivalent to uranium enriched to 70% ^{238}U. If we had at our disposal such uranium as well as uranium of various lower enrichments, let us now determine the realizable feed and separative work values for various usable product materials that can be produced by blending. The results of these assessments are presented in Tables V-VII.

The separative work destruction incurred by producing 3% enriched uranium via the blending of 70% enriched material with uranium of lower enrichment is indicated in Table V. This product enrichment was chosen for illustration since 3% is the approximate fissile content of LWR uranium fuel and would be the approximate fissile content of mixed-oxide fuel if plutonium were recycled by simply replacing some ^{235}U atoms with fissile plutonium to maintain reactivity over the same cycle length as is enjoyed with uranium fuel. The results show significant degradation losses, being greatest when blending is done with material having the lowest enrichment (in the tabulated results, depleted uranium having an enrichment of 0.3%). Specifically, the results show that blending with depleted uranium will allow realization of only 60% of the separative work content inherent in the 70% enriched fuel material. Recycling plutonium via such blending will result in the irretrievable loss of 40% of the inherent separative work value of the plutonium! Blending with natural uranium to produce the 3%

TABLE V

3% ENRICHED URANIUM PRODUCT VALUES

EFFECT OF 70% ENRICHED MATERIAL UTILIZATION ON ITS REALIZABLE FEED AND SEPARATIVE WORK CONTENT
(Based on 1 kg of each listed material; 0.3% tails assay)

	Feed Materials				
	0.3% En-riched U	Natural U	2.8% Enrich. U	3.0% Enrich. U	70% Enrich. U
Fissile Content, g	3.00	7.11	28.0	30.0	700.0
Feed Content, kg U	0	1.00	6.08	6.57	169.59
SW Content, kg SW	0	0	3.07	3.42	147.59

	Product Materials			
	3% Enrich. From Cascade	3% Enrich.U 0.3%U+70%U	3% Enrich.U Natural U+70%U	3% Enrich.U 2.8%U + 70% U
Low-enriched Feed Required to make 1 kg Product, g:				
Low enriched U	—	961.26	996.96	997.02
Fissile Content, g		2.88	6.87	27.92
Low-enrich.U	—	0	0	0
Feed Content, kg U		0	0.97	6.06
Low-enrich. U	—	0	0	0
SW Content, kg SW		0	0	3.06

Table V (Continued)

	Product Materials			
	3% Enriched From Cascade	3% Enrich.U 0.3%+70% U	3% Enrich.U Natural U+70%U	3% Enriched U 2.8%U + 70% U
70% U Required to make 1 kg Product, g	–	38.74	33.04	2.98
70% U Fissile content, g	–	27.12	23.13	2.08
70% U Feed content, kg U	–	6.57	5.60	0.51
70% U SW Content, kg SW	–	5.72	4.88	0.44
Values Used to Make 1 kg of Product:				
Fissile Content, g	30.00	30.00	30.00	30.00
Feed Values, kg U	6.57	6.57	6.57	6.57
SW Values, kg SW	3.43	5.72	4.88	3.50
SW Lost per kg of Product Produced, kg SW	–	2.29	1.45	0.07
SW Lost per kg 70% Enriched U used, kg SW	–	59.11	43.89	23.49
(SW Lost/SW Value) of 70% Enriched U, %	–	40.0	29.7	15.9

TABLE VI
15% ENRICHED URANIUM PRODUCT VALUES

EFFECT OF 70% ENRICHED MATERIAL UTILIZATION ON ITS REALIZABLE FEED AND SEPARATIVE WORK CONTENT

(Based on 1 kg of each listed material; 0.3% Tails Assay)

Feed Materials

	0.3% Enrich.U	Natural U	10% Enrich.U	15% Enrich.U	70% Enrich.U
Fissile Content, g	3.00	7.11	100.0	150.0	700.0
Feed Content, kg	0	1.00	23.60	35.76	169.59
SW Content, kg SW	0	0	17.28	27.72	147.59

Product Materials

	15% Enrich.U From Cascade	15% Enrich.U 0.3% U+70% U	15% Enrich.U Nat.U+70% U	15% Enrich.U 10%U+70% U
Low-enriched Feed Required to make 1 kg Product, g	—	789.10	793.78	916.67
Low-Enrich.U Fissile Content, g	—	2.37	5.64	91.67
Low-Enrich.U Feed Content, kg U	—	0	0.79	21.63
Low-Enrich.U SW Content, kg SW	—	0	0	15.84
70% U Required to make 1 kg Product, g	—	210.90	206.22	83.33
70% U Fissile Content, g	—	147.63	144.36	58.33
70% U Feed Content, kg	—	35.76	34.97	14.13
70% U SW Content, kg SW	—	31.13	30.44	12.30

Table VI (Continued)

	Product Materials			
	15% Enrich.U From Cascade	15% Enrich.U 0.3% U+70% U	15% Enrich.U Nat.U+70% U	15% Enrich.U 10%U+70% U
Values used to Make 1 kg Product:				
Fissile Content, g	150.00	150.00	150.00	150.00
Feed Values, kg U	35.76	35.76	35.76	35.76
SW Values, kg SW	27.72	31.13	30.44	28.14
SW Lost per kg of Product Produced, kg SW	—	3.41	2.72	0.42
SW Lost per kg 70% Enriched U used, kg SW	—	16.17	13.19	5.05
(SW Lost/SW Value) of 70% Enriched U, %	—	11.0	8.9	3.4

TABLE VII

8% ENRICHED URANIUM PRODUCT VALUES

EFFECT OF 70% ENRICHED MATERIAL UTILIZATION ON ITS REALIZABLE FEED AND SEPARATIVE WORK CONTENT

(Based on 1 kg of each listed material; 0.3% tails assay)

	Feed Materials			
	Natural U.	7.0% Enriched U.	8.0% Enriched U.	70% Enriched U.
Fissile Content, g	7.11	70.0	80.0	700.0
Feed Content, kg U	1.00	16.30	18.73	169.59
SW Content, kg SW	0	11.16	13.19	147.59

	Product Materials		
	8% Enrich.U From Cascade	8% Enrich. U Nat.U+70% U	8% Enrich.U 7.0%U+70% U
Low-enriched Feed Required to make 1 kg Product, g	—	894.80	984.13
Low-Enrich.U Fissile Content, g	—	6.36	68.89
Low-Enrich.U Feed Content, kg U	—	0.89	16.04
Low-Enrich.U SW Content, kg SW	—	0	10.99
70% U Required to make 1 kg Product, g	—	105.20	15.87
70% U Fissile Content, g	—	73.64	11.11
70% U Feed Content, kg U	—	17.84	2.69
70% U SW Content, kg SW	—	15.53	2.34

Table VII (Continued)

	Product Materials		
Values used to make 1 kg Product:			
Fissile Content, g	80.00	80.00	80.00
Feed Values, kg U	18.73	18.73	18.73
SW Values, kg SW	13.19	15.53	13.33
SW Lost per kg of Product Produced, kg SW	—	2.34	0.14
SW Lost per kg 70% Enriched U used, kg SW	—	22.27	9.01
(SW Lost/SW Value) of 70% Enriched U, %	—	15.1	6.1

product material is somewhat more efficient and results in
a 30% loss of inherent separative work. Maximum utilization
of inherent separative work is achievable by blending with
uranium that has enrichment not much lower than the desired
product enrichment; the tabulated example shows that blending with 2.8% enriched uranium will result in a separative
work loss of only 16%. In short, the results show that the
realizable separative work value of the recovered plutonium
can be increased by 20% if such were blended with 2.8% enriched uranium rather than with natural uranium.

The magnitude of this irreversible loss of a resource can
be readily quantified by noting (in Table V) that one kilo-
gram of 70% enriched material has inherent separative work
value of 147.59 kg SW. By blending with natural uranium,
43.83 kg SW are lost for each kg of total plutonium utilized
or 62.62 kg SW are lost for each fissile kg of plutonium
utilized. Under the Low Growth, No Plutonium Recycle pro-
jection of installed MWe capacity for the United States included in the GESMO Final Environmental Statement (30), it is
indicated that the cumulative fissile plutonium production
to the year 2003 is expected to be 897 tonne. If all this
plutonium were blended with natural uranium to produce the
necessary mixed-oxide, the separative work degradation loss
would be 56,170 tonne. At a round number separative work
cost of $100/kg SW, the total monetary loss would amount to
$5.6 billion. On the other hand, if the plutonium were blen-
ded with uranium enriched to 2.8% in order to produce the 3%
fissile product, $2.6 billion of this could be saved. More
precisely stated: if blending were carried out with uranium
of the higher enrichment, the separative work required to
enrich the uranium needed in conjunction with plutonium recycle would be $2.6 billion less than would be required if
the diluent were natural uranium.

Similar results for a product fuel material having 15%
fissile content are presented in Table VI. Fuel of this
enrichment would be suitable for fast breeder application.
Therein it is seen that the degradation losses are significantly lower, amounting to only 11% when material enriched
to 70% is blended with depleted uranium. Blending with
natural uranium further reduces the degradation loss to 8.9%.
Blending with 10% enriched uranium reduces the loss to only
3.4%; however, such blending for fast breeder application

is really not of interest since ^{235}U is not equivalent to
fissile plutonium for this end use. Consequently, comparison of the two tables gives indication that approximately
90% of the inherent value of plutonium can be effectively
utilized in fast breeder application while only 70% can be
utilized in recycle in LWRs if natural uranium is used as
the diluent; however, if 2.8% (or thereabouts) enriched
uranium is used as a diluent in the recycle application,
effective use can be made of 84% of the inherent separative
work. In short, for LWR recycle application, by blending
the plutonium with uranium having enrichment close to the
desired product enrichment of 3%, one can come to within 10%
of utilizing as much of the separative work as can be effectively utilized in the fast breeder application.

In Table VII, we show a similar analysis for 8% enriched
product fuel. If such could be used for LWR fueling, the
effective utilization of separative work would be well above
90% if blending were carried out with uranium enriched to
almost 8% (in the tabulation, blending with 7% enriched
uranium results in only a 6.1% degradation loss). Use of
fuel having such high fissile content has been suggested (11)
in order to increase the length of the fuel cycle.

Obviously, based on this separative work analysis, utilization of fuel having relatively high fissile content (and
deriving benefit therefrom) would be the most efficient way
to realize the separative work resource inherent in plutonium. This is contrary to the recent suggestion that plutonium might be used as a fuel in CANDU reactors (36). For such
application, the plutonium would be blended with natural or
depleted uranium to yield a fuel material having about 1%
fissile content. The resulting separative work degradation
loss would amount to at least 56% without commensurate benefit.

Because of its unique nuclear properties in the fast neutron spectrum, it is certainly desirable to use plutonium as
the fuel in fast breeder reactors; however, since more plutonium is produced than can be effectively utilized in this
manner (30), most of the produced plutonium must be recycled
in thermal reactors. Accordingly, it is suggested that
practical utilization of plutonium should be as efficient
as possible.

<u>Safeguards Considerations</u>. The present philosophy
relative to handling of dangerous material is to keep it in
concentrated form so that a minimum quantity of matter need
be handled, processed and guarded. Obviously, this minimizes handling and processing costs and simplifies surveillance and control; however, it must be recognized that if
control is lost and upsets occur, the consequences to people
and the environment could be severe.

This philosophy generally has been applied to toxicants
including radioactive materials. It also can be applied to
other "dangerous" materials, plutonium being one, since it
can be used to make a nuclear explosive device.

An alternate philosophy that merits consideration states
that toxic or dangerous material should, as soon as practicable after its production, be diluted with innocuous material, to the extent practicable, so that if control is lost
and upsets occur, the consequences to people and the environment will be minimal.

It recently has been suggested that this latter philosophy be considered for handling plutonium (35) and nuclear
waste (37). Relative to plutonium, practical application of
this philosophy would dictate that plutonium be diluted with
uranium at the reprocessing plants where the plutonium is recovered, with the amount of dilution such that the product
mixture could still be used as a power reactor fuel. The
effect of application of this philosophy on the safeguards
issue can be readily evaluated.

The critical mass data presented in Table II gives indication that a crude nuclear device probably can be constructed with a plutonium mass of about 20 kilograms. This quantity would be chosen as a minimum by any terrorist who had
only this freely-published critical mass data available to
him. His choice would be based on the assumption that implosion of plutonium oxide powder would probably result in
achieving near-theoretical oxide density (11.46 Mg/m^3) and
therefore, that the associated critical mass would be
approximately the 12.7 kg indicated for oxide having a density of 10 Mg/m^3. His choice of 20 kg would include a reasonable safety factor. If he were metallurgically more
sophisticated, he could plan on using plutonium metal,
thereby reducing his assessment of the plutonium requirement

down to about 10 kg. He might wonder whether implosion
could conceivably result in densities greater than theoretical, but without firm knowledge and detailed experimental data on implosion characteristics, he could not risk
the use of a lesser quantity of plutonium, at least not if
he seriously desired to have reasonable assurance that the
device would work. From this assessment, we conclude that
a terrorist would probably attempt to divert at least 20 kg
of plutonium or 23 kg of plutonium oxide.

If pure plutonium oxide were the material of commerce,
diversion of 23 kilograms would suffice; however, if dilution with uranium were immediately effected at the reprocessing plants, significantly larger quantities of the mixture would have to be stolen. If dilution were such as to
produce a mixture suitable for fast breeder fueling (6 to 1
dilution), approximately 150 kg would be required. If dilution were with natural uranium to produce a mixture suitable for thermal reactor recycle (say 3% fissile content),
approximately 700 kg would be required. If dilution were
with 2.8% enriched uranium to produce a mixture suitable
for thermal reactor recycle, approximately 8,000 kg would
be required.

Certainly, it would be appreciably more difficult to
divert 8 tonnes of material than 23 kilograms. Further,
after such diversion, the plutonium would have to be chemically separated from the uranium in order to obtain the
pure plutonium compound required for device construction.
The quantities requiring such chemical treatment (dissol-
ution and ion exchange, selective precipitation, or solvent
extraction) would dictate large equipment and/or a long
processing time. Both these factors increase the probability of detection.

Of course, dilution at the reprocessing plants would
imply that greater quantities of material containing plutonium would have to be shipped. This is not a horrendous
amount of material since similar quantities of uranium would
require shipment from enrichment plants to fuel fabrication
plants if the plutonium were not diluted at the reprocessing
plants. Obviously, the plutonium-bearing mixture would require more sophisticated transport because of its greater
toxicity and its potential attraction to terrorists. This
added cost must be weighed against the safeguards advantage

just described. Quantitative balancing is complex since it
is difficult to estimate the reduced probability of successful diversion and device manufacture that is a consequence
of early dilution.

B.   Minimum Perturbation of the Uranium Fuel Cycle

As previously indicated, initial investigation relative
to plutonium utilization in thermal reactors took an open-
minded approach and contemplated optimization of core design (fuel rod diameter and lattice spacing), fuel burnup,
and fuel-fabrication techniques; however, because of costly
delays experienced throughout the nuclear industry, the
industry rapidly became conservative and discarded those
options that deviated from standardized uranium practice.
Development rapidly narrowed to the recycle scheme that most
closely paralleled uranium technology and whose institution
would have the least effect on uranium operations. Accordingly, the scheme that received development impetus was
the one that allowed substitution of a few uranium fuel elements with plutonium-bearing fuel elements without alteration of the fuel burnup level and the reactor refueling
schedule. To keep the number of plutonium-bearing fuel
elements to a minimum dictated blending of the plutonium
with natural (or near-natural) uranium. Paralleling uranium
fuel fabrication technology dictated the pellet-in-tube concept.

Even though plutonium recycle is not yet being employed
to any significant extent, in the ensuing presentation we
shall call this the "conventional" method of recycle. Only
recently has the validity of this conservative approach been
questioned (35,38). The alternatives to the conventional
approach will be addressed later; the following presentation
deals exclusively with the conventional recycle scheme.

Uranium Blending Options. If plutonium is to be used
in only a few fuel assemblies and if it is deemed desirable
to minimize the amount of mixed-oxide fuel that must be fabricated, then plutonium should be blended with uranium that
has little fissile content. This implies that it should be
blended with depleted uranium; however, evaluation of separative work utilization, as presented in Table V indicates
that blending with depleted uranium is economically unat-
tractive. Approximately 40% of the separative work inherent

in plutonium would be lost upon blending with depleted uranium while only about 30% would be lost upon blending with natural uranium. Further, conversion of UF_6 (the chemical form of stored depleted uranium) to fuel grade oxide is more costly than conversion of virgin yellow-cake. Based upon these considerations, it can be readily concluded that blending with natural uranium is economically more desirable than blending with depleted uranium.

Another, alternate blending scheme is to mix the plutonium with uranium recovered from spent fuel. This uranium has residual enrichment of about 0.85%. Analysis identical to that performed in the development of Table V indicates that approximately 28% of the inherent separative work value of the plutonium would be lost by such blending. The additional 2% separative work saving is inadequate to cover the current conversion cost differential. Besides, such blending would require that a specific amount of available plutonium be spread through a somewhat greater quantity of fuel and this would result in somewhat greater fuel fabrication expenditure. On balance, the overall cost differential is small, and there is little difference between blending with natural uranium and blending with residual uranium.

Reactivity Value of Plutonium. It is both difficult and potentially misleading to quantify the reactivity value of fissile plutonium relative to ^{235}U. Such value depends upon the plutonium isotopic composition utilized, the location of the plutonium-bearing fuel in the core, and the condition of the surrounding fuel material. Further, these interrelationships vary as the reactor is operated. The complexity of the problem is compounded because of the complex cross-section behavior of the plutonium isotopes (Figures 1 and 2) which makes reactivity worth particularly sensitive to the local neutron temperature.

To allow meaningful discussion, consider a plutonium-bearing fuel assembly that can be used to replace a uranium fuel assembly without altering the reactivity life of the core configuration. Generally, such a replacement assembly will have lower k_∞ and higher blackness than the uranium assembly. Substitution will, therefore, tend to decrease neutron absorptions in the surrounding materials; substitution will decrease the worth of surrounding fuel and the worth of surrounding control elements. Overall core reactivity, therefore, could be either increased or decreased

depending upon the nature of the surroundings. Since decrease of control rod worth is generally not desirable, in
practical application, one can say that the surroundings
will be predominantly fuel material; hence, the overall
core reactivity probably will be decreased. Based upon
this starting conclusion, similar reasoning will indicate
that the overall core reactivity will tend to drop less as
a function of core burnup; however, one must guard against
equating such reduced reactivity variation to the adequacy
of control capability as a function of time. As the reactor
operates, uranium fuel located between the plutonium-bearing
element and the control rods is burned, and accordingly with
time, the relatively black plutonium-bearing fuel element
may have increasingly more effect on the worth of the control
rod. Consequently, the worth of neighboring control rods may
be reduced as a function of time. Obviously, quantitative
generalization is impossible; for each specific application,
detailed analysis is required to evaluate the effect of substitution of plutonium-bearing fuel elements.

Such detailed evaluations have been made for a number
of specific reactors (33,39-42); in addition, the reactor
vendors have made evaluations to support their current offerings (44-46). Much useful information also is being generated under the EPRI plutonium recycle program (47).

The report on the SENA reactor (33) has particularly in
teresting information relative to the reactivity value of
plutonium of isotopic composition expected to be generated
in that reactor. Therein, it is indicated that when such
plutonium is blended with natural uranium to make a mixed-
oxide equivalent to that required to sustain operations with
stainless steel cladding (equivalent to 4% enriched uranium),
1.24 atoms of fissile plutonium will be required to replace
one atom of ^{235}U. If blending were carried out with uranium enriched to approximately 1.8%, it is indicated that
only 1.18 atoms of fissile plutonium would be required to
replace one atom of ^{235}U. With Zircaloy cladding (3.35%
^{235}U enrichment), corresponding derived values are 1.28 and
1.17, respectively. These specific reported values for
plutonium equivalence when natural uranium is used as the
diluent (1.24 and 1.28) compare favorably with recent vendor
offerings for large LWRs (apparently this value is not par-
particularly sensitive to reactor type or plutonium isotopic
composition) and with the value (1.25) assumed by the AEC
(and now ERDA) in its forecasting activities (48).

Fuel Design Features. Fuel design analysis is primarily concerned with making plutonium-bearing fuel elements compatible with surrounding uranium fuel elements and in maintaining a relatively flat power distribution throughout the plutonium-bearing assemblies during fuel burnup. In general, this requires that fuel assemblies contain 3-5 different types of fuel rods. These may all be mixed-oxide rods having different plutonium contents (generally, for PWR use), and in addition, some may be uranium rods with different enrichments (generally, for BWR use). In addition, burnable poisons (e.g., Gd_2O_3) may be used to, in effect, neutronically isolate the plutonium-bearing assemblies from the surrounding uranium elements. Specific design configurations are presented in the above-referenced documents.

Few generalizations can be made; detailed analysis is required for each situation and the relative location of the rod types and their acceptable fissile content are dependent upon the isotopic composition of plutonium that is available at each refueling.

Fuel Fabrication Options. Once the decision is made that mixed-oxide will be fabricated by the pellet-in-tube process, process variability is severely restricted. Primarily, process alternatives exist at the head end. The mixed-oxide can be produced either by coprecipitation or by the mechanical blending of the component oxide powders. If the latter, then there is choice of equipment for powder preconditioning to adjust particle size and surface activity; for example, ball-milling or jet-milling can be used. Precise blending, generally, is not a simple operation since component weights must be accurately determined with corrections made for moisture content and since an acceptable degree of homogeneity is necessary. Product variability as a function of processing operations is described in the literature (13,49).

The conventional method of plutonium recycle creates significant fabrication problems because of the multiple fuel compositions involved. This implies small batch operations and the need for strict accountability. One can reasonably question whether it is possible to guarantee that the proper compositions always will be provided in the proper locations. Once placed in a reactor, one pellet of higher fissile content can result in overheating and rapid cladding failure. Consequently, heavy reliance must be placed on quality assurance

procedures. Small batch operations, strict accountability,
and quality control to assure powder homogeneity and pellet
composition result in inordinately high fabrication cost.

Fuel Performance. There is ample evidence to indicate
that mixed-oxide fuel fabricated with the standard pellet-
in-tube process performs at least as well as comparable
uranium fuel (50,51). Irradiations have been carried out at
levels of up to 68 kW/m peak power and up to 51 MW-day/kg
HM burnup levels. Although numbers are still statistically
small relative to uranium fuel experience, to date, no fuel
failures have been observed in mixed-oxide.

C.   Possibility of Extended Fuel Cycles

The separative work analysis results presented in Table
VII indicate that the separative work inherent in plutonium
can be most efficiently employed if fuel having relatively
high fissile content is utilized. The tabulated results
show that 85% of the separative work value inherent in plutonium can be recovered if a product fuel material with 8%
fissile content is produced by blending the plutonium with
natural uranium. Utilization of such high fissile content
fuel in thermal reactors will allow almost as much separative work to be recovered as is possible with fast breeder
fueling (Table VI, 89% recoverable when plutonium is blended
with depleted uranium, as contemplated).

In spite of this significant conservation advantage,
use of high fissile content fuel has not been investigated
in any great detail, since, as pointed out previously, the
industry has chosen to concentrate on the plutonium recycle
alternative that gives minimum perturbation of uranium operations.

If high fissile concentration fuel were employed, fuel
cycles must be extended (burnup levels must be increased)
to gain practical benefit from recovery of the high separative work values. This implies the necessity of having
fuel elements that can operate satisfactorily to the exposure limits that are possible, based upon reactivity considerations. Current LWR fuels have not been pushed to these
limits, although there are indications that they may be
satisfactory. At best, significant proof-testing will be
required.

Also, current fuel cycle accounting practices mitigate against employment of fuel having such high inventory value. Under conventional accounting, in-core inventory carrying charges would more than offset the separative work savings. While one can argue that the separative work savings are true resource savings (primarily energy savings; energy that otherwise must be used for isotope separation) while the carrying charges are not true savings but account only for the time value of money, it would be difficult to obtain a consensus of opinion that different accounting evaluations should be employed. It is pointed out that the breeder reactor suffers from the balancing of the same economic factors. In that case, the inventory-carrying charge is made acceptable by going to higher-power density, i.e., by making more effective use of the inventory. The Naval reactors program suffers similarly; in that case, the longer core life is given a high military value so as to counterbalance the high inventory-carrying charge.

This plutonium recycle scheme also has an additional economic advantage in that more plutonium is concentrated into a lesser amount of fuel material, thereby minimizing the amount of fuel that must be fabricated with use of the high-cost plutonium technology; however, even this incentive is inadequate to initiate the necessary development work.

Even though it appears that this method of plutonium recycle will not be soon accepted, a brief discussion of pertinent factors is warranted. It is inevitable that resource conservation will be taken seriously some day, at least more seriously than the level of the prevailing interest rate.

Reactivity Considerations. Effective utilization of the separative work inherent in plutonium is feasible because k_∞ of the fuel, as a function of the plutonium concentration in mixed-oxide, tends to peak at a relatively low concentration (a few percent, a value that varies with fuel rod size and lattice spacing) and then tends to decrease as the concentration is further increased (11). At concentrations greater than the peak, reactivity therefore tends to increase with fuel burnup (plutonium depletion) and this tendency substantially counterbalances the negative effect of fission product buildup. Extended core life is feasible because the reactivity swing over the burnup range is not large and can be controlled.

With conventional LWR reload practices (1/4 - 1/3 core removal at each refueling), fuel exposure to about 60 MW-days/kg HM is possible, from the reactivity standpoint, with the use of mixed-oxide fuel having a fissile content of about 8%. High fissile content fuel can be used in LWRs fueled only with such material (plutonium burners), or it can be used in conjunction with uranium fuel in a scatter-pattern loading.

Use in plutonium burners has the advantage that the water-to-fuel-ratio can be optimized; however, this generally implies a looser lattice and the reduced specific power results in higher capital cost per unit power output. Fuel cost benefits derived from optimization of the moderation must be balanced against this contrary cost increase.

Use in conjunction with uranium fuel is not simple; power peaking and controllability must be satisfactorily resolved as discussed above for conventional recycle. Luckily, solution to these problems does not become much more difficult as the fissile content is increased. Even for conventional recycle, the plutonium-bearing fuel assemblies are relatively black, and adding additional plutonium has little effect on the surrounding neutron flux. Accordingly, solutions offered for 3% fissile content fuel are equally effective for fuel having greater fissile content.

Obviously, when all fuel has high plutonium content, fuel life can be generally extended, and the time between refueling outages can be increased. Such will not be possible with mixed uranium and plutonium-bearing fuel assemblies. For such utilization, refueling still will be required on approximately an annual basis, but fewer fuel assemblies would have to be removed at each outage; the plutonium-bearing assemblies could remain in the core for approximately twice as many cycles as the uranium assemblies.

Materials Considerations. Irradiation of LWR pellet fuel to an average exposure of about 60 MW-day/kg HM, while high, is not completely outside the range of feasibility. As previously indicated, Zircaloy-clad mixed-oxide has been irradiated to 51 MW-day/kg HM without failure. If fuel/clad interaction should prove to be unacceptable with Zircaloy, at least for PWR application, one always has the option of using stainless steel cladding. Since plutonium has higher

cross-sections than uranium, neutron absorption in the steel
would be much reduced from that experienced in the early PWR
cores wherein such cladding was satisfactorily utilized with
uranium fuel.

D.   Low Concentration Plutonium Recycle

The previously-presented analyses and discussions indicate that the conventional method of plutonium recycle is
far from ideal.

Such utilization of plutonium results in a deplorable
waste of a valuable resource (approximately 30% of the separative work inherent in plutonium is lost by blending of
the high fissile content plutonium with natural uranium).
Also, use of plutonium blended with natural uranium results
in substantial hardening of the local neutron spectrum and
thereby gives plutonium relatively low reactivity equivalence compared with uranium. As indicated in calculations
applicable to the SENA reactor (33), blending the plutonium
with uranium enriched to 1.8% rather than 0.71% would result
in an 8% increase in the reactivity value of the plutonium.
Use of such blending with 1.8% enriched uranium would "make
the plutonium go farther"; natural uranium and separative
work savings could be increased by an additional 8%, a considerable saving of valuable resources.

Further, it was indicated that concentration of plutonium results in excessively high fabrication cost due to
small batch sizes and the need to simultaneously process
mixed-oxide having a variety of plutonium concentrations.

Effective use of the inherent separative work content
of the plutonium can be made by utilizing mixed-oxide having
substantially greater fissile content than contemplated for
the conventional method or recycle; however, current economic assessment militates against utilization of such high
inventory values. Also, reliable operation of conventional
LWR fuel has not been demonstrated out to the burnup levels
(about 60 MW-day/kg HM) required to derive maximum benefit
from the separative work contained in the plutonium.

An alternative is available at the other end of the
concentration range (35). If the plutonium were blended with
uranium enriched almost to the fissile content required for

LWR operation, much of the wastefulness and many of the
difficulties associated with the conventional method of
recycle could be ameliorated. Of course, use of such low
concentration plutonium fuel has disadvantages as well as
advantages and these must be balanced. The following presentation concentrates on the important consequences that
would result from the use of low plutonium concentrations
in mixed oxide.

    In addition to the separative work and neutronic sav-
ings already considered, sizable savings of virgin uranium
would also result. This comes about because more residual
uranium would be available for recycle than is possible
under the conventional recycle scheme. In conventional recycle, as indicated in Table IV, about 27% of all fuel in a
reactor eventually becomes mixed-oxide. This 27% of the
core loading is, upon insertion, natural uranium blended
with plutonium. Upon discharge, this uranium is well depleted. Based upon economic considerations, it should be
discarded following discharge. If such were done, the
amount of uranium recovered from each discharge batch would
then be 73% of the uranium that would be recovered if plutonium were not recycled since with recycling, the only uranium
that would be recovered would be that contained in the uranium assemblies. (In practice during reprocessing, it is not
now contemplated that the mixed-oxide fuel will be segre-
gated from the uranium fuel. Accordingly, the depleted
uranium from the mixed-oxide will be blended with the
slightly-enriched uranium recovered from the uranium fuel.
The result will be that almost as much uranium will be recovered under plutonium recycle operation as with uranium-
only operation; however, the recovered uranium enrichment
will be reduced by more than 0.07%. Under this method of
reprocessing plant operation, the uranium loss will not be
quite so great but the separative work degradation loss will
more than negate the monetary savings associated with the
additional uranium recovery.) Under the low concentration
method of recycle, the uranium contained in the mixed-oxide
would, upon insertion, have enrichment of a few tenths of
a percent less than the uranium contained in uranium-only
assemblies. Since even the small amount of plutonium would
offer the contained ^{235}U significant competition for neutrons, the uranium enrichment upon discharge would not be
essentially different from the discharge uranium enrichment
of the uranium-only assemblies; certainly, the difference

upon discharge would be less than the difference upon insertion. Accordingly, most of the 27% uranium recycle loss sustained with conventional recycle would not be experienced with
low concentration recycle.

Comparison of the Low Growth MWe capacity projections
presented in the final GESMO Environmental Statement (30),
indicates that uranium recycle with plutonium recycle can
save approximately 200,000 Mg U_3O_8 between now and the year
2000. About one-quarter of this saving, or 50,000 Mg U_3O_8
could not be realized if plutonium were recycled in the conventional manner; however, it can be essentially fully realized under the low concentration recycle scheme. At an
average cost of about \$90/kg U_3O_8, this amounts to a monetary
saving of about \$4 billion.

Monetary savings that can be anticipated by application
of the low concentration recycle scheme therefore amount to
this \$4 billion plus the \$2.6 billion saving in separative
work discussed previously. Augmentation of these values due
to the increased neutronic efficiency brings the total realizable saving (to the year 2000) up to about \$8 billion.
Since, as was previously indicated, it is projected that
approximately 900 Mg of fissile plutonium will be recovered
from LWR operation over this time span, it can be concluded
that employment of low concentration recycle will add almost
\$10/g to the monetary value of fissile plutonium over that
which can be realized by the conventional method of recycle.
In addition, as discussed herein, since concentrated plutonium would not be the available material of commerce if blending were carried out at the reprocessing plants, there would
be added safeguards advantage.

These potential monetary benefits must be weighed
against added complexity in the areas of safeguards and fuel
fabrication. For the evaluation of both these factors, let
us make pessimistic assumptions.

If it is assumed that identical safeguards provisions
will be applied to the low concentration mixed-oxide as to
pure plutonium oxide, the scale of safeguards operations
will obviously be increased. To obtain indication of the
magnitude of such increased activity, it is again worthwhile
to base such assessment on the Low Growth projections contained in the final GESMO Environmental Statement (30).

These projections, under the plutonium recycle option, indicate that a maximum of 91 Mg of fissile plutonium will be
recovered in any one year prior to the year 2000. In conjunction with such recovery, a maximum of 2900 Mg of mixed-
oxide fuel will require fabrication in any one year under
conventional recycle.

Presumably, such fabrication will be carried out in
facilities similar to those proposed by Westinghouse (27)
having stretch production capability of 400 Mg per year HM.
Accordingly, eight fabrication plants will have to be safe-
guarded under the conventional method of recycle. Under the
low concentration recycle scheme, if mixed-oxide having 0.5%
fissile plutonium content were used rather than 3% as in
conventional recycle, six times as much mixed-oxide fuel
would have to be fabricated. Since production demand would
exist, such fabrication would be undertaken in significantly
larger plants. Since 1500 Mg/year facilities are used to
fabricate uranium fuel, it is reasonable to expect that similar sized plants would be constructed to fabricate low-concentration mixed-oxide fuel. Accordingly, 12 such plants
would be required. In summary, conventional recycle will
require safeguarding of about eight fuel fabrication plants
by the year 2000, while low concentration recycle will require
safeguarding of about 12 significantly larger plants. Without
getting into detailed safeguards measures, it does not appear
that plant safeguarding would present an insurmountable technological or economic barrier to low-concentration recycle.

The shipping differential can be similarly estimated.
In the final GESMO Environmental Statement, it is indicated
that shipping of PuO_2 in specially-designed trucks with maximum builtin safeguard protection (0.5 Mg PuO_2/shipment)
would cost between two and six cents per gram of plutonium.
If the plutonium were diluted at the reprocessing plants to
a concentration in mixed-oxide of 0.5% (200/1 dilution), the
mixture would be no more reactive than slightly enriched
uranium, and therefore, could be shipped with approximately
15 Mg per truckload as is the practice with uranium. Since
200 times as much material would have to be shipped following dilution at the reprocessing plants, but since each shipment could contain 30 times as much material, it is seen that
implementation of low-concentration recycle would require
approximately seven times as many shipments. Taking the NRC

transportation cost estimate, it is concluded that fully-safeguarded transportation would detract at most, $.42/g from the plutonium value, a significant number, but certainly not appreciable when compared to the $10/g potential cost saving indicated above.

Assessment of the fuel fabrication penalty is a much more difficult undertaking. As introduction, let us first examine the radiological aspects. Then let us develop fabrication cost estimates based upon conceptual plant design and contemplated operations.

Radiological Considerations. Relative to setting design criteria for a fuel fabrication plant, a number of radiological aspects must be considered. Workers must be protected during normal operations, and the population at large and the environment must not suffer unduly in case of major upsets.

The upset problem can be expeditiously resolved by requiring natural phenomena protection on those portions of the plant wherein substantial quantities of mixed-oxide will be processed. In short, for the processing of low concentration mixed-oxide, plant construction will be assumed to be similar to that now required for the processing of conventional mixed-oxide.

Setting plant design criteria for worker protection requires some analysis since the lower radiation levels associated with low-concentration mixed-oxide can result in significant plant design and operational changes.

Historically, when working with pure compounds of plutonium, it soon became apparent that the maximum permissible air concentration limits (Table I) could not be met if plutonium were handled out in the open as was the custom with uranium. Mixed-oxide, however, is not a pure plutonium compound; its toxicity, obviously, approaches that of uranium as the plutonium concentration approaches zero. To allow comparison, relative toxicities of various materials are tabulated in Table VIII under the assumption that the compounds are insoluble (e.g., the oxides) and that the plutonium has the isotopic composition of LWR-produced material. It is readily seen that any mixture that contains even a small amount of plutonium is significantly more toxic than uranium. The least toxic mixture tabulated (that which contains 0.5% total plutonium, or 0.35% fissile plutonium) is

about 700 times more toxic than the 3% enriched uranium that
is the standard LWR fuel material. Highly enriched uranium
is about 26 times more toxic than 3% enriched uranium, and
the low-concentration mixture is 28 times more toxic than
highly enriched uranium. When placed in this latter perspective, it is seen that the toxicity of low-concentration
mixed-oxide is not extremely different from the toxicity of
uranium that is processed without elaborate containment provisions. Accordingly, it is conceivable that low-concentration mixed-oxide can be safely processed with containment
being provided only around dusty operations (the head-end
powder handling areas); however, in order to make a conservative assessment of the fabrication cost, the following
developed fabrication plant concept is based upon full containment provision for all mixed-oxide processing.

Another radiological aspect that must be considered for
plant design is the magnitude of the external radiation level
since mixed-oxide emits both neutrons and gamma rays. Based
upon shielding assessments made for the processing of conventional mixed-oxide (27) (4-5% Pu content) it appears that
little or no shielding will be required for processing
mixed-oxide that has only about one-tenth of the plutonium

TABLE VIII

RELATIVE TOXICITIES OF URANIUM AND PLUTONIUM
(Based upon Inhalation of Insoluble Compounds)

Material	Relative Toxicity
Natural Uranium	1.000
3% Enriched Uranium	2.9
95% Enriched Uranium	75
Plutonium	430,000
5% Pu - 95% U Mixture	21,000
0.5% Pu - 99.5% U Mixture	2,100

content. This can be readily verified by numerical assessment. The dose rate of neutrons at the surface of a
large amount of material will be about 40 µSv/hr. Geometric attenuation afforded by material-to-operator distance (usually at least one foot for glove box operation)
will reduce this to about 10 µSv/hr at the operator
location. Similarly, the surface gamma dosage rate is estimated to be about 4 mSv/hr; however, since this is primarily
due to soft gamma rays from ^{239}Pu and ^{241}Am, much attenuation is afforded by the glove box walls, windows and gloves.
A 5 mm thickness of steel or a 1 mm thickness of lead
will result in gamma attenuation of about a factor of 100.
Accordingly, it is confirmed that little or no gamma
shielding is required. If plant design necessitates that
large sources be located near operator stations, shielding
can be readily provided with a minimal thickness of material. Based upon these considerations, it is concluded
that glove box operation is entirely practical and safe
for the processing of low-concentration mixed-oxide; canyon-
type facilities as contemplated for the processing of conventional mixed-oxide (27) are not necessary.

Fuel Fabrication Plant Design and Cost. A conceptual
design of a low-concentration mixed-oxide fabrication plant
was developed by extrapolation from comparable uranium and
conventional mixed-oxide fuel fabrication facilities. Sufficient background information and rationale will be presented to justify the developed concept and the associated
estimated costs.

The design bases used for this extrapolation are presented in Figures 4 and 5. Figure 4 shows the layout of a
typical uranium fuel fabrication plant having capacity of
1500 Mg/yr U. Figure 5 shows a comparable layout for a
conventional mixed-oxide fuel fabrication plant having capacity of 300 Mg/yr HM. Also shown on this figure is the
outline of the area that is hardened for natural phenomena
protection.

The estimated cost of the uranium plant is presented in
Table IX. Our total estimated capital cost of $51 million
(for a 1500 Mg/yr U plant) appears reasonable; it is somewhat less than a recently-publicized industry figure of $50
million for a 1250 Mg/yr U plant (52). A comparable cost
breakdown for the conventional mixed-oxide fabrication plant

is presented in Table X. Again, the total estimated cost of
$67.5 million for a 300 Mg/yr HM plant is in good agreement
with a recently-published industry figure of $90 million for
a 400 Mg/yr HM plant (52).

Now, keeping in mind the radiological criteria discussed
above, a rationale can be developed for extrapolation to a
low-concentration mixed-oxide fuel fabrication plant. In
Figure 5, the various process areas located in the hardened
portion of the facility are depicted. When working with
dilute plutonium, activity levels will be so low that there
will be no need for the laboratories to be located within
the natural phenomena-hardened structure. Similar reasoning
is valid for the waste treatment area; in fact, this area
can be much reduced in size if minimal treatment and packaging were determined to be acceptable. Since sinterable,
coprecipitated mixed-oxide is expected to be received as
feed material from the reprocessing plants, there will be
no need for the powder preparation area, although a modest
area will be required for powder storage. Also, because of
the low radiation levels anticipated, there will be no need
for the hot maintenance area; equipment repair will be capable of being performed in place, through gloves. Further,
since there will be no need for massive external shielding
and the concomitant aisles, the remaining pellet and rod
manufacturing area should be capable of some reduction. In
summary, it appears that a hardened area of about 1300 m^2
would suffice for the fabrication of pellets and rods in a
1 Mg/d HM facility. Accordingly, a 5 Mg/d HM facility need
contain no more than 6500 m^2 for such activity. It is to
be noted that this is 1.56 times greater than that contemplated in a comparable uranium manufacturing plant (Figure
4). If the powder were coprecipitated in the same plant,
this scaling factor would probably hold for the powder area
also; consequently, it would appear that powder preparation
could be performed at a 5 Mg/d HM rate in a hardened area
of about 5,000 m^2. Even though it is expected that such
powder preparation will be performed at the reprocessing
plants, for purpose of assessment, it is simple to consider
such operation to be performed in the fabrication facility.
This should not affect the validity of the assessment, since
substantial overhead and laboratory facilities can be expected
to be present at either plant.

Based upon this assessment and further comparison between
Figures 4 and 5, one can lay out the design of a 1500 Mg/yr

HM, low-concentration, mixed-oxide fabrication plant. The resultant layout is shown in Figure 6. It is seen that the plant would have approximately 20,000 m^2 in unhardened facilities and 11,500 m^2 in hardened facilities. This hardened area is 1.7 times that of the hardened area in the 1 Mg/d HM plant shown in Figure 5.

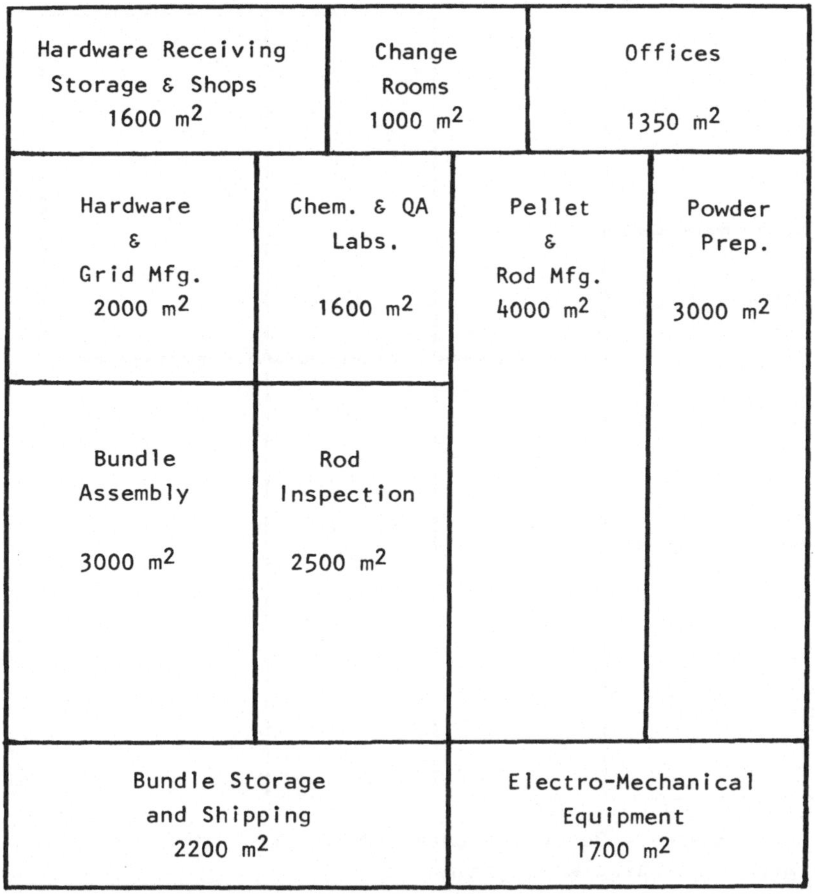

Hardware Receiving Storage & Shops 1600 m^2	Change Rooms 1000 m^2	Offices 1350 m^2	
Hardware & Grid Mfg. 2000 m^2	Chem. & QA Labs. 1600 m^2	Pellet & Rod Mfg. 4000 m^2	Powder Prep. 3000 m^2
Bundle Assembly 3000 m^2	Rod Inspection 2500 m^2		
Bundle Storage and Shipping 2200 m^2		Electro-Mechanical Equipment 1700 m^2	

Total Plant Area = 24,000 m^2

Figure 4

Typical Uranium Fuel Fabrication Plant Layout
(1500 Mg/yr U Capacity)

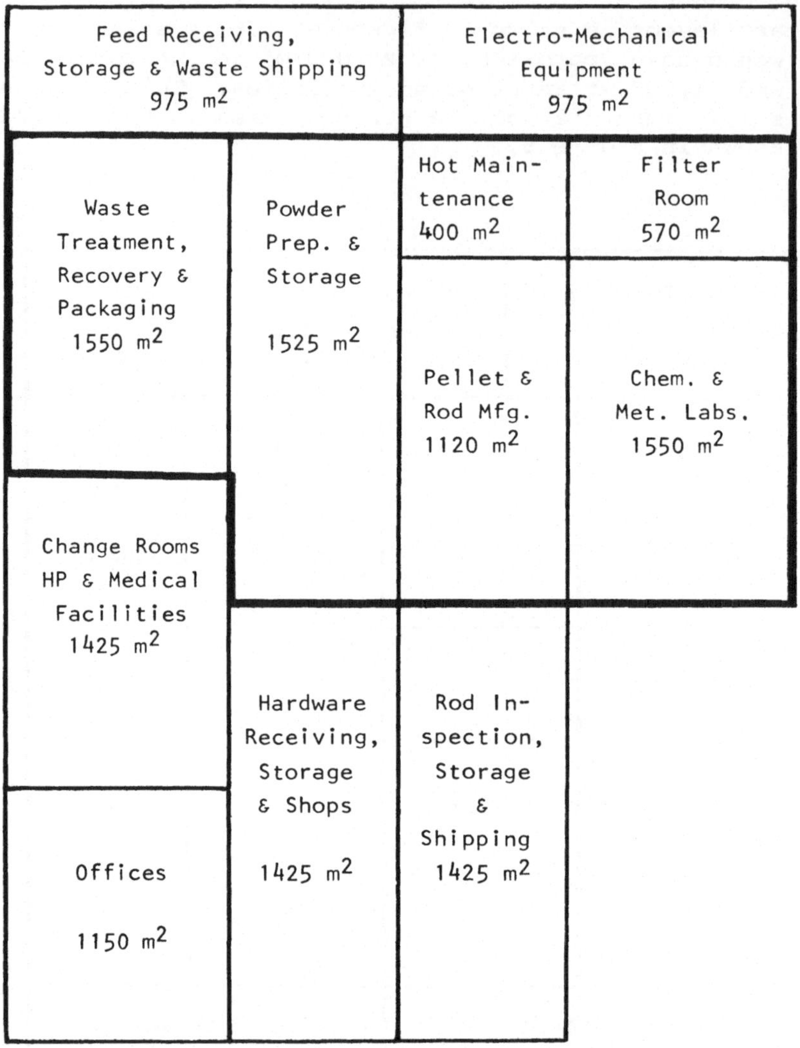

Feed Receiving, Storage & Waste Shipping 975 m²		Electro-Mechanical Equipment 975 m²	
Waste Treatment, Recovery & Packaging 1550 m²	Powder Prep. & Storage 1525 m²	Hot Maintenance 400 m²	Filter Room 570 m²
		Pellet & Rod Mfg. 1120 m²	Chem. & Met. Labs. 1550 m²
Change Rooms HP & Medical Facilities 1425 m²	Hardware Receiving, Storage & Shops 1425 m²	Rod Inspection, Storage & Shipping 1425 m²	
Offices 1150 m²			

Heavy Boundary Denotes Natural Phenomena Protection
Total Plant Area = 14,000 m²

Figure 5

Typical Conventional Mixed-Oxide Fuel Fabrication
Plant Layout (300 Mg/yr HM Capacity)

Estimation of capital cost can be done by simple ration-
ing. Since the uranium fabrication plant of 25,000 m^2 was
estimated to cost $51 million, it seems reasonable to con-
clude that an unhardened area, fully equipped, will cost
about $2,070/m^2. Since the 1 Mg/d HM conventional, mixed-
oxide fabrication plant contains 76,000 m^2 of unhardened
facilities and 7,000 m^2 of hardened facilities, it can be
concluded that the unhardened facilities contribute $15.7
million to the total estimated cost of $67.5 million (Table
X). Therefore, it is ascertained that the hardened areas
contribute about $51.8 million to the total cost. Now from
the detailed cost breakdown (Table X), it is to be noted
that $16 million of this is for containment, automation and
shielding provisions. If one simply assumes that only half
this amount would be required if dilute mixed-oxide were
processed, then the hardened facilities for such processing
can be expected to cost $43.8 million for 7,000 m^2 or 6300
m^2. Using these data, it is concluded that a 1500 Mg/yr HM
low-concentration facility will cost approximately $113.2
million ($41.1 million for unhardened areas and $72.1 million for the hardened areas).

TABLE IX

CAPITAL COST OF A URANIUM-FUEL FABRICATION PLANT
(1500 Mg/yr U Capacity)

	Millions of Dollars
Land, Site and Yardwork	2.0
Buildings	8.0
Process Equipment	15.0
Building Mechanicals	10.0
Auxiliaries	6.0
Total Direct Cost:	41.0
Design and Engineering	4.0
Indirect Cost and Contingency	6.0
Total Capital Cost:	51.0

TABLE X

CAPITAL COST OF A CONVENTIONAL MIXED-OXIDE FABRICATION PLANT
(300 Mg/yr HM Capacity)

	Millions of Dollars
Land, Site and Yardwork	3.0
Buildings	
Base Cost	5.0
Natural Phenomena Hardening	4.0
Safeguards Hardening	0.5
	9.5
Process Equipment	
Base Cost	6.0
Modification for Containment	3.0
Automation Provisions	6.0
Containment Provisions	5.0
Shielding Provisions	2.0
Natural Phenomena Hardening	4.0
	26.0
Building Mechanicals	
Base Cost	4.0
Natural Phenomena Hardening	2.0
	6.0
Auxiliaries	
	3.0
Total Direct Cost:	47.5
Design and Engineering	10.00
Indirect Cost and Contingency	10.00
Total Capital Cost:	67.5

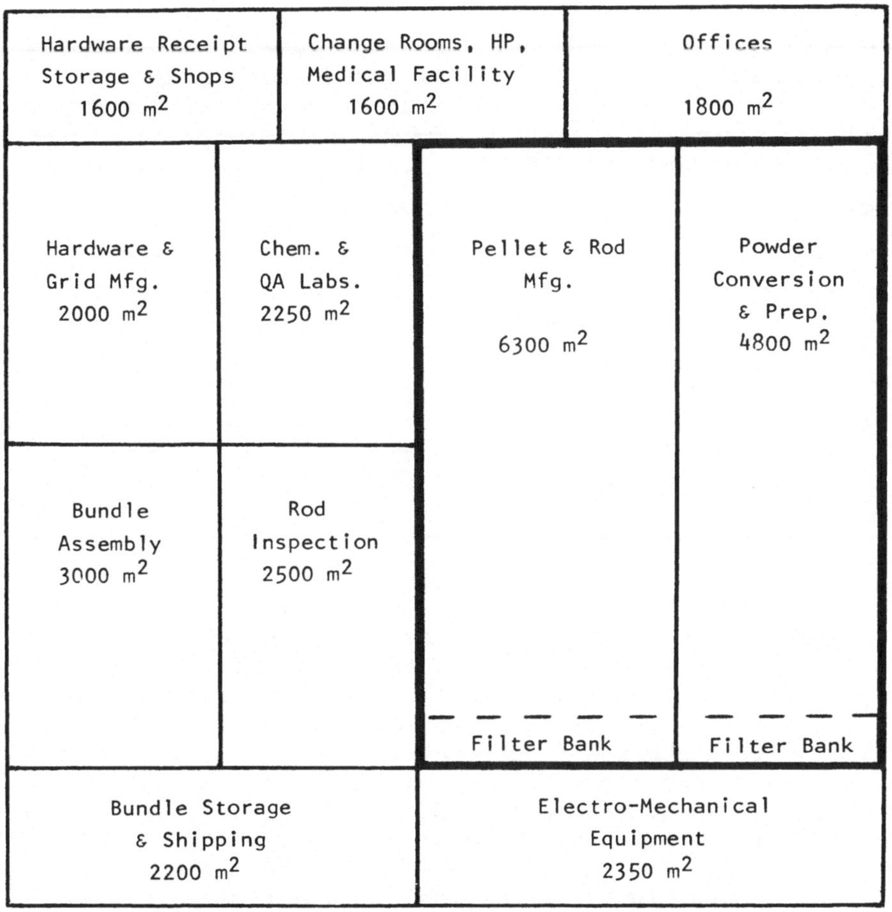

Hardware Receipt Storage & Shops 1600 m^2	Change Rooms, HP, Medical Facility 1600 m^2		Offices 1800 m^2
Hardware & Grid Mfg. 2000 m^2	Chem. & QA Labs. 2250 m^2	Pellet & Rod Mfg. 6300 m^2	Powder Conversion & Prep. 4800 m^2
Bundle Assembly 3000 m^2	Rod Inspection 2500 m^2		
		Filter Bank	Filter Bank
Bundle Storage & Shipping 2200 m^2		Electro-Mechanical Equipment 2350 m^2	

Heavy Boundary Denotes Natural Phenomena Protection
Total Plant Area = 30,500 m^2

Figure 6

Low-Concentration, Mixed-Oxide Fuel Fabrication Plant Layout
(1500 Mg/yr HM Capacity)

TABLE XI

URANIUM FUEL FABRICATION PRICE

Price Components	Total Fab.Price ($/kg U)	Rod Fab.Only ($/kg U)
Capital Retirement and Return	12.75	8.00
Hardware Components	30.00	-
Direct Labor	20.00	16.00
Variable Overhead	10.00	8.00
Fixed Overhead	15.00	11.00
Process Materials & Tooling	6.00	5.00
Warranty Provision	2.00	-
	95.75	48.00

Fuel Fabrication Price. Based upon these plant layouts and familiarity with operations, the price of fuel fabrication can be ascertained. The results are presented in Tables XI-XIII. In all three assessments, the capital contribution to the price (including return on investment) was calculated by assuming that 30% of the total capital cost would be recovered each year and that this would be prorated over 80% of the production capacity.

The uranium fuel fabrication price estimate is presented in Table XI. Since production plants call for fabrication of only mixed-oxide fuel rods in a conventional mixed-oxide fabrication plant, to allow comparison, also shown in Table XI are the uranium price components relative to such fabrication operation (without hardware). It is seen that our total estimated uranium fuel fabrication price of $95.75/kg U is reasonable, based upon current quotations (34).

These estimates for uranium fuel fabrication were considerably expanded to illustrate differences expected to be encountered with conventional mixed-oxide fuel. The re-

sults are shown in Table XII. Included in the unit price
estimate is the cost for fuel bundle assembly and the cost
of plutonium nitrate to oxide conversion. The former cost
component is not expected to be appreciably higher than for
uranium fuel bundle assembly; some added costs will be incurred because of the neutron and gamma radiation levels
that surround the mixed-oxide. In practice, some of the

TABLE XII

CONVENTIONAL MIXED-OXIDE FUEL FABRICATION PRICE

	$/kg HM
Capital Retirement and Return	84.37
Hardware Components (total bundle)	32.00
Direct Labor	20.00
Variable Overhead	
Basic	8.50
Equipment Maintenance	6.00
Containment Maintenance	5.00
Safeguards and Accountability	4.00
	23.50
Fixed Overhead	
Basic	12.50
Containment Maintenance	2.00
Safeguards and Accountability	2.00
	16.50
Process Materials and Tooling	6.00
Warranty Provision	5.00
Mixed-Oxide Rod Fabrication	187.37
Added Price for Bundle Assembly	18.00
Plutonium Nitrate-Oxide Conversion	40.00
Mixed-Oxide Fuel Assembly Fabrication Price	245.37

bundle quality assurance procedures probably will have to be automated for remote viewing. For conversion, a cost of $1.00/g of plutonium was assumed. This unit price is multiplied by 40 to account for the 40 grams of total plutonium expected to be contained, on the average, in each kilogram of mixed-oxide under conventional recycle. The total derived fabrication price of $245.37/kg HM again appears to be reasonable when compared with recent industry estimates (34).

A comparable fabrication price estimate for low concentration mixed-oxide is presented in Table XIII. The derived fabrication price of $128.50/kg HM is directly comparable

TABLE XIII

LOW-CONCENTRATION MIXED-OXIDE FUEL FABRICATION PRICE

	$/kg HM
Capital Retirement and Return	23.50
Hardware Components (total bundle)	30.00
Direct Labor	22.00
Variable Overhead	
Basic	12.00
Equipment Maintenance	2.00
Containment Maintenance	3.00
Safeguards and Accountability	4.00
Fixed Overhead	
Basic	19.00
Containment Maintenance	1.00
Safeguards and Accountability	2.00
	22.00
Process Materials and Tooling	8.00
Warranty Provision	2.00
Mixed-Oxide Fuel Assembly Fabrication Price:	128.50

to the derived prices of $95.75/kg U for uranium fabrication and $245.37/kg HM for conventional mixed-oxide fabrication.

In Table XIV we present a comparison of the average fuel fabrication price as a function of the mixed-oxide fabrication fraction under conventional recycle. These results point up two factors that merit consideration.

First, it is seen that the low-concentration fabrication price is not much greater than the average fabrication price with plutonium recycle as currently contemplated. If one takes credit for the added separative work and virgin uranium savings associated with low-concentration utilization (about $40/kg HM at a fissile plutonium concentration of four grams per kg HM) then it appears that overall fuel cycle costs may be somewhat lower for the low-concentration method of recycle over the entire range of uranium-to-mixed-oxide fabrication mixes tabulated.

Second, it is to be noted that the low-concentration scheme becomes economically more favorable as the industry fabrication mix tends toward a higher mixed-oxide fraction under conventional recycle. At a mix of 25-30%, the low-concentration scheme appears to have the possibility of a

TABLE XIV

FUEL FABRICATION PRICE COMPARISON

Conventional Mixed-Oxide Fraction of Total Fuel Fab.	Average Conventional Fuel Fab. Price ($/kg HM)	Low-Concentration Mixed-Oxide Fab. Price ($/kg HM)
0.10	110.70	128.50
0.15	118.20	128.50
0.20	125.70	128.50
0.25	133.20	128.50
0.30	140.60	128.50

decided price advantage. With the separative work and
uranium savings, this could amount to about $50/kg HM.
These results indicate that the entire industry need not
immediately convert to the use of low-concentration fuel.
Rather, the use should be limited to a segment of the industry so that mixed-oxide fabrication, if conventionally
undertaken, would constitute 20-30% of the total fabrication
requirements for that segment. In periods when new reactors
come on line or when produced plutonium is reserved for
fast breeder fueling, uranium fuel should continue to be
fabricated. The tabulated results suggest that about half
the fuel that is fabricated should continue to be uranium.
Introduction of the low-concentration recycle scheme would
not require major reorientation of the industry. Existing
uranium plants can continue to operate; however, the results
do suggest that new fabrication plants should be of the low-
concentration variety. This conclusion also can be stated
somewhat differently; namely, the results indicate that within the range of low-concentration plutonium, the plutonium
concentration should be as high as possible to obtain maximum economic benefit. The plutonium concentration must be
sufficiently low so that the fabrication economy, separative
work savings, safeguards protection, etc., can be realized,
but it need be no lower. Because of increased penetrating
radiation levels as the plutonium concentration is increased,
it appears that this maximum concentration is well below
one percent; possibly, it is as low as the 0.5% used in the
above shielding assessment. Detailed plant design will be
required to ascertain the exact breaking point.

It is pointed out that the low-concentration method of
plutonium recycle would have attraction, even if the associated fabrication prices had turned out to be significantly
greater. Fabrication costs imply labor primarily, while
virgin uranium and separative work savings imply resource
savings. While present economic analysis dictates comparison on a cost basis, common sense dictates that resource
savings are more important than savings in human labor.

IV. CURRENT STATUS AND CONCLUSIONS

Based upon recent governmental decisions (53), it does
not appear likely that plutonium recycle will be commer-
cialized in the near future; certainly, not in the United
States. Accordingly, there is time to consider all the
available options prior to initiating large-scale activities.

The viewpoint expressed herein should be given due consideration. The presented analyses certainly indicate that the use of low-concentration mixed-oxide is a more efficient use of the valuable plutonium resource material than is conventional recycle. Also, with judicious design of fabrication plants, there appears to be adequate economic incentive for taking this direction.

REFERENCES

1. "The Plutonium Economy," A Statement of Concern Submitted by the Committee of Inquiry, The Plutonium Economy to the National Council of Churches of Christ in the U. S. A., Page 13, September, 1975.

2. "Maximum Permissible Amounts of Radioisotopes in the Human Body and Maximum Permissible Concentration in Air and Water," NBS Handbook 52, March, 1953.

3. "Maximum Permissible Body Burdens and Maximum Permissible Concentration of Radionuclides in Air and In Water for Occupational Exposure," NBS Handbook 69, June, 1959.

4. Morgan, K. Z., Nuclear Engineering Handbook (Harold Etherington, Editor), PP 7-27, McGraw-Hill Book Company, New York, 1958.

5. Tamplin, Arthur R. and Cochran, Thomas B., "A Report on the Inadequacy of Existing Radiation Protection Standards Related to Internal Exposure of Man to Insoluble Particles of Plutonium and Other Alpha-Emitting Hot Particles," Natural Resources Defense Council, Washington, D. C., February, 1974.

6. Wilson, R. H., Plutonium Handbook (O. J. Wick, Editor) Vol. 2, Page 842, Gordon and Breach, Science Publishers, New York, 1967.

7. "Operational Accidents and Radiation Exposure Experience Within the United States Atomic Energy Commission," WASH-1192, Page 28, 1971.

8. Hempelmann, L. H., Richmond, C. R. and Voelz, G. L., "A Twenty-Seven Year Study of Selected Los Alamos Plutonium Workers," LA-5148-MS, January, 1973.

9.   "Plutonium Utilization in Commercial Power Reactors,"
     Nuclear Technology, Vol. 15, August, 1972.

10. "A Review of Plutonium Utilization in Thermal Reactors,"
     Nuclear Technology, Vol. 18, May, 1973.

11. Puechl, K. H., "The Potential of Plutonium as a Fuel
     in Near-Thermal Converter Reactors," Nuclear Science
     and Engineering, Vol. 12, PP 135-150, 1962.

12. Puechl, K. H., "The Potential of Plutonium as a Fuel
     in Near-Thermal Burner Reactors," Nuclear Science and
     Engineering, Vol. 12, PP 151-156, 1962.

13. Thomas, I. D., Plutonium Handbook (O. J. Wick, Editor)
     Vol. 2, PP 619-641, Gordon and Breach, Science Pub-
     lishers, New York, 1967.

14. Proceedings, Plutonium as a Power Reactor Fuel, ANS
     Topical Meeting, Richland, Washington, HW-75007, 1972.

15. Commercial Plutonium Fuels Conference, AIF, EEI, ANS
     Conference, CONF-660308, Washington, D. C., 1966.

16. Report of Committee II on Permissible Dose for Internal
     Radiation, ICRP Publication 2, Health Physics, Vol. 3,
     June, 1960.

17. Denial of Petition for Rule Making, Federal Register,
     Vol. 41, No. 71, April 12, 1976.

18. Leonard, B. R., Jr., "Thermal Cross-sections of the
     Fissile and Fertile Nuclei for ENDF/B-11, BNWL-1586
     (ENDF-153), 1971.

19. Final Generic Environmental Statement on the Use of
     Recycle Plutonium in Mixed-Oxide Fuel in Light Water
     Cooled Reactors, NUREG-0002, Vol. 3, PP IV C-35-41,
     1976.

20. "Reactor Physics Constants," ANL-5800, PP 566-567,1963.

21. Nuclear Safety Guide, TID-7016, Rev. I, Page 10, 1960.

22. Bierman, S. R. and Clayton, E. D., "Critical Experiments with Unmoderated Plutonium Oxide," Nuclear Technology, Vol. 11, No. 2, 1971.

23. Clayton, E. D. and Reardon, W. A. in Plutonium Handbook (O. J. Wick, Editor, Vol. 2, PP 875-919, Gordon and Breach, Science Publishers, New York, 1967.

24. Hansen, L. E., Clayton, E. D., Lloyd, R. C., Bierman, S. R. and Johnson, R. D., "Critical Parameters of Plutonium Systems. Part I: Analysis of Experiments," Nuclear Technology, Vol. 6, No. 4, 1969.

25. Final Environmental Statement Light Water Breeder Reactor Program, ERDA 1541, 1976.

26. Rickover, Admiral H. G., Naval Nuclear Propulsion Program--1975, Hearing before the JCAE, Page 15, 1975.

27. Recycle Fuels Plant License Application, Westinghouse Nuclear Fuel Division, 1973.

28. Smith, R. C., Faust, L. G. and Brackenbush, L. W., "Plutonium Fuel Technology Part II: Radiation Exposure from Plutonium in LWR Fuel Manufacture," Nuclear Technology, Vol. 18, PP 97-108, May, 1973.

29. Puechl, K. H., "Plutonium Recycle--A Status Report," Presented at the AIF Topical Conference Nuclear Fuel '73, Chicago, Illinois, 1973.

30. Final Generic Environmental Statement on the Use of Recycle Plutonium in Mixed-Oxide Fuel in Light Water Cooled Reactors, NUREG-0002, 1976.

31. Puechl, K. H., "The Commercial Plutonium Business," Presented at the AIF/EEI/ANS Commercial Plutonium Fuels Conference, Washington, D. C., CONF-660308, PP 185-192, 1966.

32. Burnham, J. B., Merker, L. G., Deonigi, D. E., "Comparative Costs of Oxide Fuel Elements," BNWL-273. 1966.

33. Debrue, J., Deramaix, P. and de Waegh, F., "Plutonium
     Recycle Studies for the SENA PWR Reactor," Nuclear
     Applications & Technology, Vol. 9, PP 516-527, October,
     1970.

34. Bain, E. E., Jr. and Gordon, E., Summary Report, Fuel
     Cycle Conference "75", AIF, 1975.

35. Puechl, K. H., "The Case for Low Concentration
     Plutonium Recycle," Nuclear Engineering International,
     Vol. 20, PP 687-692, September, 1975.

36. Duret, M. F., "The Use of Plutonium in CANDU Reactors,"
     ANS Transactions, Vol. 23, Page 238, 1976.

37. Puechl. K. H., "The Nuclear Waste Problem in Perspec-
     tive," Nuclear Engineering International, Vol. 20,
     PP 950-954, November, 1975.

38. Hellens, R. L., Rohr, P. C., Shapiro, N. L., "Pluton-
     ium Burners for Nuclear Energy Centers," Trans. Amer-
     ican Nuclear Society 23, PP 240-241, 1976.

39. Ariemma, A., Gualandi, M. Paoletti, Rosa, I., Zaffiro,
     B., "Evaluation of Alternate Solutions for Plutonium
     Utilization in Light-Water Reactors," Transactions
     American Nuclear Society 15, PP 112-113, 1972.

40. Deramaix, P., Andriessen, H., Bairiot, H., Charlier,
     A., Leenders, L., Mostert, P., "Irradiation of
     Plutonium Assemblies in the DODEWAARD BWR Power Plant,"
     Transactions American Nuclear Society 15, PP 113-114,
     1972.

41. Mertens, P. G., Lanning, D. D., "Analysis of a Pluton-
     ium Recycle Fuel Assembly for the Yankee (Rowe) PWR,"
     Transactions American Nuclear Society 15, PP 109-110,
     1972.

42. Grochowski, G. S., Hamilton, G. T. and Israel, S.,
     Nuclear Design, Economics and Safety Analysis of the
     Dresden 1 Mixed-Oxide Demonstration Assemblies,
     GU-5291, December, 1972.

43. Hanson, G. E., Jones, H. H., Tulenko, J. S., "Improved Plutonium Recycle Capabilities," Transactions American Nuclear Society 17, Page 292, 1973.

44. Hellens, R. L., Shapiro, N. L., "Plutonium Fuel Management Options in Large Pressurized Water Reactors," Transactions American Nuclear Society 17, Page 297, 1973.

45. Crowther, R. L., Hopkins, G. C., Lam, S. K., Sawyer, C. D., Wolters, R. A., "BWR Plutonium Recycle," Transactions American Nuclear Society 17, PP 297-298, 1973.

46. Hill, D. J., Henderson, R. R., "Plutonium Recycle in Westinghouse PWRs," Transactions American Nuclear Society 17, Page 298, 1973.

47. Zolotar, B. A., Roberts, J. T. A., "EPRI Plutonium Recycle Program," Transactions American Nuclear Society 23, PP 238-239, 1976.

48. Nuclear Power Growth 1974-2000, WASH-1139, Page 26, 1974.

49. Ross, W. J., Puechl, K. H., Caldwell, C. S., "Irradiation Testing of UO-PuO Fuels," Presented at the Third International Conference on Plutonium, Institute of Metals, London, 1965.

50. Meyer, R. O., Wood, P. M., Sheaks, O. J., "Effects of Plutonium Utilization on Reactor Performance," Trans. American Nuclear Society 21, PP 328-329, 1975.

51. Freshley, M. D., Plutonium Handbook (O. J. Wick, Ed.) Vol. 2, PP 643-706, Gordon and Breach, Science Publishers, New York, 1967.

52. The Nuclear Fuel Cycle U. S. Capital and Capacity Requirements 1975-1985, an AIF Staff Report, 1975.

53. Gapay, Les, "Policy to Limit Spread of Nuclear Arms, Delay Plutonium Use is Cleared by Ford," The Wall Street Journal, October 4, 1976.

54.   Bulman, R. A., "Development of New Chelating Agents
       for Removing Plutonium from Intracellular Sites,"
       <u>Nat. Rad. Prot. Bd/Rand Dl</u>: Annual Report, 1976.

COMPUTER ASSISTED LEARNING
IN NUCLEAR ENGINEERING

P. R. Smith

Department of Nuclear Engineering
Queen Mary College, University of London

I.   INTRODUCTION

The use of the digital computer in engineering education has increased in recent years, and there is every
indication that it will continue to make a significant and
increasing contribution in a teaching-learning role (1).
The paper surveys this area of education, with particular
reference to nuclear engineering, to demonstrate the various
roles of the computer and to establish a rationale for its
use. The discussion is restricted to digital computers,
without implication as to the relative merits of the digital
and analog computer as a teaching device. The examples that
are quoted and the projects that are described are taken
from the tertiary sector of education.

Following an analysis of teaching applications of the
computer and a description of the "laboratory" role which
is of particular interest in nuclear engineering, the use
of graphical display in an educational environment is illustrated with examples from current teaching programs. The
range of available computer-based teaching material in nuclear engineering is indicated by a summary of the work of
three current projects including brief descriptions of the
teaching programs that have been developed; this is followed
by a more detailed presentation of a selected program to provide an example of implementation.

The theme of implementation is continued in a discussion
of the problems that are met when teaching programs are
transferred between educational institutions, and a solution

to some of these problems is indicated. The computing requirements associated with computer-based teaching are considered in relation both to the type of application and to
the associated quality of service. The relationship between
cost and effectiveness is explored by considering appropriate evaluation techniques and procedures.

II.   EDUCATIONAL ROLES OF THE DIGITAL COMPUTER

The digital computer first appeared in teaching estab-
lishments as an aid to research, to perform calculations of
a complexity that could not be contemplated without this computing power. The teaching role was relatively slow to develop, partly due to limitations of computer languages and
systems, but also, perhaps principally, due to a reluctance
on the part of many academics to become involved with the
unfamiliar techniques associated with the use of a computer.
Happily, both of these inhibitions have largely disappeared
as computer systems have become more attractive for educational applications, with the advent of interactive multi-
access systems and graphical display, and as a new generation of academics has gained competence and confidence in
digital computation.

The first educational role to be established was the
obvious one of using the computer to teach about computers.
The emergence of departments of computer science brought an
increasing demand for student experience of computer systems
and software, with a parallel demand in electronic engineering departments for practical experience of computer hardware. In science and engineering, there quickly developed
a requirement for courses in computer programming; these
now represent a substantial commitment of computing resources. The acquisition of competence in programming allowed
the student to develop skills in problem-solving by computer
ranging from relatively trivial coursework exercises through
the processing of experimental data, to extensive program
writing associated with final year projects. This was the
first use of the computer in a teaching role in these discipline areas; the first use of the computer to help students
to learn about engineering or science; the first stirrings
of computer-assisted learning.

In reviewing developments from these early beginnings,
it is possible to distinguish the emergence of two distinct

roles that have been described (2) as *tutorial* and *laboratory* roles. The tutorial role, which requires direct terminal access, exploits the ability of the computer to select and display alphanumeric information, and is in the direct line of evolution of programmed learning. A page of teaching material is presented to the student, followed by a multi-choice question; depending upon his response, he is then offered either new or remedial material. This tutorial role is usually referred to as *computer-aided instruction* (CAI); most of the significant CAI developments have been in the United States where the results of their evaluation have been inconclusive.

The laboratory role exploits the computer as a calculator and is closely associated with computer modeling and simulation, and with *computer-aided design* (CAD). The undergraduate scientist or engineer needs to become familiar with processes and systems of such mathematical or physical complexity that the use of a computer is essential if numerical responses are to be obtained. There will be instances when it is possible, and perhaps desirable, for the student to write a computer program to gain this experience, but there is a danger of a confusion of objectives between learning to write a program and learning about the process or the system that is the subject of the program. In any case, limitation of time will often make it essential for a prepared program to be made available for the student, and it is then a short step to the concept of an interactive and conversational *computer-assisted learning* (CAL) package.

The description of this experience as a "laboratory" role reflects the intended relationship between this type CAL and other components in the educational environment. It will usually be associated with a lecture course in which the mathematical basis of the model or simulation is derived and discussed. The realation between the CAL experience and the lecture course is similar to the relation between conventional laboratories and lectures; it seeks to illuminate and enrich the teaching process by offering a new experience to the student. CAL complements laboratories, tutorials and lectures and rarely seeks to replace them, although as the CAL component of a course becomes significant there will be a redistribution among the three components.

A third role of the computer in education is in the

management of teaching and learning, in marking and analyzing tests, in keeping and updating records, and in routing
students through courses. It is known as *computer-managed
learning* (CML).

In nuclear engineering, the educational applications of
the digital computer have been principally in the modeling
and simulation, or laboratory role. A significant recent
development in the UK has been the formation of a National
Development Programme in Computer-Assisted Learning. This
five-year program, which ends in December, 1977, has stimulated activity in CAL and CML by setting up and supporting
some 30 projects in secondary and tertiary education involving 80 institutions, 400 academics and 100 full-time supporting staff (3). Much of the work in nuclear engineering
in the United Kingdom has resulted from the formation of two
projects; the Engineering Science Project in Computer-Assisted
Learning (4) at Queen Mary College, and the Computer-Assisted
Learning in Nuclear Science and Engineering Project at Royal
Naval College, Greenwich.

The implementation of CAL takes a variety of forms,
depending upon the nature of the teaching material and the
software and hardware of the computer system. It should be
noted, however, that the quality of the computing service
can have an important bearing upon the effectiveness of the
implementation. It is relatively easy to provide a computing service to support the batch processing of programs
written by students to solve simple coursework problems. It
is relatively difficult to provide for scheduled allocation
of a group of teaching terminals using CAL packages that
make extensive use of graphical output, with guaranteed priority and an adequate response time and reliability. CAL
requirements are tending to become more demanding, and the
provision of an adequate computing service will require increasingly, a dedicated computer system.

To be effective, CAL must be academic-initiated. The
association of a CAL package with a particular course has
little chance of success unless the teacher is committed and
enthusiastic. This means that material produced by a computer oriented service group is unlikely to be used unless
there has been consultation with prospective users on the
teaching staff during its development. There is also a much-
valued independence in the tertiary sector of education where

the presentation of a common syllabus may differ significantly in different courses; this can lead to difficulties when
implementing a CAL package that has been developed for another teacher in another institution. This is not an insuperable problem, as will be shown in the following section,
but one that must be recognized and anticipated when preparing
teaching packages.

III.   CLAIMS FOR CAL

     To many engineers and scientists who are now using com-
puter-based teaching methods in their courses, the development of CAL packages has seemed a natural, almost inevitable
consequence of the nature of the curriculum. For the uncom-
mitted, however, it is necessary to attempt to analyze this
application of computers in education and to identify some
of the advantages to be gained. The claims that can be made
for CAL fall into two categories (5); the first comprises
those claims that arise when the learning experience can be
provided only by using a computer, and where consequently,
the decisions to be made are concerned only with the educational value of the experience or the precise nature of
the student-computer interaction. The second category of
claims relate to functions that can be performed either by
computer or by some other medium; for example, tape-slide or
textbook. These are usually more difficult to establish.

     In engineering at undergraduate level, and particularly
in nuclear engineering, the student meets many examples of
systems that demand the use of a computer for evaluation.
In some of these, the use of a computer-based numerical method is essential, as in the determination of neutron flux
distributions in a multiregion system by finite differences;
in others, the defining equations, though capable of analytical solution, are too complicated to allow hand calculation for more than an occasional illustrative case, as in
the evaluation of neutron multiplication factors; there is
a third category where the investigation requires repetitive
or recursive evaluation of relatively simple equations, as
in calculations of frequency response.

     Engineers must be numerate; they need to experience the
numeric response of engineering systems. Traditionally, at
undergraduate level, it has been the custom to simplify systems so that analytical solutions can be obtained, but while

there is a place for this, there is also a need to examine
more realistic systems and to present the associated techniques that are currently in use in the engineering industry.   The use of a computer allows the student to tackle
more complex and more realistic problems, gives him access
to numerical solutions of problems that cannot be solved
analytically, and introduces him to computing techniques
that are relevant to his later career.

       Many engineering systems are multivariate, often with
multiple inputs and multiple outputs.   A great advantage of
a computer model or simulation is that it is relatively simple to design the program to display the effect upon a selected output of changes in a selected input, so that these
often complex relationships can be investigated.   If a
graphical display terminal is available, this interaction
can be shown in the form of a graph or phase plane diagram,
with immediate impact.   By providing an option to hold selected data sets, the student can build up multiple plots
relating families of systems and can be introduced to the
techniques of computer-aided design.   The use of a computer
allows the student to get a feel for magnitudes, sensitivities and the interrelationship between system parameters.

       Engineers need to get "hands-on" experience in the lab-
oratory, but there are many engineering systems that cannot
be made available to them as students.   Some systems are too
expensive, others are too dangerous.   For example, while a
few teaching institutions can provide access to a nuclear
reactor, there are many that cannot, and none can offer a
range of reactor systems for direct experimentation.   The
reactor experiments that can be performed are strictly limited by considerations of safety.   Computer simulation in
nuclear engineering can offer the indirect experience of a
variety of reactor systems and of "accident" conditions that
would not be possible in the laboratory.   The time available
for laboratory experiments is limited either by availability
of the experimental facility or by restrictions of the student timetable.   Typically, a CAL package allows a whole
family of systems to be studied in the time that would be
required for a laboratory experiment on a single system.
The use of a computer allows the student access to dangerous,
expensive and time-consuming systems that he would not otherwise experience.

The nuclear engineering industry is heavily committed
to computer modeling and computer-based design, and in the
education of a nuclear engineer, the associated techniques
assume some importance. The CAL package is usually centered
upon a model of an engineering system and often the "solution" requires the application of numerical analysis. It
is a natural development to design a package so the student
can become aware of the limitations of the model and of the
relation between parameters of the numerical method and the
speed and accuracy of computation. The use of a computer
can give the student direct experience of numerical analysis
and system modeling.

The advantages discussed so far have all been in the
first category of claims in which the computer plays a unique
and irreplaceable part in the process. If these experiences
are not provided by CAL, they will not be provided at all.
The following claims are more difficult to substantiate but
they represent the views, however, subjective, of a group of
academics with wide experience of CAL in higher education in
the United Kingdom (6).

The individualization of the student learning process
allows the student to proceed at his own pace in an imper-
sonal teaching situation wherein he can make mistakes in
private. In engineering, it has been found that students
prefer to work in pairs, and there is some indication that
this is an optimum arrangement; the speed of progress is
determined by the students, and there is usually no super-
visory intervention unless this is requested. This self-
pacing is particularly valuable for the lower ability student
who tends to get left behind in other teaching environments.
Computer-assisted learning is student-centered, with a more
active role demanded of the student who must assume responsibility for his own progress.

The emphasis upon modeling and simulation in CAL in
engineering may suggest that it is a thinly-disguised form
of computer-aided design. There are, however, very significant differences, both in program content and in presentation. CAD provides an algorithm to achieve a design objective; CAL provides insight into the physical processes
that are modeled in the program, and will display parameters
that are of interest in the learning context but not in the
design context. As a simple example, the infinite multipli-

cation factor in a Magnox lattice exhibits a maximum when
pin size is varied at a given pitch. A CAD program would
simply calculate and display the maximum; the CAL package
displays, in addition, the variations of resonance escape
probability and thermal utilization factor that give rise
to the existence of the maximum. CAL provides opportunities for the student to consolidate theoretical concepts
previously encountered in lectures.

The students appear to enjoy the CAL process, even
when it is no longer a novel experience. There is some
evidence of increased motivation as a result. Their en-
joyment is no doubt increased by the capability that the
CAL packages offer to perform extensive calculations without drudgery.

The specification of a CAL package imposes upon the
teacher a requirement to think carefully about the aims of
his course, and this can be only beneficial for the teaching process.

Finally, it should be noted that although the case for
CAL in nuclear engineering rests firmly and securely on the
first category of claims, the marginal cost of introducing
additional CAL packages once the computing system and infra-
structure has been provided will reduce sharply, and it may
well be convenient then to apply CAL techniques in curriculum areas where use of the computer is not essential.

IV.   COMPUTERS IN NUCLEAR ENGINEERING EDUCATION

In the United Kingdom, the emergence of specially designed courses for nuclear engineers has been a gradual process; in the early days of the nuclear power program, the
recruitment of professional engineers to the UK Atomic Energy
Authority and to the consortia was from the ranks of grad-
uates in electrical or mechanical engineering or applied
science, with in-house training to provide the nuclear expertise. During the 1950's, the universities began to rec-
ognize the need for special courses and occasional short
courses, and one-year postgraduate courses emerged. There
was a reluctance among academic engineers to accept the concept of nuclear engineering as a separate discipline, and
this debate still has not been fully resolved after almost
20 years.

At Queen Mary College, the department of nuclear engineering was formed in 1957 and has the distinction of being the first such department to be established in a UK university. The emphasis initially was upon research, but in 1958, a one-year postgraduate diploma course was introduced; this later developed into an M.Sc. course and is now offered jointly with Imperial College. Queen Mary College also was first in introducing a full undergraduate course in nuclear engineering in 1967; there are still only two such courses available in the United Kingdom, the second being offered by the department of nuclear engineering of the University of Manchester.

There are now five one-year postgraduate nuclear engineering courses in the United Kingdom leading to the award of M.Sc. or equivalent, and seven universities offer courses at undergraduade level as part of other engineering or applied science degree programs. Within the industry, the Central Electricity Generating Board is notable in providing a range of short courses in various aspects of nuclear engineering education.

In the United States there has been less reluctance to provide for the academic requirements of the nuclear engineering industry and some 30 universities offer full first degree programs, 47 have graduate programs, and a total of 69 institutions offer some form of nuclear engineering education (7). In 1975, the enrollment of both third and fourth year full-time undergraduates exceeded 800, and the number of full-time master's candidates was almost 1,000; 515 bachelor's, 475 master's and 101 doctoral degrees were awarded.

In parallel with these civil programs there has been a substantial military involvement in nuclear engineering education both in the United Kingdom and the United States, in association with the education and training of personnel for the respective nuclear propulsion programs.

The development of the digital computer has been stimulated by the requirements of the nuclear engineering industry which relies heavily upon the computer in design, development and operation of nuclear plants. The new graduate, on joining the industry, will be confronted with a complex of interrelated programs that model, simulate or mimic and predict or record performance of nuclear systems or processes.

He must be sufficiently familiar with the techniques of programming to be able to identify the structure and components
of the complex. This familiarity can be introduced in undergraduate programming courses and developed by suitable programming exercises. He must also be sufficiently aware of
the limitations of the models and of the accuracy of the
procedures, particularly of any numerical techniques, in
order to be able to assess the validity of the results of
the computation, and this experience is more difficult to
provide.

The teacher of nuclear engineering has a first responsibility to establish a firm basis of understanding of fundamental concepts, and it is acceptable and often desireable to
treat greatly simplified and idealized systems for this purpose. These simplified systems may yield direct analytical
relationships between parameters, and indeed, may have been
selected for this purpose. It is not, in general, a function
of nuclear education to provide skill and expertise in manip-
ulating the large computer programs that are used in the industry; the emphasis is upon principle rather than detail,
for both philosophical and pragmatic reasons. Between these
extremes there is an area in which more realistic models are
required but where the degree of complexity precludes both
analytical solution and numeric hand evaluation. It is in
this area that computer-assisted learning can play a uniquely
significant role and, at the same time, provide insight into
the limitations of computer modeling and numerical analysis.
To take an example from reactor physics, an analytical solution of a one-group neutron diffusion theory model of a
homogeneous medium can provide a valuable illustration of
concepts of thermal diffusion and criticality, but the transition to a multigroup, multiregion model requires a computer-
based approach.

Similar examples can be readily found in reactor kinetics, heat transfer, control and shielding. It also should
be noted that the education of a nuclear engineer embraces
many conventional subject areas in electrical and mechanical
engineering, and there is a case to be made equally for a
computer-oriented approach in these disciplines. Examples
of suitable curriculum areas are control theory, electronic
circuit design, waveform analysis, convective heat transfer,
turbo-compressor design and heat exchanger design (8,9,10).
An indication of the range of engineering topics in which

CAL has been applied can be obtained from the proceedings of a recent conference (11).

There is a firmly-held traditional view that an engineering student should spend a significant proportion of his supervised time in the laboratory in order to gain firsthand experience of the engineering equipment whose theoretical treatment forms the basis of his lecture courses. As engineering systems have grown in complexity and cost, it has become more difficult to meet this objective; a philosophy has been developed similar to that resulting in the simplification of lecture content. Experiments are often devised to establish basic principles using relatively simple equipment, and there is a gap between this limited laboratory experience and that of the real world of industry. In nuclear engineering, the problem is accentuated both by cost and by requirements of safety. CAL can contribute to the solution of this problem in two ways: First, by associating a CAL package with an existing laboratory experiment, the range of concepts investigated can be widened and the relationship between the computer model and the laboratory system determined; second, the CAL package can be a relatively inexpensive substitute for experiments that are too expensive or too dangerous, as is the case for certain reactor kinetics or criticality studies. In many ways, the CAL experience can be made to model the conventional laboratory experiment, as will be shown in the following section.

V.   THE ROLE OF GRAPHICS IN COMPUTER-ASSISTED
     LEARNING

The computers on which CAL is implemented range from large centralized systems providing facilities for both research and teaching for an institution or for a group of institutions, to free-standing local minicomputers offering a dedicated teaching service to a department or to a faculty. On the large central system, there usually will be some batch-oriented teaching applications in which a job is submitted as a deck of punched cards and queued, the results being produced some time later on a line printer. The development of fast turnaround systems, sometimes described as "cafeteria" systems, has made this batch operation more attractive for teaching, and in at least one CAL project (12) this is the preferred mode of operation.

The majority of CAL applications, however, are associated with interactive computing; most centralized computer
systems now provide such a service through remote terminals.
The advantage for teaching is immediately apparent; there is
now the possibility of a "conversation" between the student
user and the teaching program, the results of a computation
can be made "immediately" available at the terminal, and the decisions about subsequent calculations can be made on the spot.
The effect of this interaction is that in situations where
the central processor time (the time taken by the computer
to perform the prescribed calculations) is of the same order
as decision time (the time taken by the student to decide
what to do next), a much greater amount of work can be processed from a terminal than through a batch stream in a given
elapsed time. Since the calculations can be more selective,
the interactive process is usually more economical in central processor time, although expensive in "connect time".
For CAL, the terminal response time is all important; a requirement of a one-second response is often quoted. The
response from a given computer system is dependent upon the
number of simultaneous terminal connections and the ratio
of central processor time to terminal connect time. Unless
there is management intervention or some form of scheduling
and priority allocation, the average response time on a
system providing an interactive service for both research
and teaching tends to increase to a level which, though
still acceptable for research computing, is barely acceptable for CAL. The quality of service required for CAL will
be discussed more fully, but it is apparent that many of the
difficulties concerning scheduling, priority allocation and
response time can be solved if a dedicated computer system
is provided for teaching.

As a medium for teaching, the interactive environment
is more attractive than batch; the student is intimately involved in the computation, his interest is maintained by the
rapid response of the program to his requests and commands,
and he can progress an investigation and achieve significant
educational objectives for a modest outlay of time. The
interactive environment makes possible the scheduling of
sessions in which a group of terminals is allocated to a
class for supervised use of a CAL package or the scheduling
of CAL "experiments" in the student course-work timetable.

A variety of interactive terminals is now available.

The teletype or its silent equivalent, the visual display
unit (VDU) is still the most common. For teaching, the
teletype has the advantage that it produces a "hard copy"
record of input and output; the VDU does not. On the other
hand, the teletype is noisy and subject to mechanical failure, while the VDU is silent and more reliable. The VDU is
also capable of operation at higher speeds than the common
teletype, although faster teletypes can be obtained at a
price.

For CAL, the most significant development in terminals
in recent years has been the graphical display terminal. It
has long been possible to output results from a computer in
graphical form on an incremental plotter, but this is a
mechanical device and painfully slow in operation. The
graphical display terminal, whether of the "storage" or "refresh" type (13), has an addressable screen that allows vectors to be drawn to communicate information, either in the
form of graphs or of diagrams, at high speed. At the same
time, it retains the capability of the teletype or VDU to
display alphanumeric information.

The "refresh" graphical display consists of a cathode
ray tube, X and Y deflection amplifiers and a short decay
time phosphor on the screen which makes it necessary to refresh the display in order to prevent flickering. Refresh
rates are of the order of 30 times a second. A typical device of this kind allows vectors and characters to be generated and displayed using a 256 x 256 grid on a 7 x 8 inch
screen and has the capability of selective erasure and re-
writing for high-speed editing of the display. A cursor is
provided in the form of adjustable crosshairs operated by
thumbwheels, to allow points on the screen to be addressed.
The screen coordinates of the indicated point can then be
transmitted to the computer, thus providing the basis for
computer/screen interaction.

A "storage" display consists of a cathode ray tube with
a charged grid between the electron gun and the phosphor of
the screen. The picture is written on the grid. A secondary, low-energy beam floods the grid and passes through to
brighten the screen. This secondary beam is modified by the
stored picture on the grid so that corresponding areas on
the screen are brightened. The storage tube has the advantage of higher definition, limited only by the resolution of

the grid and the electron beam focus; the refresh display
is limited by memory size and the frequency of the refresh
cycle. A typical storage tube has a 1024 x 1024 display
matrix on a 200 x 170 m^2 screen. The primary disadvantage
is that selective erasure is no longer possible; a picture
can be modified only by erasing the whole screen and redrawing the modified picture, but the speed of display makes
this quite tolerable for most applications. The storage display suffers from a further disadvantage in that it is no
longer a simple matter to take the display signal to a repeat monitor as is the case with a refresh system. It may
be desirable to take computer output into the lecture room
for tutorial purposes, and for a class of any size, this
will require a number of monitors to repeat the displayed
picture. Closed circuit television can be used to effect
this for a storage tube, but there is inevitably a loss of
definition that makes it difficult to achieve a fully legible
repeat on the monitors. A scan converter is available commercially for the purpose, but it is expensive. Computer-
screen interaction by cursor and thumbwheels is also a feature of storage displays.

     There are several developments from the two basic types
of graphical terminal, involving more sophisticated screen
pointing devices such as the light pen and the graphical
tablet, a larger screen, or a builtin computer and memory.
A terminal with an attractive combination of storage and refresh characteristics is now available. From the CAL standpoint, however, these are all too expensive to contemplate
the provision of an effective number for teaching purposes.
The choice between the two basic devices is to some extent
subjective, but experience of both seems to indicate that
the storage tube has a significant advantage in that the
amount of information that can be displayed is considerably
greater, both in the number of lines of alphanumeric output
and in the definition of graphical or pictorial information.

     An effective "teaching station" is a combination of a
graphical display unit and a teletype. The graphical unit
is used to display descriptive information, option sets and
program output. The teletype is used both for numeric input,
so that the student has a record of the parametric values
he has chosen, and for selective tabular output of the data
contained in the displays; thus, the speed of the graphical
unit can be exploited to give rapid exploration of a
problem, while the teletype provides the necessary hard-

copy record of progress and results. Where the shape of
the display is important, a photographic hard copier can
be made available; as its name implies, this is a device
that makes a photographic reproduction of the picture displayed on the screen.

The use and advantages of graphics in CAL are now illustrated by examples taken from teaching packages used by
undergraduate nuclear engineers at Queen Mary College. The
diagrams are reproduced from the output of a Tektronix hard
copier which itself gives a photographic reproduction of the
content of the screen of a Tektronix 4010 graphical display
terminal which is a storage display.

DEPARTMENT OF NUCLEAR ENGINEERING , QUEEN MARY COLLEGE

A COMPUTER EXPERIMENT IN REACTOR PHYSICS.

NEUTRON MULTIPLICATION IN A FUEL-MODERATOR LATTICE

A FOUR FACTOR FORMULA MODEL OF THE NEUTRON CYCLE IS

USED TO CALCULATE THE INFINITE MULTIPLICATION FACTOR

FOR A SQUARE LATTICE OF CLAD CYLINDRICAL FUEL PINS

SEPARATED BY MODERATOR

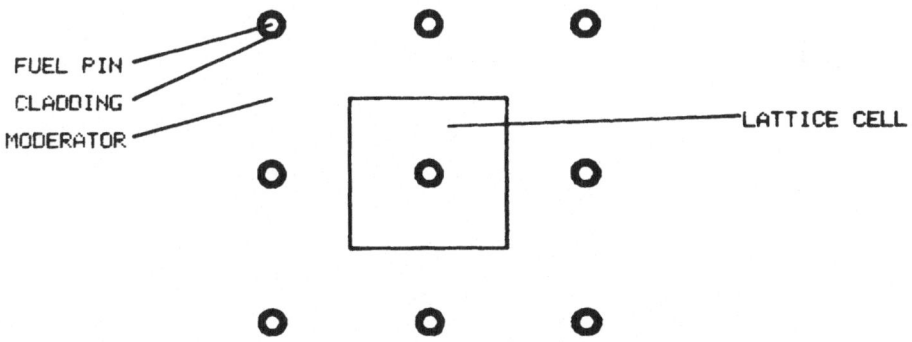

SECTION OF LATTICE SHOWING LATTICE CELL

Figure 1

Graphical Display of Introductory Information
From a Lattice Design Program

Figure 1 illustrates the use of graphics to present introductory or descriptive information using both text and pictures. This is taken from a package dealing with neutron multiplication in a fuel moderator lattice similar to Magnox. The introduction serves to remind the student of the purpose of the package and of material he will have met in lectures; the use of a diagram reduces the need for a lengthy explanation.

Figure 2 is a simple example of graphical display of output and shows the variation of neutron multiplaction factor, k_∞, with moderator-to-fuel ratio, in a homogeneous mixture of fuel and moderator. In this package, the student defines the moderator by entering numerical values for moder-

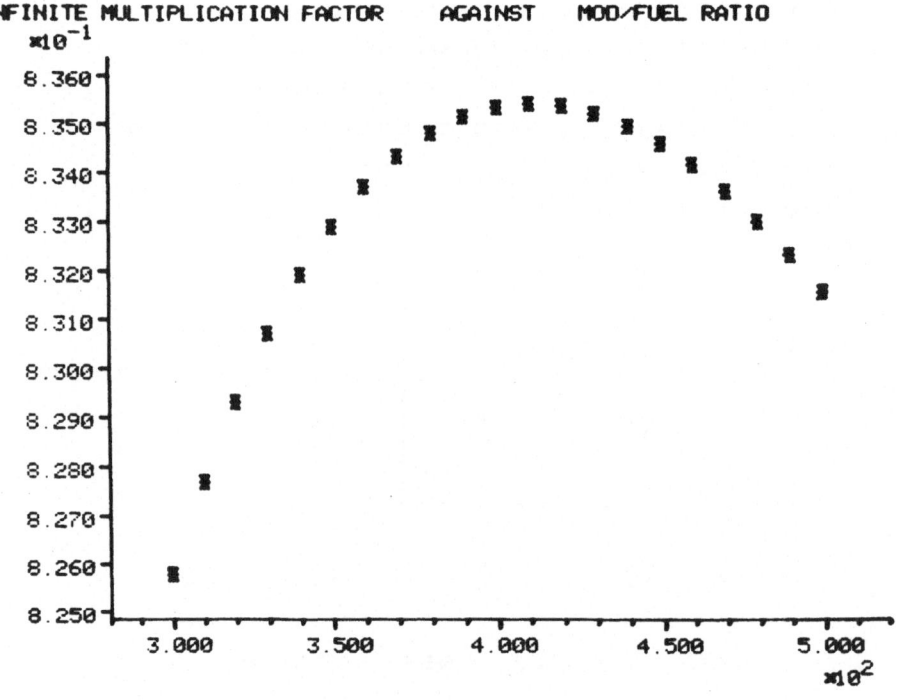

Figure 2

Graphical Display of Output from a Program
Investigating Neutron Multiplication

ator parameters, selects fuel type and enrichment, and then
defines a range of moderator-to-fuel atomic ratio. After
examination of the output, he can redefine one or more of
the input values before repeating the computation and display. The graphical display routines automatically select
and label the scales on the axes to display the data to advantage. The shape of the curve and the existence of a maximum is immediately apparent; this is not always true of the
alternate presentation of tabular data on a teletype.

Figure 3 shows a data entry technique made available by
graphics, from an electronic circuit design package that is
used in an electrical engineering course taken by nuclear
engineers. The data are entered, line by line, on an associated teletype and displayed in the table. Subsequently,

TYPE IN THE VALUES ON THE TELETYPE ONE ROW AT A TIME;
NODE NUMBERS SHOULD BE IN THE RANGE 1-11 WITH A MAX. OF 16 NODE PAIRS.
SEPARATE THE NUMBERS IN EACH ROW WITH A COMMA (","); DO NOT LEAVE ANY
TRAILING COMMAS ON THE INPUT LINE; TWO CONSECUTIVE COMMAS DEFAULT THE
CORRESPONDING VALUE TO "0" - E.G. "1,2,,1E6" SPECIFIES A
CAPACITOR (1E6 PF) BETWEEN NODES 1 AND 2.
TO TERMINATE INPUT TYPE "*".

NODE 1	NODE 2	R (K)	C (PF)	L (UH)
1	2		.1000E 07	
2	3	.1000E 02		
2	4	.1000E 02		
7	3	.3000E 01		
6	5	.8000E 00		
5	4	.3900E 01	.1000E 09	
7	8		.1000E 07	

Figure 3

Use of Graphics for Data Entry
From an Electronic Circuit Design Program

the table can be recalled and the data modified. An advantage of the selective erase facility of the refresh graphics
in this context is that individual data items can be changed
without redrawing the table. Transistors also can be included in the circuit specification using a second table.

Figure 4 shows a graphical plot of the circuit which
includes the data entries of Figure 3, providing a check on
the data entered. This combination of input and display
sequences makes the entry and validation of data rapid, attractive and simple to effect.

The most attractive features of graphics in the CAL
context, however, are found in the presentation of the
results of computation. There are many results that need
to be displayed pictorially or graphically in order to

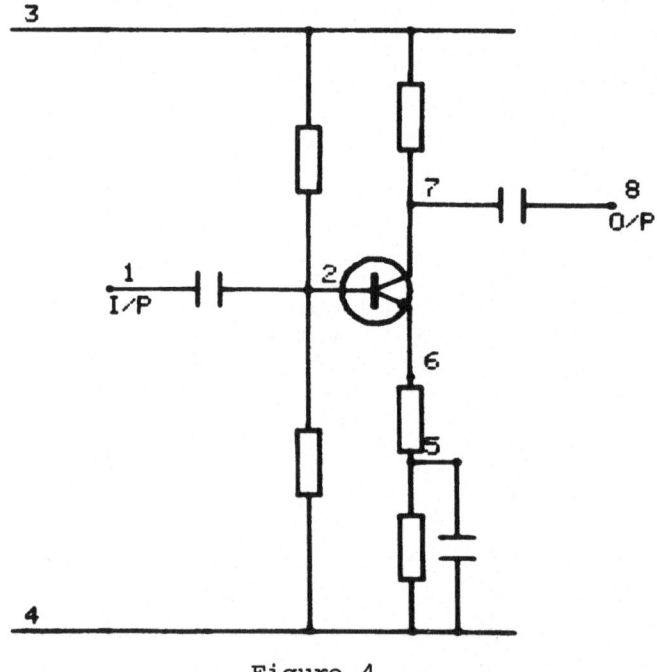

Figure 4

Graphical Display of an Electronic Circuit
As Specified in Figure 3

demonstrate a relationship that is only apparent when the shape of the plot can be observed.

A simple example of this is shown in Figure 5 which is taken from a statistics package associated with a laboratory experiment. A sequence of counts is taken from a long half-life beta source in order to investigate the statistical nature of the radioactive process. After entering the data, the student selects the size of the sample interval for the histogram (referred to in Figure 5 as the bin size) and the data are sorted and displayed, together with the equivalent gaussian distribution for the data.

Figure 6 shows a multiple plot that illustrates the use of graphics to compare the performances or character-istics of related systems. This display is from a package

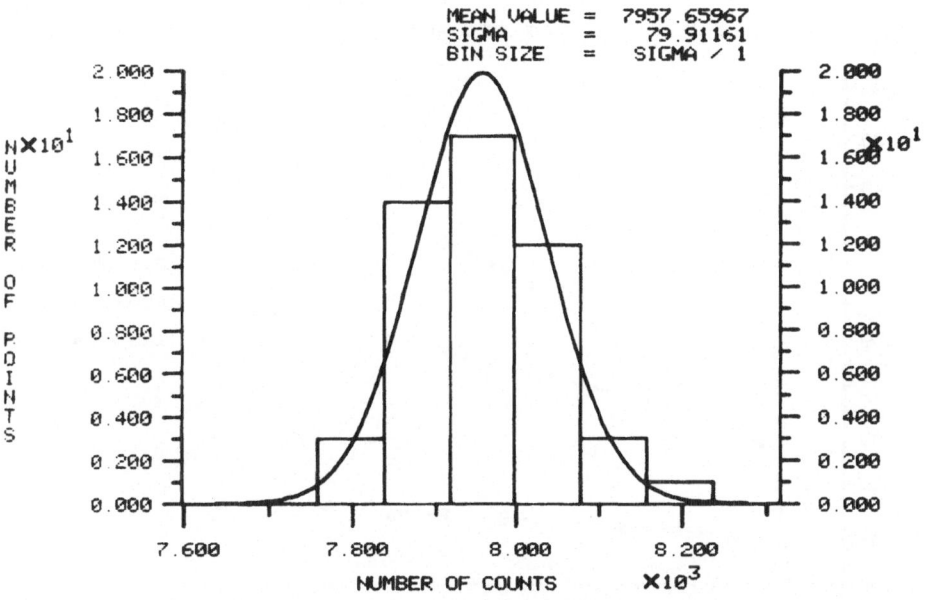

Figure 5

Display of a Count Histogram
From a Program on Statistics of Counting

that investigates the response of a critical reactor to a
step change in reactivity. Here the student can compare
the first few seconds of response, for a given reactivity
step, for three fissile materials.

The ability of the computer to hold selected data for
later comparison, and the effectiveness of graphical display in this context are illustrated in Figure 7 which is
from the same lattice design package as Figure 1. Here the
student can build up, step by step, a display that indicates
the overall optimum arrangement of the lattice for the selected materials, to give maximum neutron multiplication.

Finally, a lighter, but none the less useful application
of graphics is shown in Figure 8. In any interactive computing environment it is usual, during a lengthy computation,
to send a message to the terminal from time to time to in-

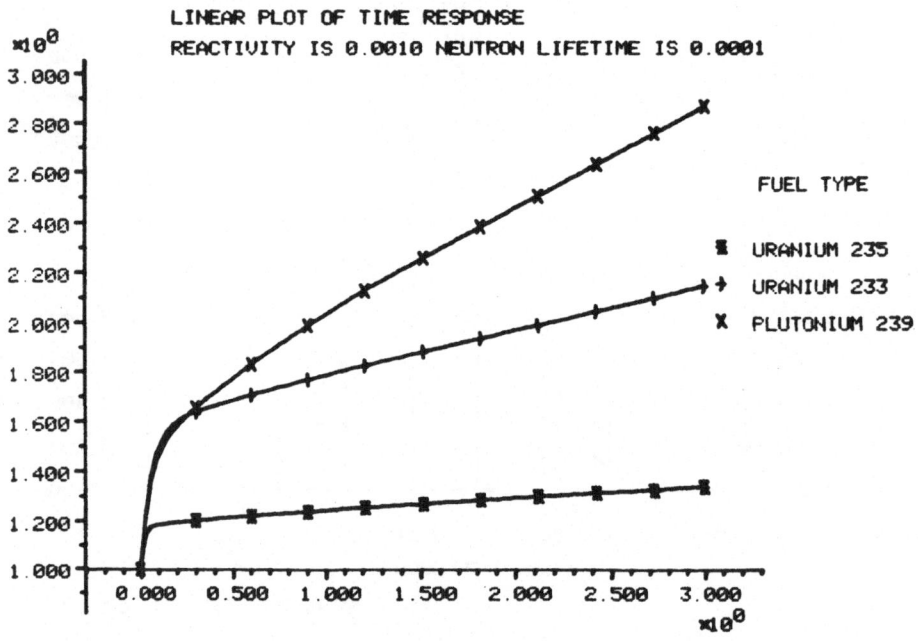

Figure 6

A Comparative Display from a Reactor Kinetics Program

dicate that all is well and that the program is still running. This is even more necessary in the CAL context, where
the student user has not been directly involved in the writing of the program and where he may not be well aware of the
length of time needed for computation. Figure 8 illustrates
a device that has been used to replace, on the graphics
screen, the usual formal indication that the system is still
active. It is output by a routine that draws a design based
upon random numbers so that a different picture is drawn on
each occasion; the progress of the drawing can be made to
indicate the progress of the computation.

    The availability of graphical display adds a new dim-
ension to the use of computers in education. Information
can be entered and output sent quickly and clearly in a
variety of forms that contribute significantly to the interpretation and understanding of the processes that are
being studied. The cheapest graphical display terminals

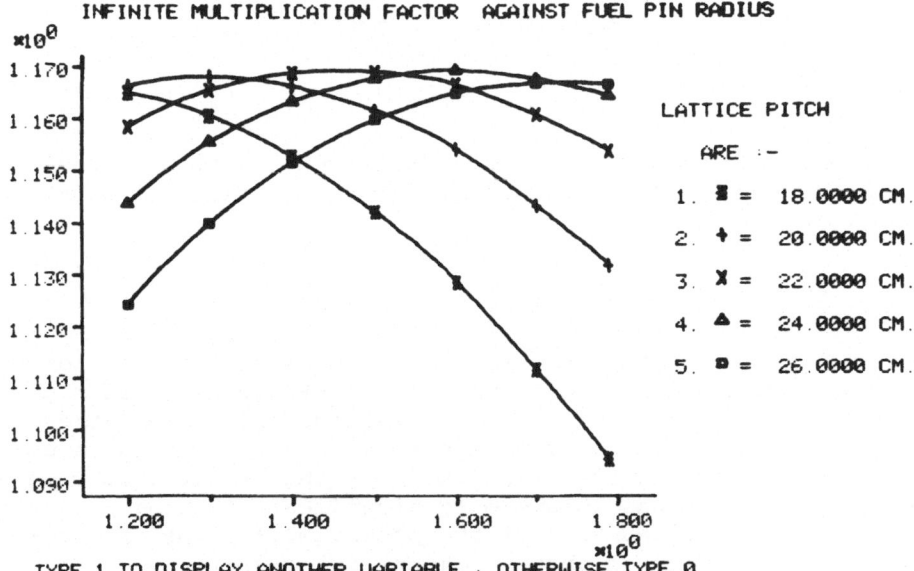

Figure 7

An Illustration of the Use of Graphics
In Lattice Optimization

Figure 8

Computing in Progress

are several times the cost of a teletype or VDU, but their
effectiveness, although much harder to quantify, is also
much greater.

VI.   CURRENT CAL PROJECTS

All institutions offering courses in nuclear engineering at undergraduate level or equivalent can also offer access to computers, and it can be assumed that in all these
courses, there will be some application of computing in nuclear studies. In some institutions, the educational role
of the computer has become more significant through projects
that have exploited some of the potentialities herein described. The work of three projects in nuclear engineering
will now be discussed; the project institutions are, in
alphabetical order: Royal Naval College Greenwich (RNC);
Queen Mary College (QMC); and Virginia Polytechnic Institute
and State University (VPI).

A.   CAL At Royal Naval College Greenwich

The Royal Naval College Greenwich was established by
order in Council in 1873 in order to provide the professional and academic education needed by Naval officers.
With the inception of the UK naval nuclear power program in
the late fifties, a new academic department was established
at the College: the Department of Nuclear Science and Technology. The purpose of this department was to provide, to
agreed standards, academic nuclear education and training
for Naval officers and civilians designated for roles in
support of the naval nuclear program. Currently, the department provides some 19 courses each year ranging from
short familiarization courses to postgraduate diploma and
M.Sc. studies.

Since 1961 the department has held a one-year Nuclear
Advanced Course that from 1967 has been recognized by the
Council for National Academic Awards for the award of the
degree of M.Sc. in Nuclear Science and Technology. In the
third term of this course, the students conduct a design
study of specified physics and engineering aspects of nuclear propulsion plants in which extensive use is made of
the department's digital computer. This work, which began
in 1961 using an IBM 1620 computer, and which has continued
using successively more sophisticated systems, can claim to
be the first effective application of computers in nuclear
engineering education in the United Kingdom (14).

In the design study work, the programs are developed
predominantly by the students, who usually are experienced
naval or civilian engineers, under staff supervision (15).
The work of a given year will usually build upon the accumulated experience and the accumulated computer programs
of previous years, resulting in the evolution of a comprehensive set of computer-based models for the assessment or
optimization of a nuclear propulsion system design. Over a
period of years, it has been found that the development and
use of these models provides an ideal vehicle for studying
not only the design process itself but also the detailed
techniques that are used in designing a nuclear propulsion
system.

To illustrate this use of the computer, the physics
modeling of a nuclear reactor core is discussed; it should
be understood, however, that the design studies cover all

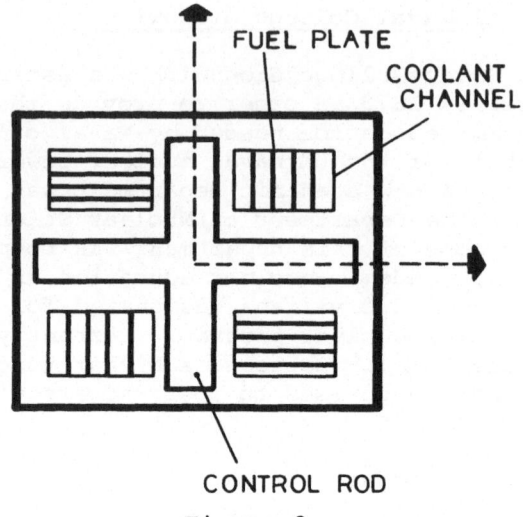

Figure 9

Propulsion Reactor Module Plan

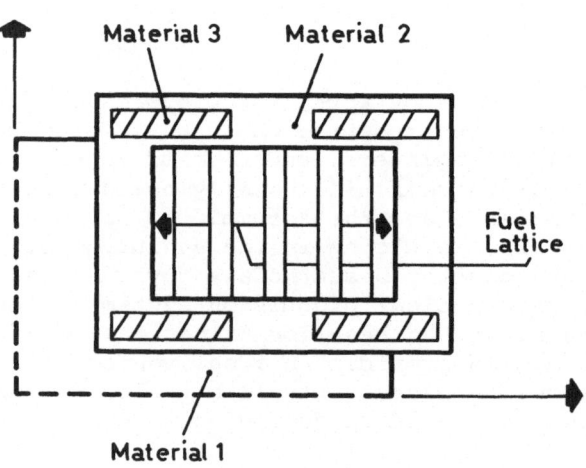

Figure 10

Quarter Module Plan

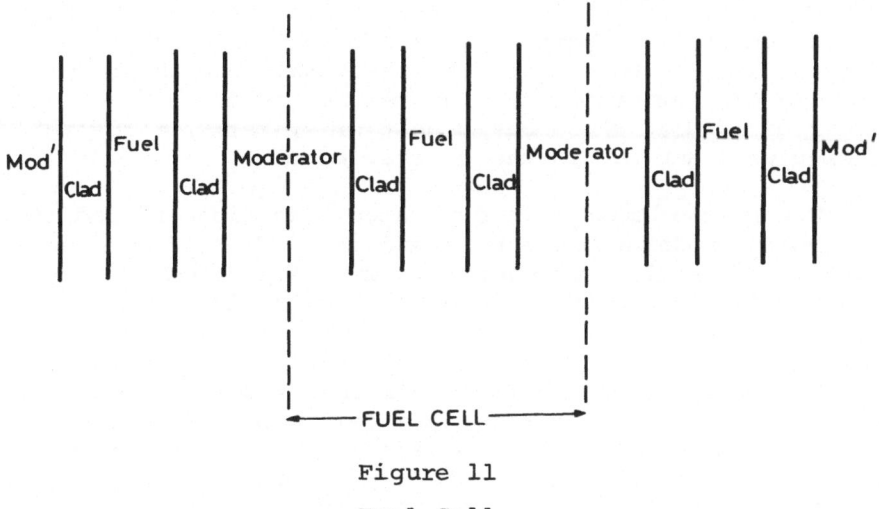

Figure 11

Fuel Cell

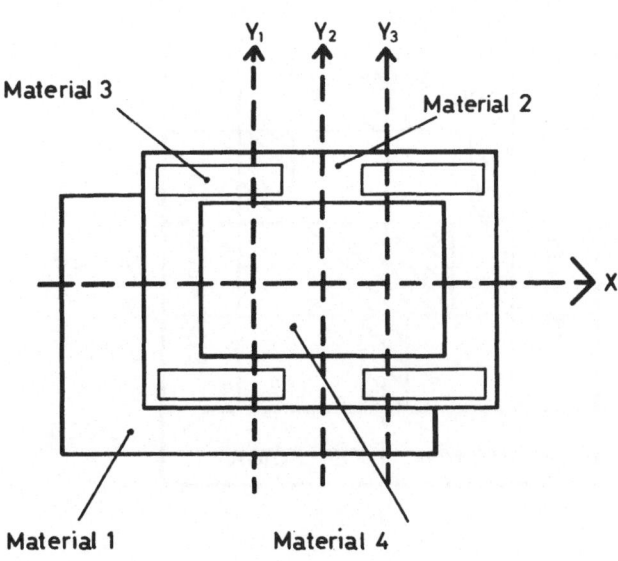

Figure 12

Quarter Module Flux Profiles

aspects of the propulsion plant of which the core is only
one component. The principal objectives of the physics
analysis are the calculation of reactivity and the determination of flux profiles. The reactor core comprises a
number of modules whose basic unit of symmetry, illustrated
in Figures 9 and 10, is the quarter module.

The quarter module incorporates a lattice of clad fuel
and coolant/moderator channels and the effect of an inserted
control rod is also represented. The lattice can be recon-
structed from a suitable number of fuel cells, shown in Figure 11.

The analysis of the fuel cell begins with a transport
theory calculation to determine a cell flux profile, which
is then used to determine equivalent homogeneous cell parameters. A further transport theory calculation is then applied to the quarter module to determine the flux profiles at

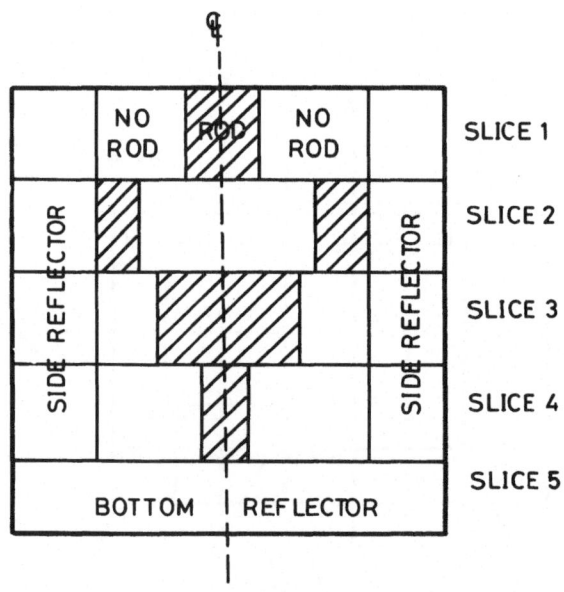

Figure 13

Axial Slices and Radial Zones

X, Y_1, Y_2 and Y_3 in Figure 12, from which a two-dimensional flux map is synthezised. An overall reactor diffusion theory calculation is then performed to determine the radial flux profiles in an arrangement of axial slices shown in Figure 13, and finally, an axial diffusion theory calculation, using homogenized slice parameters, determines the axial flux profile and reactivity.

The components of this program of analysis have been developed by successive courses of students, initially as independent models with manual editing and transfer of data, and finally, as an automated system called ORPHEUS (Overall Reactor Physics Evaluations), as is shown in Figure 14.

The model has been used by students to determine a range of reactor physics parameters including critical rod positions, temperature coefficient of reactivity, reactivity worth of poisons, control rod reactivity, core lifetime and reactor transients.

This long-standing interest in the applications of digital computers in nuclear education has led more recently to the establishment at RNC of a project under the auspices of the National Development Programme in Computer-Assisted Learning (2). This is known as the Nuclear Science and Technology Project (NSTP) and has the three broad objectives of extending the digital modeling of complex physical systems, of developing and implementing computer-based teaching packages and of on-line data analysis. The first of these objectives is related to the work already described; the second is linked with the Engineering Science Project at Queen Mary College which will be described here; the third is concerned with the on-line analysis of laboratory-generated data.

The computer-based teaching packages include a homogeneous burnable poison study that was conceived as a specification provided by RNC, given birth as an embryo CAL package at QMC and has now returned, developed but not fully matured, to RNC for further implementation. A second package, in reactor kinetics, also has been transferred from QMC and will be implemented and extended at RNC. A new package currently being specified is concerned with end-of-core lifetime analysis of fission product poison transients which are of particular significance in a propulsion reactor, whether military or civil.

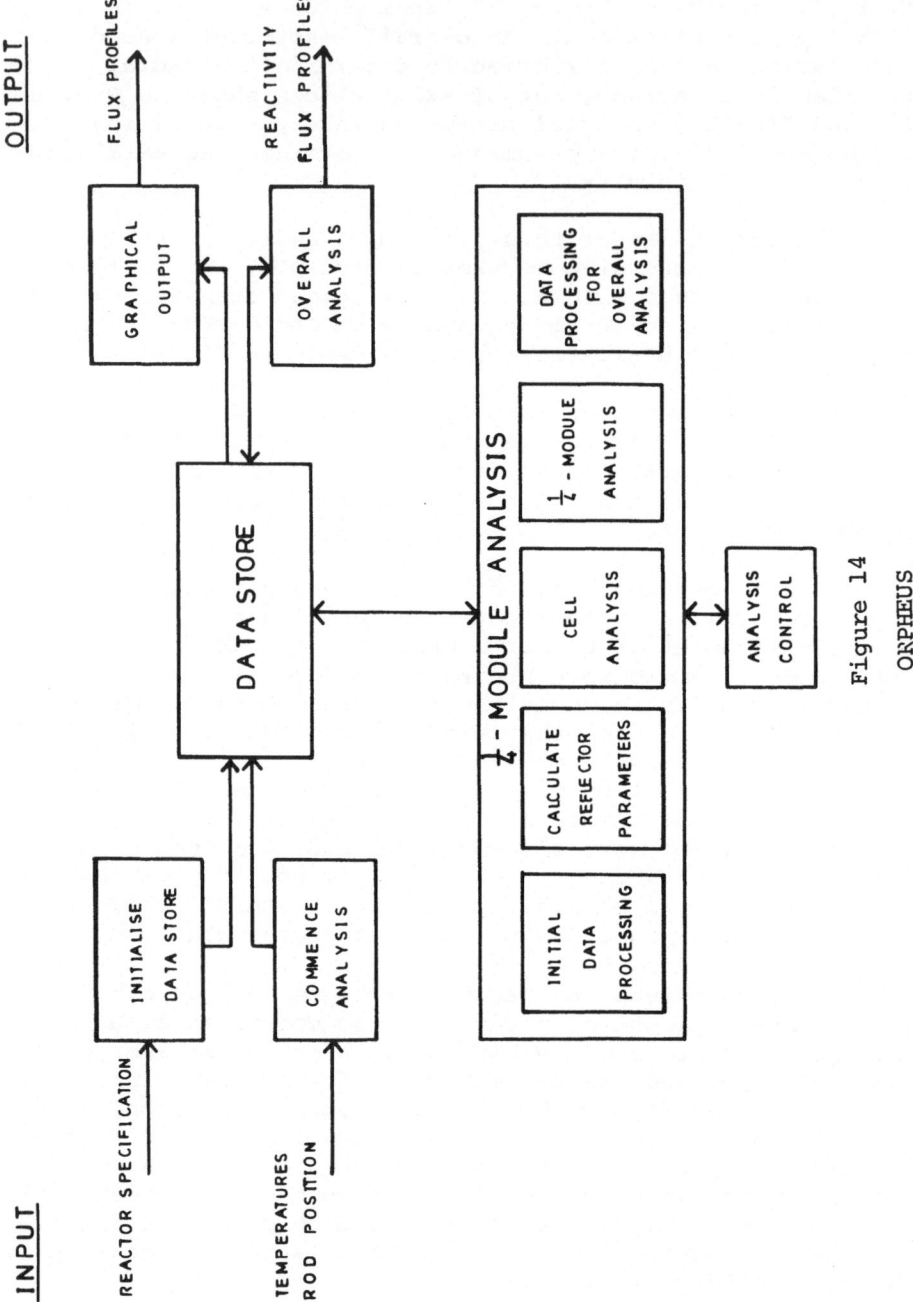

Figure 14

ORPHEUS

B.   CAL at Queen Mary College

The introduction of computer-assisted learning at Queen
Mary College coincided with the introduction of an undergraduate course in nuclear engineering in 1967. The first
interactive packages were based upon existing computer demonstrations in reactor physics and were implemented on an 8K
PDP-9 which had a single teletype for interactive input and
output (16). The backup storage was magnetic tape (Dectape)
and the operation of the overlaid programs was accompanied
by much spinning of tapes. In spite of this there was a
very positive student response to the packages, and when a
relatively inexpensive graphical display terminal (Tektronix
611) became available in 1970, it was decided to purchase
one for the teaching system. The introduction of graphics
in a teaching package in 1971 is the first recorded instance
of this in the United Kingdom. The new dimension introduced
by graphics has been discussed above; it gave added impetus
to package development and to the provision of suitable
graphical display routines. By 1972 there were four well-
developed reactor physics packages in routine use with students (17).

Newcomers to CAL who may be dismayed by the apparently
high cost can take heart from this early work where the marginal costs were remarkably small (18). Slack time on an
existing computer was used for CAL development and implementation, and most of the programming was done by a member of
the academic staff, with some assistance from students through
undergraduate projects. The additional cost to the department was minimal apart from the provision of the Tektronix
611 graphical display. Progress under these conditions was
slow, but no less sure for that. It was not without frus-
tration, mainly centered around the reliability of the computer that was one of the early rogues of a short-lived
line.

In 1973 the College agreed to purchase a PDP-11/40 system to allow this educational application of computers to
be expanded on a faculty-wide basis; at the same time, the
UK National Development Programme in Computer-Assisted Learning was getting underway. This happy coincidence resulted
in the establishment at QMC of a Computer-Assisted Teaching
Unit for the faculty of engineering as a base for a project
in Computer-Assisted Learning (19). The College provided the

computer and most of the terminals; the NDP provided supporting staff and maintenance costs. The project was and is a
cooperative venture, initially between three Colleges of
London University and later including three other institutions (see Table I). The Department of Nuclear Science and
Technology at Royal Naval College Greenwich is associated
with the project through membership of the Project Committee.

TABLE I

ENGINEERING SCIENCE PROJECT INSTITUTIONS

Queen Mary College:	Aeronautical Engineering Electrical & Electronic Engineering Mechanical Engineering Nuclear Engineering
Imperial College:	Mechanical Engineering
University College London:	Electronic & Electrical Engineering
University of Exeter:	Chemical Engineering Engineering Science
Leicester Polytechnic:	School of Electronic & Electrical Engineering
Plymouth Polytechnic:	School of Electrical Engineering

The development of CAL packages has been greatly stimulated by the availability of programming support and a
range of teaching material in nuclear engineering is now
developed and in use in timetabled teaching at QMC. A list
of current nuclear package titles is shown in Table II. The
engineering teaching packages that have been produced or are
being produced within the Engineering Science Project range

over curriculum areas in aeronautical, chemical, electrical, electronic and mechanical engineering, as well as nuclear engineering; some of the mechanical, electrical and electronic packages are used by nuclear engineering students who take the related courses. The total number of project packages currently is 67; of these, 34 are in use at QMC.

TABLE II

COMPUTER-ASSISTED LEARNING PACKAGES
IN NUCLEAR ENGINEERING AT QUEEN MARY COLLEGE

Package No.	Package Title
ESP 01	Neutron Multiplication in a Homogeneous Fuel/Moderator Mixture
ESP 02	Critical Radius of a Reflected Spherical Reactor
ESP 03	Optimization of a Fuel/Moderator Lattice
ESP 04	Variation of Radial Flux Profile
ESP 13	Temperature Distribution in a Gas-cooled Reactor Channel
ESP 15	Calculation of Critical Reactor Size
ESP 17	Reactivity Variation Due to Burnable Poisons
ESP 19	Identification of Fissile Nuclides
ESP 27	Reactor Kinetic Response
ESP 33	Shielding of Spherical Sources
ESP 34	Growth and Depletion of Heavy Elements in Nuclear Fuel
ESP 38	Neutron Distributions by Finite Differences
ESP 42	Statistics of Counting
ESP 43	Resonance Absorption of Neutrons

The nuclear engineering packages are associated with particular course units as coursework or laboratory work. Four packages are part of an introductory course in Reactor Physics that is taken by second-year nuclear engineering undergraduates:

1. ESP 01, Neutron Multiplication in a Homogeneous Fuel/Moderator Mixture uses the four-factor formula to calculate infinite multiplication factor. The student defines the moderator by entering values of the relevant parameters; he then selects fuel type and enrichment and a range of values of moderator/fuel atomic ratio. The program calculates and offers for display the variation of thermal utilization factor, resonance escape probability, reproduction factor and infinite multiplication factor over the selected range of moderator/fuel ratio. The package aims to demonstrate the existence of a maximum k_∞, the values of the maxima and of the optimum moderator/fuel ratios for different constituents and the characteristics of resonance absorption and thermal utilization of neutrons. Sensitivities and the effects of impurities also can be investigated.

2. ESP 02, Critical Radius of a Reflected Spherical Reactor. This allow the analytical determination of critical size based upon two-group diffusion theory, evaluating the critical determinant for a two-region model of a reflected homogeneous core, either step-wise or by iterative convergence. The student can investigate the zeros of the critical determinant and vary independently any one of eight system parameters to determine the effect upon critical radius.

3. ESP 03, Optimization of a Fuel/Moderator Lattice (20) models a square lattice of clad cylindrical fuel pins surrounded by moderator and calculates infinite multiplication (k_∞) by the four-factor formula. The student can define fuel, moderator and cladding by entering suitable nuclear data, and may then vary pin radius and lattice pitch, observing the corresponding changes in resonance capture, thermal utilization and k_∞. He can "design" a lattice for a system such as Magnox and investigate and compare related systems.

4. ESP 27, Reactor Kinetic Response (21). The response of a critical reactor to a step change in reactivity is in-

vestigated. The calculation is based upon a six-delayed
neutron group model and the solution of the corresponding
characteristic (in-hour) equation. The plot can be displayed, the roots and coefficients for a given reactivity
step can be determined, and the response can be shown over
a selected time on a linear or logarithmic scale. Responses can be compared for different reactivity steps, different lifetimes and different types of fuel. A one-group delayed neutron model is available for comparison with the
six-group model.

These four packages are intended to be used as "computer experiments" in timetabled laboratory sessions lasting from two to three hours.

In addition to the four computer experiments, a tutorial
package, ESP 15, Critical Reactor Size, also is associated
with the introductory reactor physics course. The critical
size of a bare homogeneous reactor is calculated using four
models: one group, two group, large reactor approximation
and age-diffusion theory. The student inputs infinite multiplication factor, age-to-thermal, and thermal-diffusion
length; the corresponding materials buckling and the dimensions of a critical minimum volume cylindrical reactor are
output for each model. This allows comparison of accuracy
of the models for different reactor types and observation
of the changes in critical size associated with changes in
reactor parameters. This exercise normally is scheduled in
a one-hour period.

Two other tutorial packages are available and both are
used in association with a first-year course, Basic Nuclear
Engineering. The first, which also serves to introduce the
students to the computer laboratory, is ESP 19, Identification of Fissile Nuclides. This uses the semi-empirical
binding energy formula to calculate binding energy for a
nucleus of specified mass number and atomic number, and thus
determines the change in binding energy associated with the
absorption of a thermal neutron. A comparison with activation energy provides a test for thermal fission. The aim
is to familiarize the student with the concept of binding
energy, the shape of the binding energy per nucleon curve,
and the release of energy in fission. He can identify ther-
mally fissile nuclides and determine the fast fission threshold for others.

The second of the first-year packages is ESP 42, Statistics of Counting. This is used to process the results obtained by the students in a laboratory experiment in which
the activity of a long half-life beta source is measured
using a geiger counter. Fifty counts are recorded; their
distribution is then investigated using the package, which
offers graphical display of the count histogram with variable size of sample interval and a superimposed gaussian
distribution.

Seven more tutorial packages either are in use or are
about to be completed for use with third-year courses in
nuclear engineering; all are full computer experiments requiring from two to three hours of terminal time. A course
in Reactor Fuel Technology, which includes lectures on in-
core reactor fuel management, has three associated packages:

1. ESP 24, Growth and Depletion of Heavy Elements
in Nuclear Fuel will calculate the changes in concentration of uranium and plutonium isotopes with irradiation.
The consequent variation in reactivity will be estimated
and displayed to allow the student to examine changes in
channel reactivity with irradiation for different enrichments.

2. ESP 17, Reactivity Variation Due to Burnable Poison
adopts a simplified reactor model to evaluate the reactivity
change with burnup for a homogeneous poison in a constant
power program. The student can investigate the shape of the
reactivity-time curve and the effect of changes in burnable
poison concentration to show the effect upon single-batch
lifetime and reactor control margin.

3. ESP 04, Variation of Radial Form Factor simulates
the absorber-flattening problem in a reflected cylindrical
reactor with two fueled zones. Absorber density can be
varied in the inner zone whose radius also can be changed,
and the effect upon reactivity and radial form factor can
be observed.

The first of the remaining four packages is associated
with a course in Convective Heat Transfer; ESP 13, Temperature Distribution in a Gas-cooled Reactor Channel is based
upon Ginns' equation and calculates the distribution of
cladding and coolant temperature along the channel. Default

values for parameters defining channel geometry and coolant properties can be varied individually, to allow the student to investigate the effect of the changes upon the temperature distribution. The second package is used for a course in Reactor Operation and Safety and is concerned with shielding; ESP 33, Shielding of Spherical Sources calculates the thickness of slab shield to attenuate monoenergetic ratiation (gamma or neutron) from a spherical source to any desired dose rate at the shield surface. Three approximations can be compared: point source, nonabsorbing source and equivalent disk source. ESP 38, Neutron Distribution by Finite Differences will be associated with a course in Nuclear Systems Design and aims to familiarize the student with the finite difference technique as applied to solutions of multiregion, multigroup diffusion theory problems. Key parameters such as mesh interval and convergence criteria can be changed and the effect upon speed and accuracy of convergence can be observed. Finally, for an advanced course in Reactor Physics, ESP 43, Resonance Absorption of Neutrons provides a model of neutron resonance capture in a fuel-moderator cell using either narrow resonance or wide resonance approximations, for either homogeneous or heterogeneous arrangements of fuel and moderator. The student can investigate the effects of geometry and temperature upon the resonance integral.

Some of the packages described are also used to provide introductory material for students taking the one-year M.Sc. course in Nuclear Engineering, offered jointly by QMC and Imperial College of Science and Technology.

New packages are under consideration in reactor physics (spectrum calculations, cell thermal utilization) in reactor design (system optimization) and in kinetics (ramp response and other transient simulations). The time devoted to timetabled CAL is still a relatively small proportion of the undergraduate nuclear engineering course, totaling no more than 50 hours over the three years.

C.   CAL at Virginia Polytechnic Institute and State University

At VPI there has been an extensive effort over the past few years to develop a series of computer modules for use in teaching nuclear engineering courses. The modules that are

now available are listed in Table III; the development program was supported by the National Science Foundation and
the modules are designed primarily for use in senior level
undergraduate courses or in first-year graduate courses.
They are intended to supplement textbooks and lecture material generally available to students in their course work.

TABLE III

NUCLEAR ENGINEERING COMPUTER MODULES
AT VIRGINIA POLYTECHNIC INSTITUTE AND STATE UNIVERSITY

Module No.	Title
RS-1	Reactor Statics (Introduction)
RS-2	Reactor Statics (One Group Diffusion)
RS-3	Reactor Statics (Neutron Slowing Down and Epithermal Group Constants)
RS-4	Reactor Statics (Resonance Absorption)
RS-5	Reactor Statics (Multigroup Constants for Fast Breeder Reactors)
RS-6	Reactor Statics (Disadvantage Factors)
RS-7	Reactor Statics (Spectra and Group Constants for Light Water Reactors)
RS-8	Reactor Statics (Three-Group Criticality)
FM-1	Fuel Management (Burnup in Thermal Reactors)
FM-2	Fuel Management (Fast Reactor Fuel Cycle)
FM-3	Fuel Management (SWU for Uranium Enrichment)
RD-1	Reactor Dynamics (Reactor Kinetics Equation)
RD-2	Reactor Dynamics (Kinetics with Feedback)
TH-1	Thermal-Hydraulics (PWR)
TH-2	Thermal-Hydraulics (LMFBR)
TH-3	Thermal-Hydraulics (HTGCR)

Each module comprises a descriptive text and a computer program with specimen run.

The reactor statics modules are designed to introduce students to the use of numerical methods and digital computers in calculating the neutron flux distribution in space and energy that are required to determine criticality, power distribution and fuel burnup in both slow neutron and fast neutron fission reactors. The diffusion equation is used for the calculation of neutron transport; collision probabilities are used to calculate the effect of lattice heterogeneity on resonance absorption and slow neutron disadvantage factor.

Module RS-2, The Numerical Solution of the One-Group Diffusion Equation introduces the concept of eigenvalues in reactor statics and the transformation of the diffusion equation to a system of linear algebraic equations; it then demonstrates the "source iteration" procedure for solving the finite difference form of the diffusion equation.

Module RS-3, Neutron Slowing Down and Epithermal Group Constants illustrates the calculation of the multigroup constants in the fast and resonance energy regions for use in the design of thermal reactors. The program is based upon a 33-group model; resonance escape probability can be computed using a correlation for total resonance integral or externally computed values can be used. The user must supply the number densities of the nuclear isotopes and the geometric buckling.

Module RS-4, Resonance Absorption calculates an effective resonance integral for each of the micro-groups defined in RS-3. The program deals with the effect of hetergeneities, uses the Wigner rational approximation for escape probability and the Dancoff correction for interactive effects.

Module RS-5, Multigroup Constants for Fast Breeder Reactors uses the formulation of RS-3 and adopts 18 microgroups over the lethargy range of 0 to 9. The user specifies isotope number densities and a range of values of buckling. For each buckling, the program outputs micro- and macro-group fluxes and one-group and three-group diffusion equation constants.

Module RS-6, Slow Neutron Disadvantage Factors uses a transport-collision probability method to calculate disadvantage factors and derive thermal utilization factors for a lattice. The lattices considered are hexagonal and square arrays; pin radius and pitch and macroscopic scattering and absorption cross-section are input. Thermal utilization is computed by integral transport or ABH method. The program also calculates cell-averaged homogeneous cross-sections and diffusion coefficient.

Module RS-7, Thermal Neutron Spectra and Thermal Group Constants for Light Water Reactors obtains the energy spectrum for slow neutron for an infinite homogeneous system, determines group constants for materials appropriate to light water reactors and proceeds to calculate the group constants for reactor lattices. A least squares fit to cross-section data is adopted. The neutron spectrum is described by the Wigner-Wilkins equation.

Module RS-8, Three-group Criticality Program introduces the student to a few-group diffusion code typical of those in use in the industry. The program solves the finite difference form of the diffusion equations using Gauss elimination; the outer iterations are accelerated using a scheme based on Chebyshev polynomials.

Module RD-1, The Reactor Kinetics Equations solves the kinetic equations numerically for one to six delayed neutron groups for time-varying reactivity insertions.

Module RD-2, Reactor Kinetics with Feedback then examimes the temperature feedback mechanism for a PWR and solves the one delayed neutron group model with temperature feedback for step and ramp insertions of reactivity.

Module FM-1, Fuel Burnup in Slow Neutron Fission Reactors uses a three-group model and average cross-sections and fluxes derived from the appropriate RP modules. The time rate of change of each fuel isotope density is given by a first-order differential equation that is solved for constant power over the burnup step. The program computes the slow macroscopic absorption cross-section required to maintain criticality, equivalent to boric acid reactivity shim in a PWR.

Module FM-2, Fast Reactor Fuel Cycle Program demonstrates the linkage of neutron spectra, diffusion and
fuel burnup codes to compute fuel cycles. One-group diffusion equation coefficients are derived from Module RS-5;
a modification of Module RS-2 calculates criticality, power
distribution and power. Finally, Module FM-1 is used to
calculate fuel burnup and breeding.

Module FM-3, Feed Material and Separative Work
Requirements for Enriching Uranium by Gaseous Diffusion is
designed to show how the cost of enriched uranium can be
calculated, taking into account current costs for natural
uranium feed material and separative work. The program calculates and prints a table of cost versus enrichment for a
specified tails assay.

Module TH-1, Pressurized Water Reactor Thermal-
Hydraulics demonstrates simplified modeling of fuel
geometry, the selection of information to characterize
thermal-hydraulic behavior, the formulation of the theoretical relationships and their computation. The module is
concerned with temperatures, heat transfer rates and coolant
pressure drop. The code solves for the radial temperature
distributions in fuel, cladding and coolant at any axial
station and then marches axially with an energy balance in
the coolant.

Module TH-2, Liquid Metal Fast Breeder Reactor Thermal-
Hydraulics provides a similar treatment of LMFBR and Module
TH-3, High-Temperature, Gas-Coooled Reactor Thermal-Hydraulics
does the same for HTGR.

Computer teaching modules also are used at VPI to allow
the student to compare experimental data with data from a
computer calculation. Junior and senior level classes of
nuclear engineering students measure prompt neutron generation time for the VPI reactor by analyzing the power trace
resulting from a shim rod scram. The power trace is compared with the prediction of the FUMOKI code of Module RS-1.
A second laboratory experiment measures material buckling
in a subcritical natural uranium graphite system, and k_∞ is
calculated from the four-factor formula. The result is compared with the predicted curve of k_∞ as a function of pitch
generated by the Queen Mary College package, ESP 03.

VII. A TYPICAL CAL PROGRAM

A more detailed description is now presented for:

ESP-17, Reactivity Variation Due to Burnable Poisons
which has been selected on the grounds that it is of general
interest. It has survived the test of transfer between
QMC and RNC and has not been reported elsewhere. The version
described is that in use at QMC where it was first implemented in timetabled teaching in 1976-1977.

Burnable poisons are used in in-core reactor fuel management to compensate for excess reactivity and to shape the
variation of channel reactivity with irradiation. For constant power operation, neglecting spectrum effects, an analytical point model solution can be derived for the reactivity
variation, as will be shown herein. While this is amenable
to hand evaluation, the level of complexity precludes the
determination of more than a few illustrative examples and
repetitive evaluations require a computer program. This,
then, can form the basis for an investigation into the effect
of burnable poison concentration upon the reactivity-time
curve, and specifically, upon mismatch and single batch
lifetime. The compensation of increased fissile content by
burnable poison can be simulated using a simplified reactivity calculation. The importance of selecting a burnable
poison with a microscopic absorption cross-section appropriate to a given batch scheme can be demonstrated, as can
the effective limitation on homogeneous burnable poison concentration due to mismatch.

The burnup model assumes a single fissile element,
U^{235}, and takes no account of the accumulation and depletion
of plutonium isotopes. While this assumption, like that of
a constant neutron energy spectrum makes the model unsuit-
able for the accurate simulation of burnup that would be
required in a design program, it is adequate to demonstrate
the qualitive effects of the burnable poison in a teaching
package. The following symbols are used in the derivation:

N_u Number of uranium atoms per cm^3

N_5 Number of U^{235} atoms per cm^3

N_B Number of burnable poison atoms per cm^3

N_M — Number of moderator atoms per cm^3

N_P — Number of equivalent permanent poison atoms per cm^3

σ_{a_5} — Thermal microscopic absorption cross-section, U^{235}

σ_{a_8} — Thermal microscopic absorption cross-section, U^{238}

σ_{a_B} — Thermal microscopic absorption cross-section, burnable poison

σ_{a_M} — Thermal microscopic absorption cross-section, moderator

σ_{f_5} — Thermal microscopic fission cross-section, U^{235}

γ_p — Fission yield of equivalent permanent poison

σ_p — Thermal microscopic absorption cross-section, equivalent permanent poison

R — Initial moderator to fuel ratio $(N_M/N_u(o))$

E — Initial enrichment $(N_5(o)/N_u(o))$

B — Initial burnable poison fraction $(N_B(o)/N_u(o))$

ϕ — Thermal neutron flux

t — Time

f — Thermal utilization factor

k — Effective multiplication factor

ρ — Reactivity

Depletion of U^{235} is described by:

$$\frac{dN_5}{dt} = -N_5(t)\ \sigma_a\ \phi(t) \tag{1}$$

For constant power operation, this yields:

$$N_5(t) = N_5(o)\ (1-b_a t) \tag{2}$$

where

$$b_a = \sigma_{a_5}\ \phi(o) \tag{3}$$

also, $$\phi(t) = \phi(o)/(1-b_a t) \tag{4}$$

Depletion of burnable poison is represented as:

$$\frac{dN_B}{dt} = -N_B(t)\ \sigma_{a_B}\ \phi(t) \tag{5}$$

when $$N_B(t) = N_B(o)\ (1-b_a t)^\alpha \tag{6}$$

where $$\alpha = \sigma_{a_B}\ /\ \sigma_{a_5} \tag{7}$$

The accumulation of stable fission product poisons is:

$$N_p(t) = N_5(o)\ b_f \gamma_p t \tag{8}$$

where

$$b_f = \sigma_{f_5}\phi(o) \tag{9}$$

The variation of reactivity with time than can be shown to be:

$$\rho(t) - \rho(o) = c\ \{\frac{(1-E)\sigma_{a_8} + R\sigma_{a_M}}{E\sigma_{a_5}}\ [1-(1-b_a t)^{-1}]$$

$$+ \frac{B\sigma_{a_B}}{E\sigma_{a_5}}\ [1-(1-b_a t)^{\alpha-1}] + \frac{b_f \gamma_p \sigma_p t}{\sigma_{a_5}\ (1-b_a t)}\ \} \tag{10}$$

where

$$c = \frac{E\sigma_{a_5}}{E\sigma_{a_5} + (1-E)\sigma_{a_8} + R\sigma_{a_M} + B\sigma_{a_B}} [1-\rho(o)] \qquad (11)$$

The three terms on the righthand side of Equation (10) can be identified as changes in reactivity due to burnup of U^{235}, burnout of burnable poison and accumulation of permanent fission product poison, respectively, and these components can be calculated individually for display.

The flow diagram for the program is shown in Figure 15. The initial input sequence requests values, in turn, for $\phi(o)$, E, R, σ_{a_M}, B and σ_{a_B}. Initial reactivity $\rho(o)$ is then estimated using arbitrary values of resonance escape probability and fast and thermal leakage, and an initial thermal utilization factor, f_o, is calculated:

$$f_o = \frac{E\sigma_{a_5}}{E\sigma_{a_5} + (1-E)\sigma_{a_8} + R\sigma_{a_M} + B\sigma_{a_B}} \qquad (12)$$

Option sets A to E are given in Table IV. If Option E is exercised to change any one of E, R, σ_{a_M}, B or σ_{a_B}, the corresponding change in reactivity is calculated using

$$\Delta\rho \simeq \frac{f_1 - f_o}{f_1} \qquad (13)$$

where f_1 is the modified thermal utilization factor determined from Equation (12) using the modified parameter. Thus, the parameter set can be adjusted to provide or maintain a given initial reactivity; for example, enrichment might be increased and the burnable poison concentration then changed to return the initial reactivity to its original value.

The program as defined by Equations (1) to (13) in combination with the flow diagram of Figure 15 and the option sets of Table IV, provides a model for the investigation of the characteristics of homogeneous burnable poisons that is readily accessible to the student and which presents the results of the computation in an easily assimilable form.

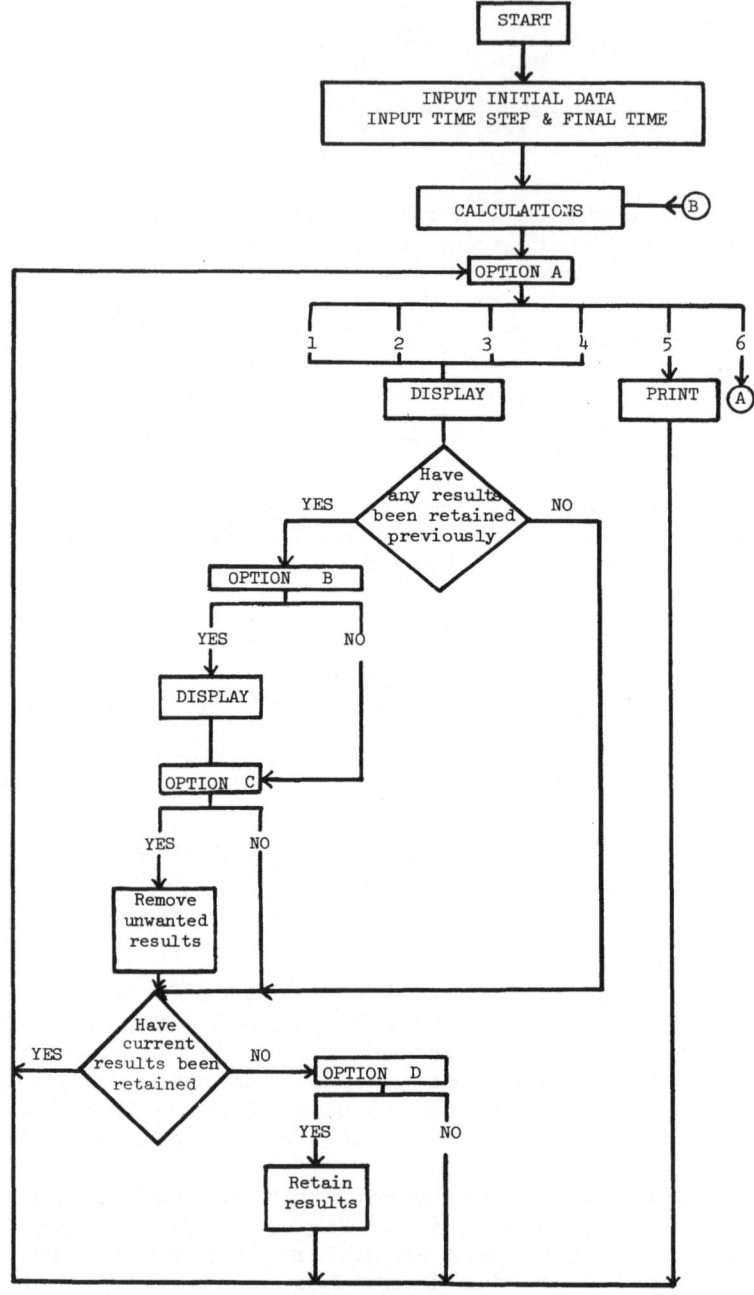

Figure 15A

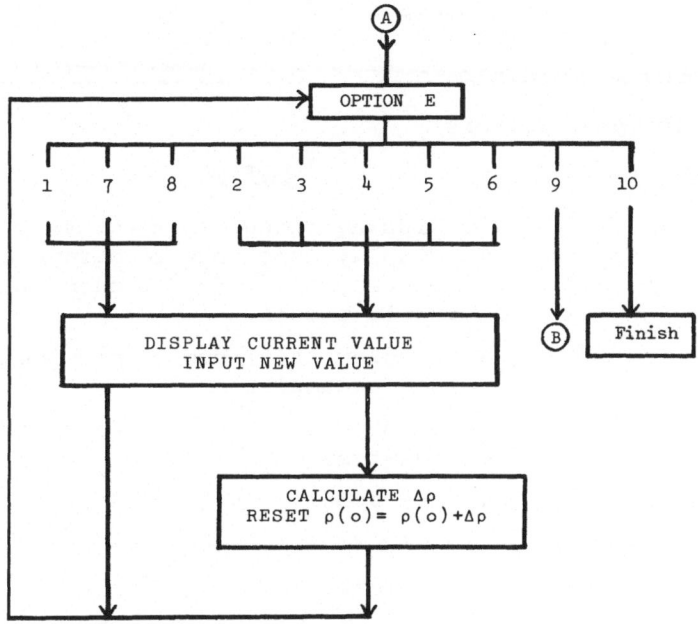

Figure 15B

Figure 15

Flow Diagram for ESP 17
Reactivity Variation Due to Burnable Poisons

TABLE IV

OPTION SETS FOR ESP 17

OPTION A (Display Options)

Type	Option
1	To display change in reactivity $\rho(t)$
2	To display $\Delta\rho(t)$ due to burnup of U^{235}
3	To display $\Delta\rho(t)$ due to burnout of burnable poison
4	To display $\Delta\rho(t)$ due to permanent poison accumulation
5	To print current set of results on teletype
6	To carry on

OPTION B (Multiple Graph Option)

Do you want a multiple graph?   (Yes or No)

OPTION C (Option to Remove Unwanted Sets of Results)

Do you want to remove any of the results retained so far? (Yes or No); if Yes,

Type the number corresponding to the unwanted set of results

OPTION D (Option to Retain the Current Results)

Do you want to retain the current results? (Yes or No)

OPTION E (To Reset Values of Parameters)

Type	Option
1	To alter initial flux
2	To alter initial enrichment

Table IV (Continued)

3	To alter initial moderator-to-fuel ratio
4	To alter moderator microscopic cross-section
5	To alter initial burnable poison-to-fuel ratio
6	To alter burnable poison microscopic absorption cross-section
7	To alter initial reactivity
8	To carry on
9	To finish

The use that is made of the model is, to a large extent, independent of the program and can be defined in the accompanying laboratory sheet. This is an advantage when the program is transferred to another institution, since it allows the detailed application to be varied to meet the specific needs of the teacher who acquires it. An example of a procedure is shown in Table V which is an extract from the laboratory sheet currently used at QMC. The laboratory sheet also provides data for moderators and burnable poisons, an appendix detailing the mathematical model and a standard note on data entry.

The graphical routines allow the student to build up a plot of a family of curves, an example of which is shown in Figure 16. Here, Set 1 refers to an unpoisoned mixture and corresponds to Item 2 of the suggested procedure of Table V. Sets 2 to 5 show the effect of introducing increasing proportions of boron as suggested in Items 3 to 5. The reactivity change associated with the depletion of burnable poison is shown in Figure 17.

TABLE V

EXTRACT FROM LABORATORY SHEET FOR ESP 17

SUGGESTED PROCEDURE

1. Find a mixture of uranium and light water which has a
    reactivity of about 20%. Use a moderator/fuel ratio
    of 50 and vary the enrichment until the required
    reactivity is obtained. Enter a flux of 10^{14}
    neutrons/cm^{2}s.

2. Note this value of initial reactivity and assume it
    is the control margin for the reactor. Use the graph-
    ical display facility to estimate the single batch
    lifetime for this mixture which has no burnable poison.
    Retain the data for comparison.

3. Introduce sufficient burnable poison (boron) to reduce
    the initial reactivity to a value slightly above zero.
    Note the mismatch and single batch lifetime. Compare
    the graph with the 'no poison' case.

4. Further increase the burnable poison fraction; compen-
    sate by increasing the fissile content to maintain
    approximately the same initial reactivity. Note the
    new values of mismatch and single batch lifetime.

5. Repeat the procedure to determine the maximum amount
    of burnable poison that can be added before the mis-
    match exceeds the control margin. Note the corres-
    ponding single batch lifetime.

6. Data are also provided for gadolinia, an alternate
    burnable poison. Repeat the procedures above, using
    gadolinia. What conclusions can be drawn about the
    selection of a suitable burnable poison?

7. As time permits, the properties of other combinations
    of fuel, moderator and burnable poison may be inves-
    tigated.

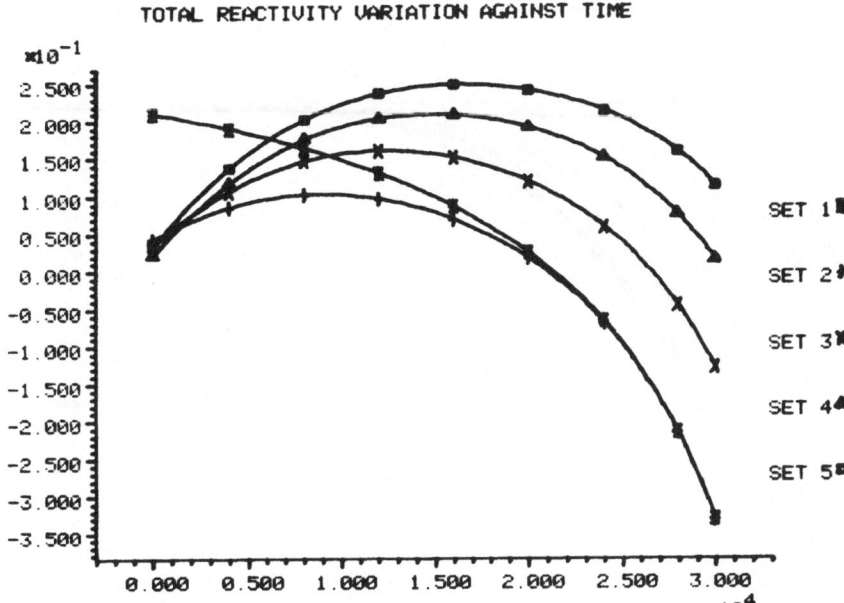

Figure 16

Output from ESP 17

Reactivity Variation due to Burnable Poisons

At QMC, the use of this program is scheduled as a full
computer experiment in a two and one-half to three hour
laboratory session. The students normally work in pairs
under the general direction of a member of academic staff,
assisted by a postgraduate student supervisor. A record of
the procedures and results are kept by each student in a
laboratory notebook and a formal laboratory report may be
called for. The notebook record and the formal report,
where appropriate, are assessed and make a contribution to
the coursework marks for the associated lecture course.

VIII.   TRANSFERRING CAL MATERIALS

The cost of CAL depends upon a number of factors. One
is the cost of the supporting computing service which is

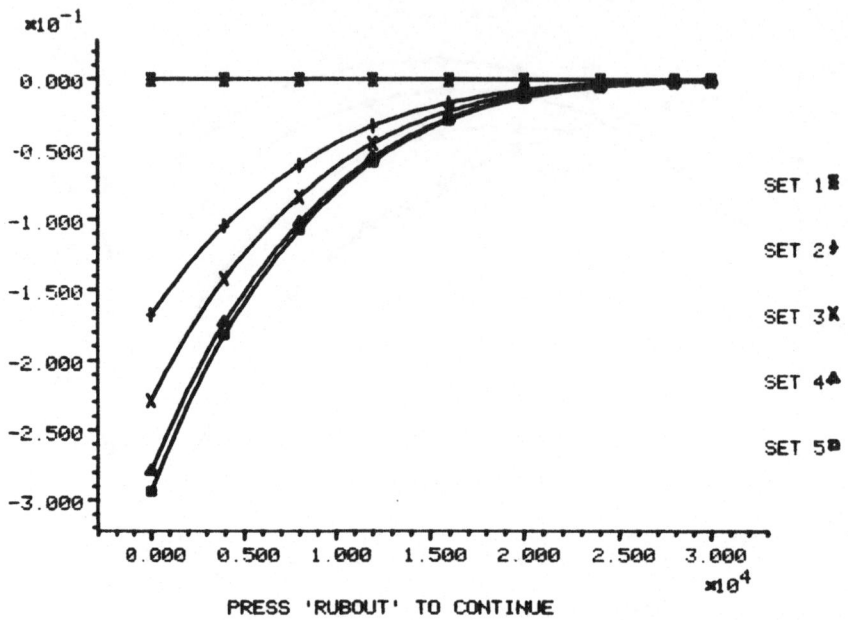

Figure 17

Display from ESP 17
Showing the Depletion of Burnable Poison

strongly dependent upon the quality of service required,
and this, in turn, depends upon the form to be taken by the
student-computer interaction. A second factor is the cost
of acquisition of the CAL teaching material, and this can
be reduced if CAL packages can be readily and effectively
transferred between institutions and implemented without
too much difficulty. Quite apart from a saving of development costs, the ready transfer of CAL materials will facilitate dissemination into institutions where it will not be
considered practicable to set up a program of CAL development.

The documentation of a CAL package should first provide information on which a teacher can base a decision

concerning the relevance and value of the package for the
course he teaches. The essential ingredients for this purpose are clear statements defining the model that has been
adopted in the program and the educational aims as seen by
the originator. It must be recognized that there is a reluctance among academics to accept teaching materials developed elsewhere, and this makes it desirable to design CAL
packages in as flexible a manner as possible so the educational aims can be varied without too much difficulty.
In the work at QMC, the program provides a model or simulation and its use by students is indicated by a suggested
procedure in a laboratory sheet. It is then a relatively
simple matter to modify the laboratory sheet procedure when
the package is used for another course in another institution.

The second purpose of the documentation is to provide
all relevant technical information about the package that
will be of use when it is implemented on another computer.
While the first section is aimed at a member of academic
staff who may have little interest or skill in computing,
this information is directed at a computer specialist and
includes a discussion of program structures, overlay arrangements and nonstandard features. It also includes detailed
listings that can be made much easier to understand by the
liberal use of comments. The task of providing this information is greatly simplified if the programs are writted to
recognized standards, or at least if there are only minor
departures from these standards that need to be documented.

The first and probably the most important standard to
be considered is the standard of language. The language of
engineering computing, and especially, of nuclear engineering computing, is FORTRAN; this is the language that has
been adopted in all the projects described here, apart from
a few tutorial packages in BASIC at QMC. There is a recognized international standard for FORTRAN, ANSI FORTRAN IV (22)
that has been adopted by the Engineering Science Project.
In most CAL applications, the student does not 'see' the
language in which the package is written, and it might be
argued that a more powerful language should be adopted.
Experience, however, shows that a major contribution to
package development comes from the academic staff who, in
engineering, will usually have a competence in FORTRAN.

The student with an inquiring mind will often ask to see
a listing, and he, as a nuclear engineer, also will be
trained in FORTRAN programming. Perhaps the most compelling
reason for adopting FORTRAN, however, is that it is widely
available on all computer systems, large and small. For
the few nuclear packages in BASIC, there has been no accepted international standard, but an ANSI standard is about
to emerge and should then be adopted.

The second standard concerns the use of graphical display. There exist a variety of graphics software packages
and locally developed graphics routines, and without some
form of standardization, the transfer of a CAL package that
makes use of graphical display can involve considerable re-
writing and restructuring of these routines. There are two
levels at which standardization is possible. At the high
level, a standard set of call sequences can be adopted, each
performing a graphics function such as drawing an axis or
plotting a graph. At the low level, the standard will consist of a small number of basic functions that are common
to all graphics devices, such as routines to move from one
screen point to another, either drawing or not drawing a
vector. In considering the transfer of CAL packages, it
would seem extremely laborious to work to the lower standard,
involving as it would, the inclusion of highly detailed routines in the package, and yet this is the only level at
which any guarantee of compatibility can be assured. The
approach to this problem at QMC has been to adopt a well-
tried and developed set of high-level call sequences, the
GINO-F system (23) that has been produced by the Cambridge
Computer Aided Design Centre, and to use, as far as possible,
only calls from that system in transferred packages. This
means that any institution that has the Gino-F system installed will have no difficulty in implementing the packages,
and it is a system that is becoming widely available. The
Computer Assisted Teaching Unit at QMC does not, in fact,
have GINO-F and consequently, has developed a set of routines that emulate Gino-F by implementing the standard calls
in terms of the lower standard of basic functions.

A third standard, which is less clearly defined, concerns program structure. CAL is often implemented on minicomputers with limited core capacity; consequently, overlay
is necessary for many of the more elaborate packages. While
it is not possible to guarantee that a particular program

structure will overlay* on any computer, it is possible to guarantee that it will not, if a large program is written as a single continuous routine. It is good programming practice to break up such a sequence into linked subroutines to facilitate validation, and this will also facilitate transfer. It has not, however, proved possible to find an agreed standard maximum overlay size or subroutine size, or indeed, to agree upon the units in which it should be measured.

Finally, in engineering and science, the development of models and simulations makes extensive use of mathematical routines, and while many basic mathematical functions are to be found in any subroutine library, there are others which, although frequently used, are not standard in this sense. Integration routines or fast Fourier transform routines are examples. Program transfer is greatly simplified if these routines are standardized. A suitable library that is widely available on a variety of computer systems is the Oxford Numerical Algorithms Group (NAG) library (24). These routines currently are being implemented at QMC.

IX. IMPLEMENTATION

There are various ways in which the variety of computing requirements for CAL can be met, but they essentially divide into two groups, the first using large central institutional or regional computing facilities and the second using a smaller local, probably departmental computer. The central facility may be dedicated to offering an educational service, but it will more often have a first duty to provide computing for research; it will usually cater for both batch and interactive computing. If there is a significant program of educational computing using a local system, usually a minicomputer, there must be a first priority to provide this service, and if the local system also provides computing for research, this must take second place. The local service is likely to be interactive.

---

*   The term "overlay" refers to a process that is commonly used to run a computer program when there is insufficient core available to accommodate it. The program is divided into segments, one of which resides in core; the remaining segments are held in a backing store and brought into core, either singly or in combination, under control of the resident segment.

The quality of the service needed for educational computing is also varied depending upon the learning content, the programming content, the terminal type and terminal access, the response time required at the terminal and the programming support. The learning content is probably least when the computer is simply used as a calculator to work class examples; it increases in problem solving and model building exercises and is greatest in computer-aided design or CAL modeling and simulation packages. The programming content here means programming by the student, who may write his own program, modify or combine existing programs or use a prepared teaching package without modification. The terminal requirement for batch is card reader/line printer; for interactive operation, it may be teletype, visual display unit, graphical display unit or a combination of these. Access refers to scheduling of terminals and priority assignment between terminals; this ranges from an 'open house' system with no scheduling or priority to a system with group scheduling of terminals with variable priority to meet the varying demands of different teaching packages. The terminal response time must be low; a response time that is adequate for research computing may not be at all satisfactory for CAL. The programming support ranges from an advisory service for students to a package development service for academics.

Some of these requirements can be relatively easily met from the resources that are to be found in most tertiary teaching institutions, but others require special provision. The variation can be illustrated by taking two extremes; it is relatively simple to cater for the use of batch stream by students who write their own programs to solve set problems, with advice from a central advisory service; it is relatively difficult to provide supervised, timetabled laboratory sessions using combined graphics/teletype terminals for CAL packages that have been developed in-house.

The Computer-Assisted Teaching Unit at QMC offers the high level service of a dedicated laboratory for the Faculty of Engineering, and is an example of a special facility. The laboratory houses a minicomputer and 12 teaching terminals; the hardware configuration is detailed in Table VI. The terminals are grouped in pairs to form six teaching stations, each having a graphical display terminal and a teletype. This arrangement combines the speed of response of the graphical display unit and its ability to present output in an

attractive form with the permanent record, or hard copy, of
input and output that the teletype provides. For particular
applications, the 12 terminals can, of course, be used independently. An alternate hard copy device is available in
the form of a photographic hard copier that can be switched
to any one of the six graphical display terminals. The
students are allowed only restricted use of this facility,
first because it is expensive, and second, because a too-
easy access to output in graphical form may result in a
transfer from computer to laboratory notebook without much
thought on the part of the student. Lengthy tabulations of
output data, which tend to affect adversely the general response of the system, can be diverted via a disc file to a
fast printer.

The operating system normally in use supports FORTRAN,
the language of the majority of the teaching programs; an
alternate operating system is used for BASIC programs. A
later version of operating system, now available, will support both languages and allow a mix of programs to be run

TABLE VI

QUEEN MARY COLLEGE COMPUTER-ASSISTED TEACHING UNIT EQUIPMENT

DEC   PDP   1/40

60K   core

Two   RKO5   2.4 megabyte discs

Papertape reader and punch

Decwriter

Six   KSR33 teletypes

Six   Tektronix 4010 graphical display terminals

Tektronix hard copier

Terminette 1200 printer

---

concurrently. The two 2.4 megabyte* discs provide for bulk
storage, swapping and overlay. Experience has shown that
if a satisfactory terminal response is to be maintained,
the division of the available core between programs should
allow a maximum of two programs swapping in each division.
When operating system overheads and libraries have been
allowed for, this gives four divisions each, of about 9,000
words* and so defines the overlay size. To avoid system
holdup for periodic shuffling when incore programs are re-
arranged to make consecutive core space available for a
waiting program, all QMC teaching programs are fitted to
precisely the same overlay size, so that an incoming program will always fit into the space vacated by the outgoing
program.

The laboratory is run by a staff of two who under academic direction, are responsible for the operation of the
computer system and the availability of teaching packages;
they also provide a package development service for the
Faculty of Engineering. A typical QMC package comprises
between 1,000 and 2,000 lines of FORTRAN, and requires about
two programmer months to develop from the initial specification to a working version that has been tested in use with
students. Some members of academic staff prefer to write
their own programs, or at least to program the central model
themselves. Some packages are based upon programs that have
already run on other computers, perhaps in batch mode; in
such cases, the development time is reduced, although it
should be noted that the restructuring and modification necessary to produce an interactive program with graphics takes
up an appreciable proportion of the total development time.

The operation of a teaching service requires high reliability and availability of the computer-terminal system,
and the associated cost of maintenance to ensure that this
is a significant proportion of the total running cost of the
laboratory. Manufacturer's maintenance is used for the computer and a service agency contract for the terminals.

---

*   The *byte* is a convenient unit of core storage for
      machine language instructions; the *word* is a con-
      venient unit for storing numerical data and is a
      small integral number of bytes. For the system
      described here, one word = two bytes.

The costing of CAL for the UK National Development Programme has been the concern of an independent agency (25,26); generalizations in this area are fraught with danger of misinterpretation. However, in round figures, the capital cost of the equipment in the QMC laboratory is $87,500, and the current annual cost of providing a CAL service is $35,000. The capacity of the laboratory for timetabled CAL probably is 5,000 student hours per annum, allowing for an effective twenty-week laboratory teaching year, giving a marginal cost of $7 per student hour, which is within the range of conventional engineering laboratory costs.

The computer-assisted teaching laboratory at QMC functions in much the same way as any other engineering laboratory except for the versatility of the "experiments" it can offer. The usual laboratory terminology is used; experiments are scheduled as part of the coursework timetable; the student spends two to three hours on each experiment, keeps a record in a laboratory notebook, and may be required to write it up as a formal report. This indicates the view that is taken of the relationship between this activity and other aspects of the educational process; it relates to a course of lectures in the same way as a conventional laboratory experiment, and is equally seen as complementing and enriching the teaching process (27,28). A computer experiment does not, in general, attempt to replace lectures or conventional laboratories, although it can make a contribution to tutorials and may have a remedial role. Sometimes, however, it will be more convenient and less expensive to introduce a computer experiment rather than to purchase the equipment needed to provide a new laboratory rig; the marginal cost of an additional computer experiment often will be small.

The laboratory was established in 1974 and was able to offer a limited CAL service in the academic year, 1974-1975. By the following academic year, demand was growing and a total of about 1,500 student hours of timetabled coursework was recorded. In 1976-1977, the total was in excess of 2,000 student hours and the demand continues to increase as new packages are specified and developed. In addition to the formal computer experiments, the laboratory is used for student project work at both undergraduate and postgraduate level. The departments that make most use of the facilities are nuclear and electrical engineering, but there is a

growing interest from the aeronautical and mechanical engineering departments.

The most common use of computer teaching packages at
QMC is as timetabled computer experiments, but other modes
are of interest. It has been explained herein that with
graphical display terminals of the storage tube variety,
the output cannot be directed to repeat monitors for lecture
room display, but this can be achieved by use of closed-
circuit television. Several lecturers have used this facility, either to illustrate a topic in the course of a lecture
or as a basis for tutorial discussion. The lecturer is
equipped with a radio-microphone through which he cues the
computer operator in the laboratory. For some of the more
complex packages, this has been found to be a valuable introduction for the students before attempting a computer
experiment based on the package.

In another application, a stereo tape recorder is combined with a graphical display terminal for recording and
playback of graphical output. A special interface is required for the terminal but any quality tape cassette recorder is adequate. The "picture" signal is recorded on
one of the stereo tracks and a commentary can be recorded
on the other; the playback can then be used off-line for
demonstrations or as a revision or remedial facility.

Finally, a package may be directly associated with
conventional laboratory work to process and display experimental data, to relate the observations made on an experimental rig to a wider family of systems or to design a
system that is then built and tested in the laboratory.

X.   EVALUATION

However convincing the arguments for CAL in engineering
may be, the decision to implement such techniques must be
based upon an assessment both of cost and of effectiveness.
An attempt has been made to indicate the cost of implementation; in this section, the determination of effectiveness
is considered.

It should be said at once that the educational evaluation of CAL presents difficulties, especially as it applies
to CAL for engineers based upon modeling and simulation programs. The commonly-used techniques of control groups and

tests on objectives are not clear indicators when class
sizes are small, or when an important aim is either to consolidate points of theory that have been presented previously
in lectures or to provide experiences that cannot be made
available by any other means. It is difficult to formulate
pre-and post-CAL test combinations that do not interfere
with the process being evaluated. The computer can be used
to monitor student performance, but the design of routines
to analyze routes taken by the student in a simulation exercise is a formidable task that itself has not been considered cost-effective.

CAL practitioners in engineering and science have little
doubt about effectiveness; this positive view is based upon
observation of students at terminals, discussion with students, and a conviction that the quality of their teaching
has been improved by CAL. One of the first students to experience CAL in nuclear engineering recently has recorded (29)
his recollection of the "moment of discovery" some 10 years
earlier, of the existence of an optimum fuel-moderator mixture. It is significant that the independent evaluation
group who were appointed to monitor the work of projects
within the UK National Development Programme in CAL, have
accepted that evaluation evidence, at least in this curriculum area, will be essentially qualitative rather than
quantitative (30).

The procedures that have been adopted at QMC are centered upon student questionnaires, interviews with staff and
students, and studies of the reactions of students to individual packages.

The questionnaires are tickoff sheets that are given
to the students during laboratory sessions; completion is
optional and anonymous. The questions and format have been
changed twice during the period of evaluation. An analysis
of returns completed in the academic year 1975-1976 is shown
in Table VII; it is clear that a high proportion of the
students found the CAL experience rewarding and wanted the
CAL content of their courses to be maintained or to increase.
The third version of the questionnaire has not yet yielded
significant returns for analysis; the questions on effectiveness have been rephrased and a range of answers allowed
to obtain a more informative response.

TABLE VII

ANALYSIS OF QUESTIONNAIRES, QMC, 1975-1976

Old Format	
Number of returns	214
Number wanting MORE CAL	190
to replace tutorials	48
to replace laboratories	72
Preferred group size of one	34
Preferred group size of two	97
Preferred group size of three	8

New Format	
Number of returns	137
In favor of having demonstrator	104
Instruction sheet adequate	91
Instruction sheet inadequate	39
Effectiveness:	
clarified concepts	96
gave feel for numbers	76
confused	3
produced no change in understanding	12
was worthwhile	111

An end-of-course questionnaire has been completed by a
group of 22 nuclear engineering students who had experienced
an average of 15 hours of CAL; the results are shown in
Figures 18 and 19. Figure 18 indicates a very positive response to the simple question: Would you have liked more
or less CAL material?", with almost three quarters of the
group indicating that they would have welcomed more CAL.
Figure 19 shows the response to a number of questions concerning the achievement of aims of the CAL packages; the
questions are shown in an extract from the questionnaire,
in Table VIII. The averaged results presented again indicate a very positive response; the detailed results are
more positive for the first three questions and less so for
Questions (iv) and (v). Similar results have been obtained
by other groups within the Engineering Science Project (31).

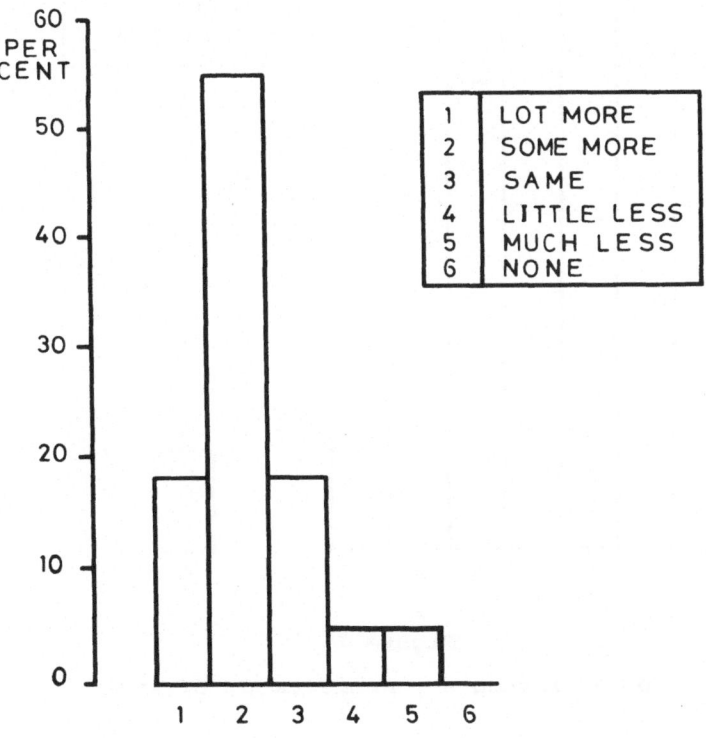

Figure 18

Student Response - More or Less CAL

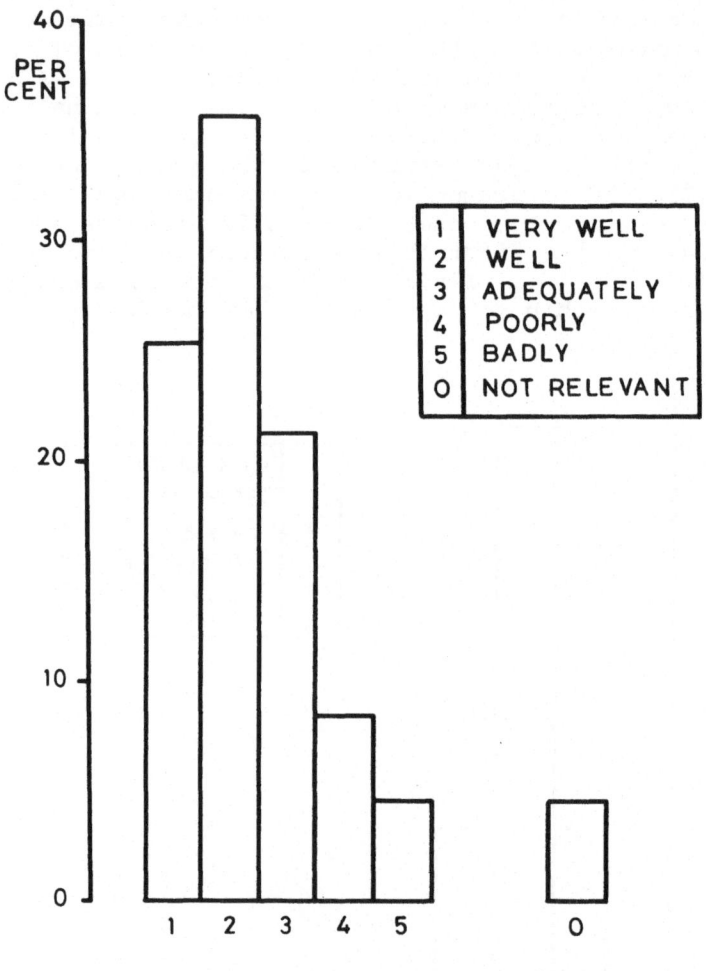

Figure 19

Student Response - Achievement of Aims

TABLE VIII

EXTRACT FROM END-OF-COURSE QUESTIONNAIRE
QMC 1976

Design Aims

5. Experiments are designed with a number of aims in mind.
    Please rate how well the following design aims are met,
    using the scale 1 - 5 below (put 0 if you feel the aim
    was not relevant to the experiments you did)

    1 = very well, 2 = well, 3 = adequately, 4 = poorly,
    5 = badly

    (i) Consolidates theoretical concepts introduced
        in lectures

    (ii) Gives a 'feel' for magnitudes of system
         parameters

    (iii) Helps to understand the relation between
          input and output in multivariate systems

    (iv) Provides insight into the use of com-
         puters in design

    (v) Provides professionally valuable exper-
        ience in using computer models of
        engineering systems

    (vi) Assists understanding of the use of
         numerical methods of analysis

    (vii) Develops awareness of the limits of
          validity of computer models of engin-
          eering systems

Two series of interviews with students have taken
place at QMC. In 1975, a total of 14 third-year students
in nuclear and electrical engineering were interviewed in
individual half-hour sesseions by David Tawney, a member of
the independent evaluation group UNCAL, who also observed
groups of students in the computer laboratory. In his report to the project evaluation meeting (32), Tawney comments
on the reaction of the nine nuclear engineering students he
interviewed as follows:

> "The majority of the students were enthusiastic,
> some very. Comments were favorable: 'very inter-
> esting'; 'could have stopped on'; 'you just don't
> seem to notice the time'; 'it made things a lot
> easier, clearer'; 'it didn't mean anything until
> I did the programs'; 'very good'; 'very, very
> dynamic'. These opinions confirm observations
> that students find the CAL packages very motivating.
> Two students who had reservations were the ones
> who tended to value lectures as an efficient way
> of covering the ground, but one exponent of this
> view admitted that CAL packages 'can impress
> something on one's mind in a lasting way!'"

The interviews also showed an appreciation of the
immediacy of the CAL experience in contrast with much conventional laboratory work, of the active involvement forced
upon the student and of the value of having graphical display.

More recently, a second series of interviews has been
conducted by an independent interviewer to provide material
toward an M.Ed. thesis; although the report is not yet available, it has been indicated informally that the student reactions were equally positive.

XI.   FUTURE PROSPECTS FOR CAL IN NUCLEAR ENGINEERING

The computer is well-established in institutions of
higher education and its continued use is assured, so that
some form of CAL in nuclear engineering education can be
confidently expected to continue. The question is not so
much whether CAL will continue as what form it will take.
It is, however, useful to identify a number of factors that
will tend to make CAL more attractive in the future, at

least in the United Kingdom; similar factors elsewhere may
be more or less evident. External pressures, both political
and financial, will create a demand for more active learning,
either for individuals or for small groups, using a range
of educational resources including CAL. This will be associated with more practical and relevant curricula and an increasing degree of continuous assessment; both of these requirements can be met through CAL. The tertiary educational
system will have to deal with a broader range of ability on
entry and with a more demanding student population. Teachers
will have more students to teach and less assistance in
teaching them. This is an environment in which individ-
ualized learning through CAL will become increasingly attractive (6).

        Costs will continue to be a principal factor, and dev-
elopment in hardware and software will critically affect
the future forms of CAL implementation. The continuing reduction in cost of computer core has already given minicomputers a capacity comparable to that of the 'large',
mainframe computers of a few years ago. The combination of
this capacity with low access time disc storage can provide
the basis for a viable and attractive local CAL service.

        The advent of low-cost video discs for local storage,
and the availability of low-cost microprocessors will have
a strong influence on terminal development; computers will
be attached to terminals instead of terminals to computers.
Free-standing graphical display terminals with 4K to 36K
core are already available at a modest cost, and these may
well prove to be as cost-effective for the more sophisticated forms of CAL as the dedicated minicomputer system,
with the added advantage that they can be moved freely from
place to place to meet changing educational needs. Television receivers can be used for graphical display including color display, and inexpensive keyboards can be provided, so it will soon be possible to convert a domestic
television into an inexpensive CAL device.

        The nuclear engineering curriculum abounds in examples
of systems and processes where computer modeling and simulation can bring light to bear upon mathematical and physical fundamentals, and give access to the student for the
purpose of evaluation, manipulation and experimentation that
will provide insight into and familiarity with the character-

istics of these systems and processes. As the requirement
of the nuclear industry for graduate nuclear engineers increases, as it must, there will be a corresponding increase
in the use of computers in nuclear engineering education.

XII.   SUMMARY AND CONCLUSIONS

The digital computer can play many roles in the education and training of a nuclear engineer; the most significant is to provide models and simulations in curriculum areas where the use of a computer is essential, where
the student can gain experience by this means or not at all.
There are especially compelling arguments for CAL in nuclear
engineering, to offer to the student experience of costly or
dangerous systems that he can investigate in no other way.

The use of graphical display adds a new dimension to
CAL by increasing the speed at which information can be displayed and assimilated, but above all, by making the learning process more attractive and capturing the interest and
imagination of the student.

Significant development projects are underway in CAL
in engineering, and an increasing range of computer-based
teaching material in nuclear engineering is now available.
The transfer of CAL programs between dissimilar systems
presents problems, but these are not insuperable, and
appreciable savings can be made by such transfers.

The implementation of CAL must depend upon available
resources of computing and personnel, but some computer-
based teaching is always possible. A dedicated CAL laboratory has advantages in terms of system response, terminal
scheduling and priority allocation and reliability.

CAL occupies a position in the educational system comparable with that of engineering laboratories in that it
complements and supplements lecture material; the cost of
CAL at its most expensive is comparable to the cost of the
more expensive forms of laboratory coursework.

There are exciting developments in terminals and computer hardware that will affect the direction of CAL development, but the continued use of computers in nuclear engineering education is assured.

REFERENCES

1. Hooper, R. and Toye, I. (Editors), "Computer-Assisted
    Learning in the UK - Some Case Studies," Council for
    Educational Technology, 1975.

2. Hooper, R., The National Development Programme in
    Computer-Assisted Learning - Origins and Starting
    Point, National Development Programme in Computer-
    Assisted Learning Technical Report TR2, Council
    for Educational Technology, 1974.

3. Hooper, R., Two Years on The National Development
    Programme in Computer-Assisted Learning, Council for
    Educational Technology ISBN 0 902204 55 6, 1975.

4. Cheesewright, R., Evans, F. J., Smith, P. R., Launder,
    B. E., Davies, O. J., "A Computer-Assisted Learning
    Project in Engineering Science," International Journal
    of Mathematical Education in Science and Technology,
    5, PP 595-599, 1974.

5. Hooper, R., "Making Claims for Computers," National
    Development Programme in Computer-Assisted Learning
    Technical Report TR3, Council for Educational Tech-
    nology, 1974.

6. Hooper, R., Editor, "CAL in Higher Education - The Next
    Ten Years," National Development Programme in Computer-
    Assisted Learning Future Study Report TR14, Council for
    Educational Technology, 1977.

7. Nuclear Engineering Enrollments and Degrees, 1975,
    Energy Research and Development Agency Report,
    ERDA-76-102, 1976.

8. Cuthbert, L. G., Stafford, B. J., "Development of a
    Package for Computer-Assisted Teaching of Electronic
    Circuits," in Proceedings of Conference on the Teach-
    ing of Electronic Engineering in Degree Courses,
    University of Hull, April, 1976.

9. Wild, R., McQuade, E., "A Case Study in the Design and
    Evaluation of a Unit on Waveform Analysis," ETIC 77,
    International APLET Conference, 1977.

10. Smith, P. R., "Computers in Engineering Education in
     the UK," Computers and Education 1, PP 13-21, 1976.

11. Proceedings of Conference on Computers in Higher
     Education, International Journal of Mathematical
     Education in Science and Technology, 5, 3 and 4, 1974.

12. Gosman, A. D., Launder, B. E., Lockwood, F. C., Reece,
     G. J., "A CAL Course in Fluid Mechanics and Heat Trans-
     fer, Ibid, 7, Page 4, 1976.

13. Scott, B., "An Introduction to Computer Graphics,"
     National Development Programme in Computer-Assisted
     Learning Technical Report TR6, Council for Educational
     Technology, 1974.

14. Smith, P. R., "The Digital Computer in Nuclear Train-
     ing at Royal Naval College, Greenwich," Journal of
     the Royal Naval Scientific Service 20, Page 3, May,
     1965.

15. Rothman, M., "The Use of Computers by Students for
     Modeling of Nuclear Propulsion Plant," Conference on
     Computers in Higher Education, Loughborough University,
     1976.

16. Smith, P. R., "Conversational Computing - A Teaching
     Application," Nuclear Energy, November/December, PP
     170-172, 1969.

17. Smith, P. R., "Four Computer Teaching Experiments in
     Reactor Physics," Queen Mary College Report QMC-EP
     6006, 1973.

18. Smith, P. R., "A Computer-Assisted Learning Project
     Using a Small Computer," DECUS Europe Ninth Seminar,
     London, 1973.

19. National Development Programme in Computer-Assisted
     Learning, Information Leaflet DP 1/02A.

20. Smith, P. R. and Curtis, R. H., "An Application of
     Dedicated Conversational Teaching Programs in Nuclear
     Engineering Design," Journal Institute Nuclear Engin-
     eers 14, 3, PP 74-77, 1973.

21.   Smith, P. R., "A Computer-Assisted Learning Package
       in Reactor Kinetics," Journal Institute Nuclear Engin-
       eers, 17,   6, PP 147-149, 1976.

22.   Standard FORTRAN Programming Manual, National Computing
       Center Limited, SBN 85012 063 2, 1972.

23.   GINO-F, Computer-Aided Design Center, Cambridge, Tech-
       nical Information Booklet, Issue 2, 1975.

24.   NAG Mini Manual for the NAG ICL 1900 Library, Numer-
       ical Algorithms Group, Oxford, 1974.

25.   Fielden, J., "The Financial Evaluation of Computer-
       Assisted Learning Projects, International Journal of
       Mathematical Education in Science and Technology 5, 4,
       PP 625-630, 1974.

26.   Fielden, J., "An Approach to Measuring the Cost of
       Computer-Assisted Learning," National Development
       Programme in Computer-Assisted Learning Technical Report
       TR9, Council for Educational Technology, 1975.

27.   Matthews, J. and Buckingham, D., "Resource-Based
       Learning: A Pragmatical Approach," Studies in Higher
       Education 1, 2, PP 159-169, 1976.

28.   Tawney, D. A., "Simulation and Modeling in Science
       Computer-Assisted Learning," National Development Pro-
       gramme in Computer-Assisted Learning Technical Report
       TR11, Council for Educational Technology, 1976.

29.   Dodd, B., "Computer-Assisted Learning in Nuclear Engin-
       eering," Journal Institute Nuclear Engineers, 17, 5,
       PP 117-119, 1976.

30.   MacDonald, B., Jenkins, D., Kemmis, S., Tawney, D.,
       "The Program at Two," Center for Applied Research in
       Education, University of East Anglia, 1975.

31.   Engineering Science Project in Computer-Assisted
       Learning, Annual Report, Queen Mary College Report
       QMC-ER 6012, 1976.

32.   Tawney, D. A., Engineering Science Project, Report
       for the End-of-Step Evaluation Meeting, 18-19 March,
       1975, unpublished.

NUCLEAR ENERGY CENTERS

Malcolm J. McNelly

Manager, Advanced Reactor Studies
Nuclear Energy Programs Division
General Electric Company

I. INTRODUCTION

U. S. Energy Reorganization Act of 1974, Section 207 (a) (2): "For the purposes of this section the terms 'nuclear energy center site' means any site, including a site not restricted to land, large enough to support utility operations or other elements of the total nuclear fuel cycle, or both, including, if appropriate, nuclear fuel reprocessing facilities, nuclear fuel fabrication plants, retrievable nuclear waste storage facilities and uranium enrichment facilities."

A. Background

In 1973, the oil embargo reminded the world that its petroleum fuel sources were indeed not limitless. It helped trigger numerous actions throughout many sectors of the U. S. energy field. Two such actions of particular relevance to the nuclear energy center concept were:

1. an Executive Office request of Dr. Dixie Lee Ray, then chairman of the USAEC, to prepare a program for U. S. energy independence

2.   Congressional action on the matter of an Energy
       Reorganization Act of 1974 under which two new
       agencies were created; NRC to regulate the
       nuclear industry, and ERDA to conduct research
       and development across the broad range of energy
       fields of potential interest to the nation

Dr. Ray's resulting $10 billion R & D plan toward
energy independence (1) carried with it an identification of
some eight specific demonstration projects, each of which
it was felt, would help focus national capabilities into
channels that support efforts toward energy independence.

Item 4 of that listing was a proposal to construct one
or more demonstration nuclear energy centers.

The Energy Reorganization Act (2) Section 207 specifically called on the newly-established Nuclear Regulatory
Commission to conduct a nationwide site survey for possible
nuclear energy centers and to report back to Congress on the
survey by October, 1975.

As a result of these two related actions, the concept
of energy centers became a focal point of significant efforts
in 1974/1975. This does not mean that the subject had not
been under study prior to that period; indeed, there is a
prior history stretching back five years or more. These
various analysis efforts (3,4,5,6,7) cover many facets of the
possible worth of aggregating generation facilities at a
center or park.*

In general, the earlier studies addressed possible improvements regarding:

       land use planning
       construction economy
       industry colocation

By contrast, the more recent focus on nuclear energy
centers, while not overlooking these points, also identifies
the possible worth with regard to:

_____

*   The two terms 'center' and 'park' are used interchange-
    ably in the referenced literature.

                    fuel cycle integration
                    special nuclear materials safeguards
                    public acceptance

     At the present time, all of these can be considered
significant issues for which the nuclear energy center concept may prove advantageous.

     Such a potential advancement in nuclear installation
concepts is at this point, however, still subject to major
uncertainties. It also should be recognized that there are
now (as of 1976) commitments for some 200 separate dispersed
nuclear generation sites across the United States. Thus,
while potentially lessening the future rate of increase of
dispersed nuclear facilities, the nuclear energy centers
and their associated fuel services will need to be developed
in conjunction with these existing dispersed installations.
Furthermore, any such trend toward nuclear energy centers
carries with it implications of potential change in the
structure of several aspects of electric utility industry
activities and thus, some major uncertainties.

     The uncertainties associated with introduction of the
concept of nuclear energy centers include:

1. meteorological effects from a large
   concentration of reject heat

2. the potential for cost economy through
   multiple unit construction

3. the requirements on the transmission
   system resulting from large, localized
   blocks of generation of electric power

4. the social/political implications of
   large and continuing aggregations of
   construction labor and the subsequent
   operators of power generation and support
   facilities

5. ownership, planning, financing

     The intent of this chapter is to provide a descriptive
overview of a number of significant technology aspects of

nuclear center development and then to identify two or three
of the more important of these for somewhat more detailed
consideration. In particular, consideration of fuel cycle
integration, modular construction and heat rejection have
been selected as among the more significant challenges
facing the industry in developing the nuclear energy center
concept.

For the purpose of this overview and in particular for
the investigation of the selected three areas, a number of
alternate scenarios might be considered. For simplicity,
we have chosen to conduct the various analyses on the basis
that at least the next phase of U. S. nuclear industry installation additions beyond present commitments might be in
the form of the start on a series of nuclear energy center
additions. These might possibly cover most, if not all,
baseload nuclear developments for the period 1985-2005. It
is fully recognized that many other alternatives are possible including a continuing full reliance on dispersed
plant site development. A likely outcome, however, is some
mix between these limiting cases. Furthermore, an overall
trend with time could well be toward more universal adoption of such aggregated facilities or nuclear energy centers.

The nearer term situation in common with virtually all
aspects of nuclear industry development in the United States
and in many other western nations is, however, one of extensive public scrutiny and a hesitance on widespread planning
and commitment of additional nuclear facilities. The aforementioned NRC report to Congress (NECSS-75) has now been
published, it indicates the full range of energy center and
dispersed siting alternatives available to the nation, it
identifies the many socio-political considerations that need
attention, but does not specify a preferred direction for
such development. Thus, the nuclear industry is in the position of having a substantial readiness to serve capably
from a technology standpoint but faces a significant growth
constraint as a result of a lack of necessary, energetic
and nationwide public enthusiasm based on energy cost economics now strongly favorable to its further broad implementation. Concerns over power plant siting, environmental
effects and special nuclear material safeguards are believed
to be among the leading sources of this public hesitance and
they are all elements to which the nuclear energy center
concept is addressed. Nevertheless, it takes a bold and

large-scale commitment by sectors of the utility industry
for such nuclear energy centers to be implemented. It is
potentially beyond the range of feasibility for many of the
smaller and medium-sized utility companies to implement
alone, with maximum effectiveness, and indeed, this size
and integration factor itself may imply some conflicts with
certain aspects of present industry structure. Thus, the
integration of nuclear generation facilities at energy centers, while offering potential improvement to the public's
view of nuclear development on the one hand, raises fundamental questions on the other to some sectors of the utility
industry. These questions are largely socio-political in
nature, and while not the subject of this chapter, are well
recognized to represent both a near-term impediment and a
major opportunity in the longer term for energetic development of nuclear energy centers in the United States. Ultimate solution or alternately, industry evolution away from
these socio-political problems is judged inevitable. The
basic systems concepts and technology for large-scale nuclear energy center development with its numerous options
stand available. Nuclear energy centers represent a major
opportunity for nuclear industry advancement, possibly with
some changes in its structure. Some of the more significant of the technology ingredients for it are surveyed here.

B.   Nuclear Energy Center Configurations and Alternatives

     The concept of nuclear energy centers has evolved as a
possible development beyond existing dispersed mode power
generation. Nuclear energy centers may develop as:

1.   large aggregations, with time, of current
      types of nuclear power generation install-
      ations clustered at especially large sites
      or centers

2.   special standardized design nuclear power
      generation units arranged for an optimum
      sequence of installation such as to permit
      significant construction (and operational)
      cost economies

3.   more extensive arrangements of nuclear
      generation and supporting fuel cycle
      facilities intended additionally to enhance
      the safeuarding of special nuclear materials

The evolution of thinking on these various alternatives
for nuclear energy centers and indeed, on the further alternative of large Integrated Fuel Cycle Facilities (IFCF's)
entirely separate from power generation facilities, is not
complete. Each alternative carries with it its own set of
potential problems and thus, overall trade-off analysis is
required. Figure 1 is illustrative of the many considerations and their typical relationship in such a trade-off
analysis. At the top of the figure, the basic question is
posed of dispersed versus energy center generation. It is
noted that this question must be looked at in terms of the
present basis or trends of the industry, uncertainties on
future growth predictions, the array of alternate types of
nuclear installations and the many components of the fuel
cycle, the environmental considerations and the overall
regulatory scene. On the righthand side of the figure, the
current dispersed mode of generation is characterized; the
installations are primarily single and dual unit installations. The installations, dispersed at more than a hundred sites across the United States, serve localized loads,
face problems of site availability and adequate cooling
water supply, and some significant problems on various aspects of the fuel cycle, particularly safeguards and transportation and what is frequently termed "closing of the fuel
cycle back end". As a result of these problems of dispersed
sites, one might look to the potential of energy centers for
the provision of some solutions. This is signified on the
lefthand side of Figure 1.

Potential exists for closing the fuel cycle on site and
for certain other cost economies. Such energy centers do,
however, pose a new set of problems, in many cases, associated with the size of the installation. These problems range
from institutional and social considerations of large aggregations of power generation and its associated labor force,
on through to the financial considerations of underwriting
such a venture and to the technological considerations of
handling large blocks of power generation that must now serve
more than local loads. Considerations of land availability,
of thermal discharges and of waste management are also significant and are really not helped by a progression to the
larger size installations implied by an energy center. Consideration of these energy center problems leads in turn to
thoughts of possible alternatives; alternatives in terms of
the size of the installations, of the cooling mode adopted,

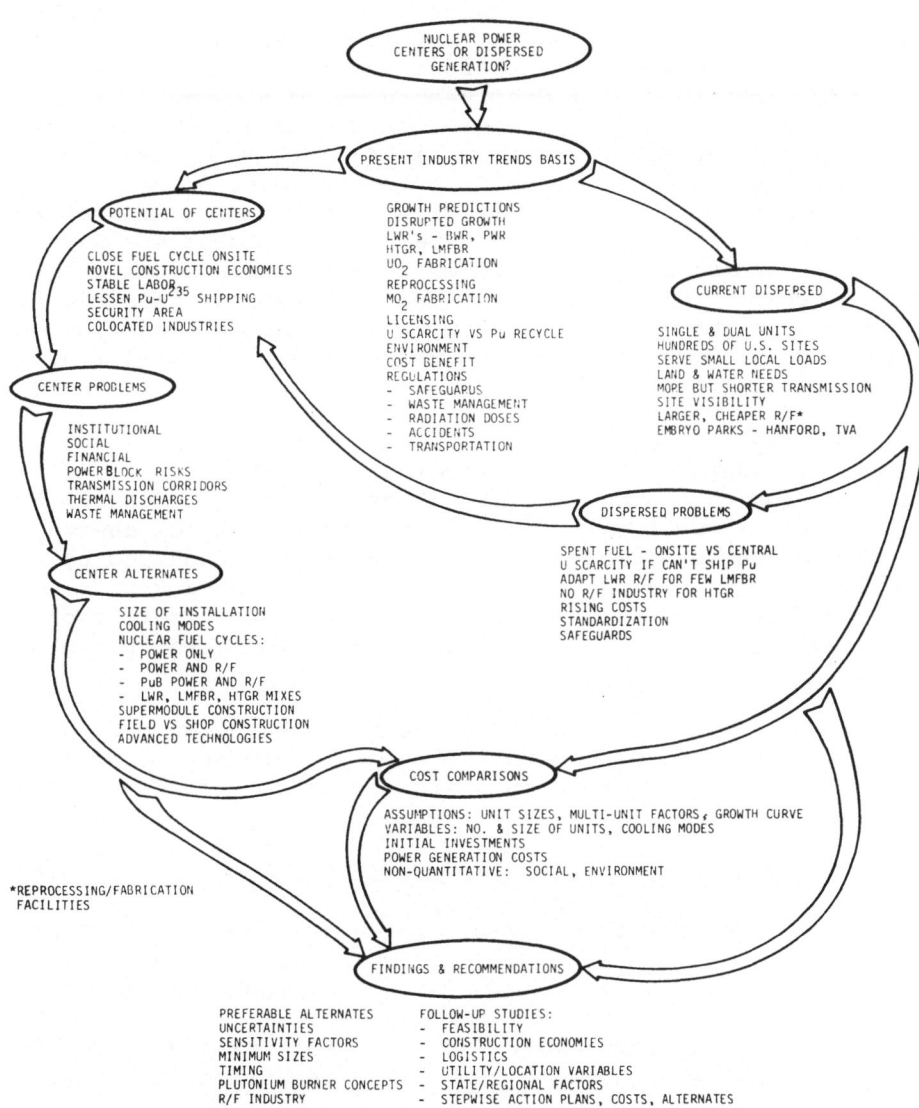

Figure 1

Typical Trade-off Analysis
Nuclear Energy Centers Versus Dispersed Sites

of the configuration of the fuel cycle, and of possible impact of other technology advances. Comparisons then can be
made between the energy center opportunities and those of
continued development of dispersed installations. This
sort of comparison has indeed been done on a generic basis
for a given set of assumptions, regarding size and facility
growth rate, and for various types of cooling, in order to
develop estimates of initial investment, power generation
cost and some of the nonqualitative social and environmental
costs. While findings can be developed from such analysis on
a generic basis; and have been in recent studies such as
"The Assessment of Energy Parks Versus Dispersed Electric
Power Generating Facilities" supported by the National Science
Foundation(5) more specific analyses for actual facility
locations are required.

      Representative data for a twenty-unit nuclear energy
center are shown in Table I, together with the equivalent
data for two unit dispersed site installations. These
data were derived from the above reference study of energy
centers or parks versus dispersed generation conducted by
General Electric Company for the National Science Foundation.
Further details of this twenty-unit center and excerpts of
comparable data for two other studies(4) are shown in Table
II.

      In principle, the nuclear energy center and dispersed
site data per unit of generating capacity are very comparable. The only four areas in which significant differences
show up are:

Generation plant site area:	3.8 km^2 vs 250/unit
Transmission R. O. W.:	14.0 km^2 vs 1050/unit where 7X transmission distance is assumed.
Construction labor and duration:	Potentially up to 25% reduction for center.
Fuel shipments offsite:	Potentially all Pluton- ium stays on site.

      An artist's rendition of the overall appearance of such
a 20-unit installation, utilizing wet cooling towers for heat
rejection, is shown in Figure 2.

TABLE I

SUMMARY COMPARISON

NUCLEAR CENTER VERSUS EQUIVALENT DISPERSED SITE GENERATION

ITEM	UNIT	CENTER	DISPERSED	
			EACH	TOTAL
Generating Units	No	20	2	20
LWR Unit Size	MWe	1300	1300	---
Total Generating Capacity	MWe	26000	2600	26000
Land Requirements:				
Generation Plant	km²	8	2	21
Heat Rejection System				
Cooling Towers	km²	17	2	17
Transmission Right of Way	km²	277	9	86
Labor Requirements				
Construction	10 Man Hours	234	31	312
Operation & Maintenance	No People	1400	175	1750
Unit Construction Time	Months	51	66	---
Transmission Voltage	kV	765	345-765	---
Average Transmission Distance	Miles	175	25	---

TABLE II

CHARACTERISTICS OF NEC'S

	GE-NSF Center Study	Hanford-Normal Case	River Bend
1. Technical:			
Electric Power, MWe	26,000 (by 2004)	26,000 by 1990	36,000 by 1992
Number of Units	20	21	28
Reactor Type	LWR-1300 MWe	Mostly LWR-1300 MWe	LWR-1300 MWe
Cooling Systems	Wet natural draft towers	Combination of all Types	Wet natural draft towers
Transmission on Voltage (kV-AC)	765	500 to 1100	500
Fuel Cycle	1-UO$_2$ Fuel Fab Plant 1-MOX Fuel Fab Plant 1-Reprocessing Plant	1-Gaseous Diffusion 8-Low Level Waste Treatment 1-UO Fuel Fab 1-MOX Fuel Fab	1-UO$_2$ Fuel Fab Plant 1-Reprocessing Plant
2. Environmental:			
Land Area, km^2	26	155 km^2 (10% of Hanford Project	44
Water Consumption, m^3/min	5662	Not available	10,324
Thermal Discharge, MWt	APX 50,000	70,000	76,000
Transmission Lines (Single circuit)	10	Not available	12
Transmission Corridor Area, km^2	129 (1.2 AC/MW @ 100 MI)		Not estimated

Table II, Continued

3. Economic (1974 Dollars):
    Capital Cost, Billion $

Generating	10.5	11.7	12.4
Transmission	0.8 ($30/MW @ 150 MI)	0.7	0.25
Fuel Cycle	0.50	1.45	0.16
Annual Operating Cost, 10^6	680	Not available	150
Cost of Electricity Mils/KW hr	14.8	Not available	12.4

4. Socioeconomic:

Construction Manpower	3,600*	20,700 Peak of which 11,200 is for fuel cycle	16,800 Peak
Operation Manpower	1,200	4,000 Peak	1,700 Peak
New Town Population	40,000-60,000	105,000 Peak	Private

---

*   Includes 30% labor reduction through modular construction.

**   1973 dollars.

*** Power only.

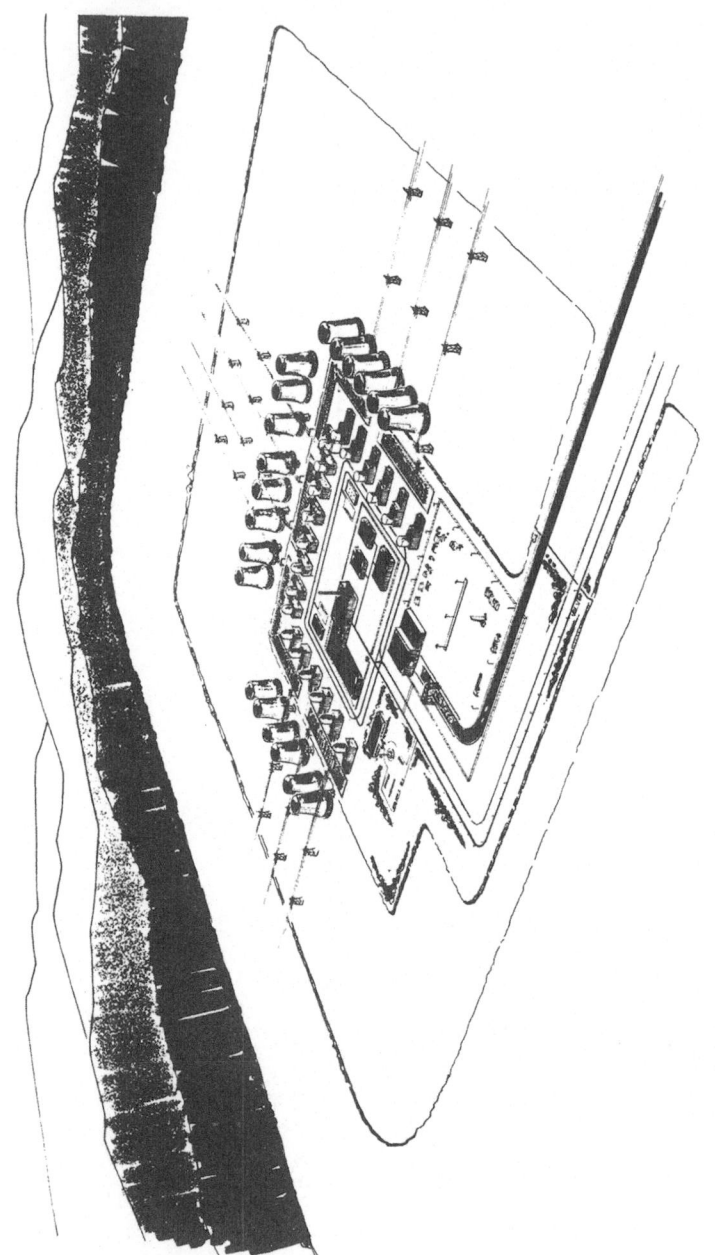

Figure 2

Nuclear Fueled Energy Center with Cooling Towers

More recent NRC evaluation work under the Nuclear Energy Center Survey (NECSS) activity (7) has focused on a range of alternatives; i.e.:

a. a large nuclear power center (48 GW)*
b. a small nuclear power center (12 GW)*
c. a regional integrated fuel cycle facility
d. a combined power generation and fuel cycle facility

Some of the key characteristics envisaged for these energy center alternatives are shown in Table III.

Growth of Nuclear Installations and the Potential Role of Energy Centers. U. S. commercial nuclear installations represent nearly 40 GW of operating power plants as of early 1976. Some 180 GW additional units are on order and are scheduled for commercial operating dates extending into the mid-80's and beyond. All of these units are presently operating or scheduled to operate at dispersed sites and will rely on fuel cycle support facilities located at other dispersed sites, in many cases yet to be decided.

The potential role of energy center developments thus is:

1. for power generation needs, primarily in the late 80's and beyond.

2. for fuel cycle support facilities, for existing nuclear units and possibly to be coupled to additional on-site nuclear generation.

In the early 70's typical projections of total growth trends to the year 2000 for energy consumption ($\sim$ 4%/year) for electrical energy (6-7%/year), and for nuclear plant installations are as shown in Figure 3. It would suggest that in addition to the 220 GW nuclear generation already committed to dispersed sites, some 800 GW plus, of new in-

---

* Power will be quoted in gross electrical output unless the context shows otherwise. Thermal power will be several thermal times larger, depending on thermal efficiency.

TABLE III

NECSS CONCEPTUAL DATA

Defining Characteristics of NEC'S	Large Power Center	Small Power Center	Regional Integrated Fuel Cycle Facility (IFCF)	Combined Facilities Larger Power Center and Small IFCF
1. Technical:				
Electric Power, MWe	48,000 by 2018	12,000 by 1995	None	48,000
Number of Reactors	40	10	None	40
Reactor Type	LWR-HTGR-LMFBR	LWR-HTGR-LMFBR	None	LWR-HTGR-LMFBR
Cooling Systems	Combination of all types	Combination of all types	No power units	Combination of all types
Transmission Voltage, (kV-AC)	500, 765 1100	500, 765, 1100	None	500, 765, 1100
Fuel Cycle	None	None	3-6 Reprocessing plants 6-10 MOX Fuel fab. 1-Low Level waste management 1-High Level waste management	Fuel Reprocessing, Fabrication, Waste Management
2. Environmental:				
Land Area, km²	194*	49.2*	21 - 65	207
Water Consumption, m³/min	530,000	130,000	Small amount	530,000
Thermal Discharge, MWt	379	95	0.76	379
Transmission Lines (single circuit)	11 - 15	4 - 7	Not applicable	11 - 15
Transmission Corridor (area, km²)	390 - 650	103 - 181	Not applicable	390 - 650

Table III, Continued

3. Economic (1974 Dollars):				
Capital Cost, Billion $				
Generating	30-40	8-10	Not Applicable	30-40
Transmission	1-1.5	0.3-0.5	Not Applicable	1-1.5
Fuel Cycle	None	None	4-5	1.0-2.0
Annual Operating Cost, 10	1,200-1,400	400-500	Not available	1,500-2,000
Cost of Electricity, Mils/KW hr 1/	20-25	20-25	20-25	20-25
4. Socioeconomic:				
Construction Manpower	10,000 Peak**	5,000 Peak***	4,000 Peak	11,000 Peak
Operation Manpower	3,500 Peak	900 Peak	4,000-5,000	4,500 Peak
New Town Population	50,000 Peak	25,000 Peak	40,000 Peak	60,000 Peak
Ownership	Public-Private	Public-Private	Public-Private	Public-Private

* Based on 1.0 Acre/MWe.
** Based on 2 LWR's per year.
*** Based on 1 LWR per year.
Source: Staff Estimate.
1/ = 62% capacity factor - 15% F. C. rate.

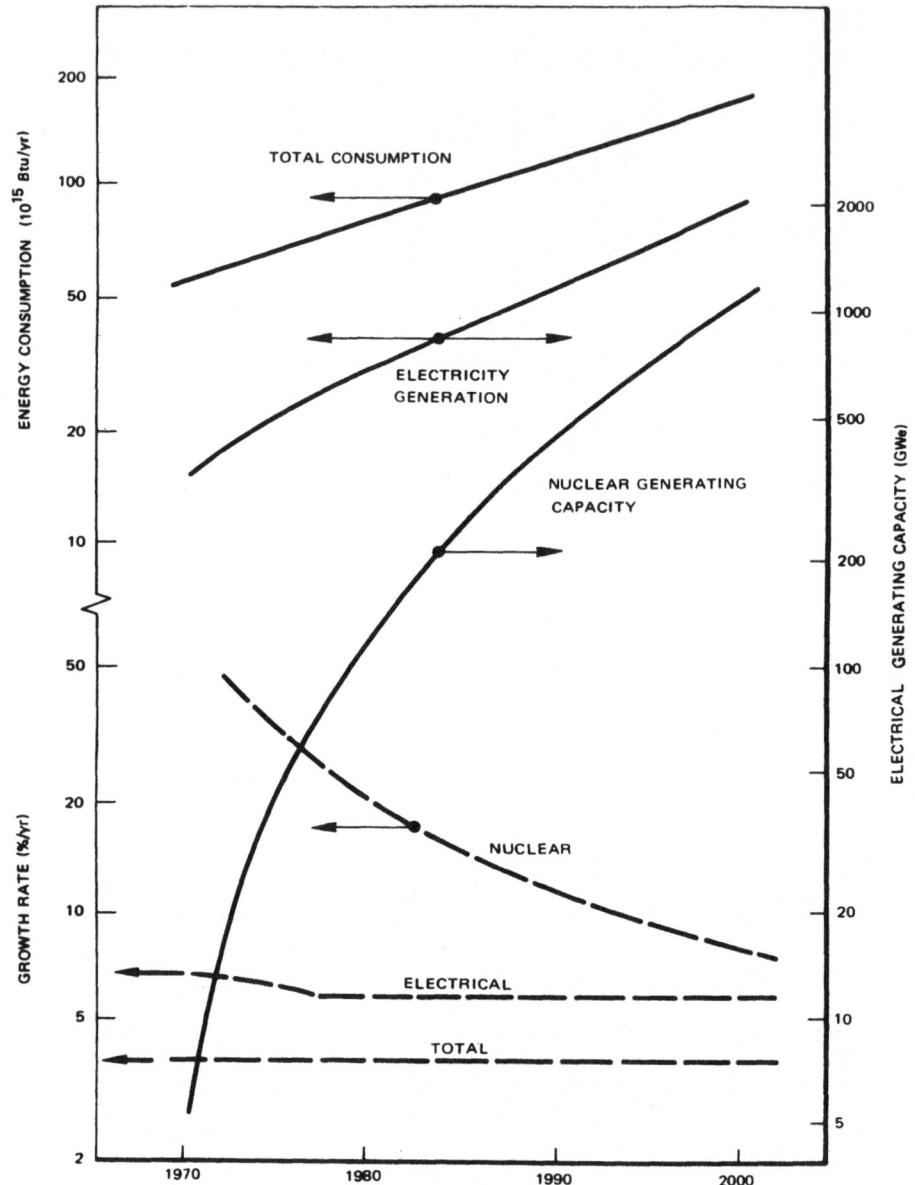

Figure 3

Pre-1975 Projections of U. S. Energy Consumption
and Electricity Generating Capacity

stallations would be committed, constructed, and placed in
operation by the end of the century. These nuclear data
are representative of that referenced in the Project Independence report by the FEA (8). The magnitude of this suggested expansion of nuclear facilities is estimated to represent well over $1,000 billion. It compares with cumulative
commercial nuclear industry facility investments through
1975 in the order of $20 billion. It would call for the
equivalent of one new 1 GW nuclear generating unit coming
on line every nine days from 1975 on through to the turn of
the century. Commissioning of new fuel cycle support facilities would be in addition to this. Together they imply an
average level of construction activity ($40 billion per year),
well in excess of the pace of all U. S. industrial and public
works construction as recently as the mid-60's. Thus, the
idea that such a burgeoning industry might be concentrated
at a series of designated energy centers is:

a. appealing in that it promises to become
a very major national investment and
potentially one in which substantial
economies should be possible through
appropriate planning and integration.

b. a source of some question in that such a
large proportion of total anticipated
construction industry activities is
potentially involved that localization
of these activities at designated energy
centers suggest some change in character
of large portions of the construction
industry.

Certain more recent forecasts (9,10) have suggested substantial reductions in the growth rate of energy consumption,
electric power capacity and potentially, though not necessar-
ily in direct response, of nuclear power plant installations.
A compilation of some of these data is shown in Figure 4. It
is based primarily on trends of ERDA 48 Scenario V data upon
which information from a late '75 electrical industry publication has been superimposed. The data shown in Figure 4
refer to three basic classes of nuclear power plants:

1. the light water reactors - PWRs and BWRs, that
constitute all operational U. S. commercial
nuclear power plants as of the end of 1975.

2. the Liquid Metal Fast Breeder Reactors -
    LMFBRs that represent a major development
    goal for the nuclear industry in its efforts
    to provide greater nuclear fuel self-
    sufficiency.

3. the High Temperature Gas Reactor - HTGR that
    had indicated signs of playing a significant
    intermediate nuclear role as a result of its
    potential for higher thermal efficiency oper-
    ation. It is now questionable whether some
    such intermediate system offering higher
    efficiency or improved fuel utilization will
    be present in future U. S. nuclear development.

The trend of Figure 4 is unmistakable; it is lower and
is linear, as opposed to exponential; it is probably more
consistent with overall thinking on future growth trends of
the United States; it has been adopted as the basis for
analyses in this review of potential nuclear energy center
development. In order to provide a helpful time horizon,
the forecast trends have been extrapolated to the year 2025.
This is believed to represent a minimum time horizon for
nuclear center study activities in which 20 or more nuclear
installations may be under study for a single center and
their rate of installation may be one per year, starting in
1985. Units would then come on line annually over a period
extending well into the twenty-first century. Indeed, the
operating lifetime of such a center would be expected to
extend to 2050 at a minimum. Ultimately, the question arises
as to the likelihood of indefinite continuance of nuclear
additions to U. S. generating capacity. While this is considered to lie beyond the selected time horizon of 2025, it
is nevertheless conceivable that ultimately, energy center
rejuvenation through gradual replacement of facilities at a
given site would become the more significant activity and
should be considered at the time of more detailed facility
design.

Two boundary values for the possible proportions of nuclear energy centers in the future growth of the nuclear industry can be suggested.

A first is that rate of installation that would be
necessary if essentially all of the plutonium produced in

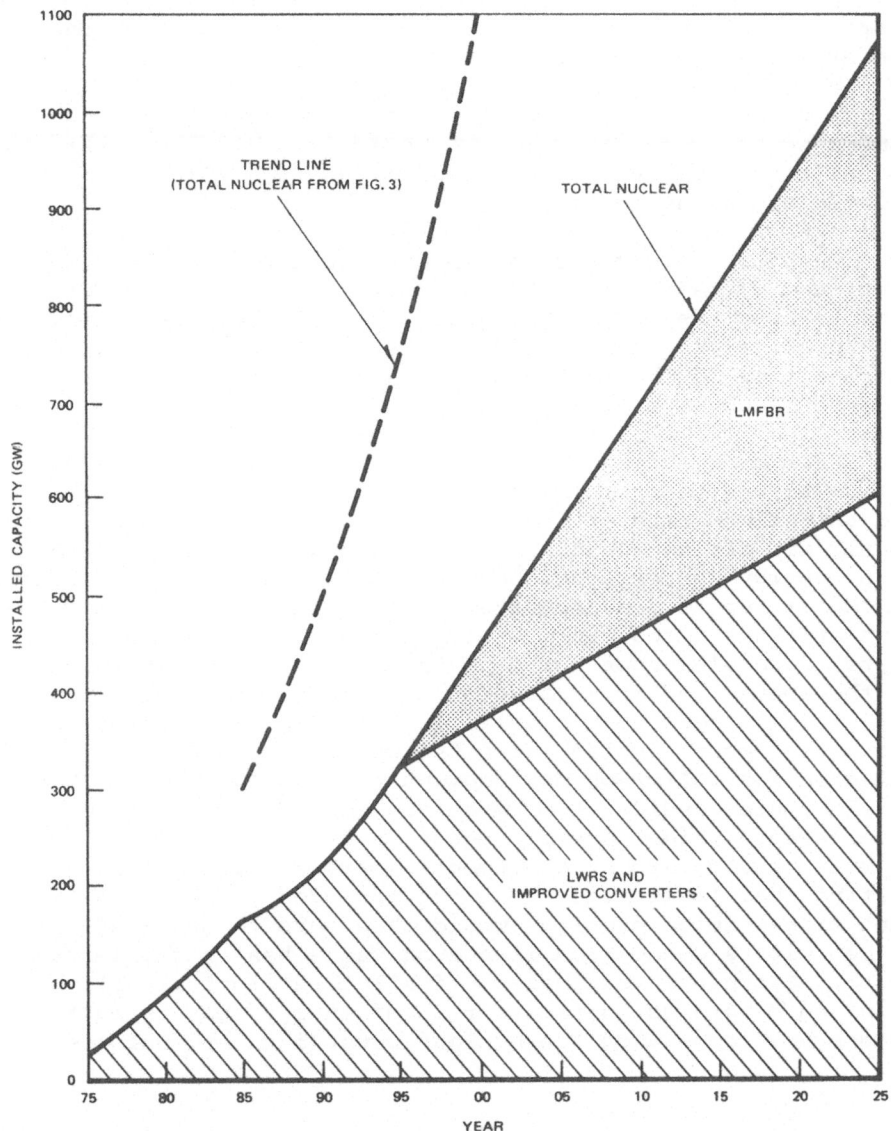

Figure 4

Revised Assumptions - Growth of Nuclear Installations

operating LWRs were to be utilized in LWRs at nuclear energy
center sites over the next few decades.

A second and near maximum role for energy center development would be one in which essentially all new nuclear plant
installations beyond those units already committed were to be
located at energy centers.

Based on this reasoning in conjunction with the aforementioned nuclear forecasts, a range of energy center developments of between 2 GW per year and 10 GW per year, rising
toward 26 GW per year around the turn of the century with the
introduction of the LMFBR appears appropriate for overview
analyses such as intended here.

Presuming average nuclear plant sizes to not grow significantly beyond the present approximately one GW level,
and an optimum rate of addition of nuclear units at an energy
center of near one unit per year, then a possible range of
U. S. energy center developments for the next several decades
may be at two to ten separate sites. Near parallel development at these sites could evolve for a period upward of 15
years, depending upon how many units were to be constructed
at the energy center. Based on this reasoning, and given
as little as four to six units for some initial energy center development, subsequent sites would then need to be
selected for follow-on development four to six years later.
An alternate consideration might be toward larger, more
durable site selections for energy centers. These would,
in turn, imply larger commitments, particularly if supporting fuel facilities are to be included. More significant transmission problems would then have to be faced, together with a greater challenge on waste heat rejection and
so on. Further investigation of all these factors will be
required before the ultimate role of energy centers in the
future development of the nuclear power industry unfolds.

II. FUEL CYCLE AND FACILITY INTEGRATION

A. Facility Arrangements

The location of the existing fuel cycle facilities, or fuel cycle-related facilities is shown in Figure 5, super-imposed on a chart that locates committed nuclear power generation facilities across the United States. The sig-nificant points are:

1. The power plant sites far outnumber the fuel facilities; hundreds versus tens; and they inherently will, on into the future.

2. Both the power plants and the fuel cycle facilities are well dispersed across the United States. This, as noted previously, is characteristic of the circumstance under which the nuclear industry has evolved to date; i.e., one in which transportation costs and constraints have been of minimal economic significance.

The present status and capabilities of these facilities are approximately as follows:

1. Fuel Fabrication, sufficient to furnish reload cores for at least the 40 GW of installed nuclear generation, plus annual additions of first cores of over 10 GW per year. Together these correspond to a capability potentially in excess of 22.7×10^5 kg per year of uranium fuel throughput.

2. Fuel Reprocessing. Facilities have been constructed with the capability to handle approximately 18.1×10^5 kg per year of spent light water reactor fuels. Operation of these facilities, as of year end '75, has been very limited. Only the West Valley facility of Nuclear Fuel Services has oper-ated. It had a design capacity of 30 MO^4 kg

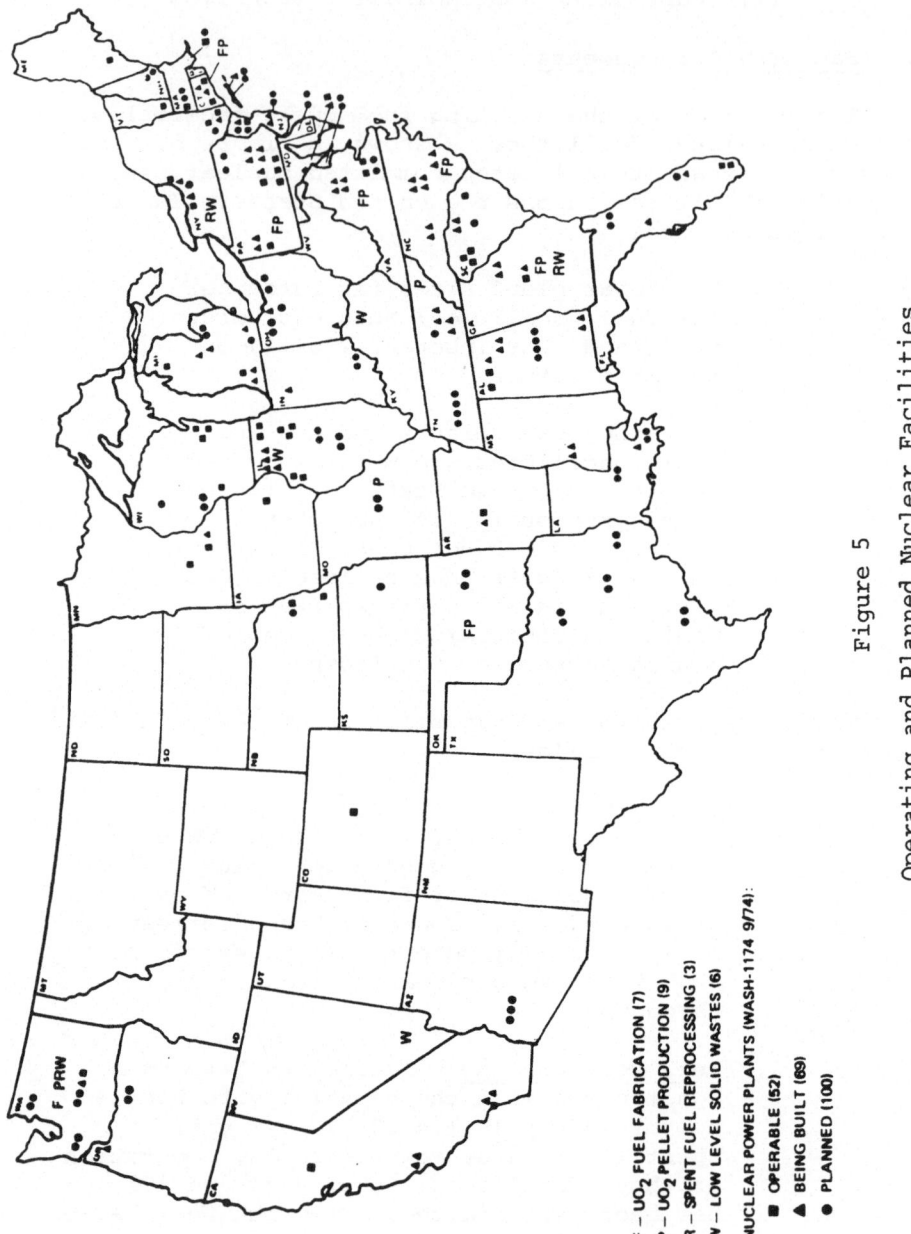

F – UO₂ FUEL FABRICATION (7)
P – UO₂ PELLET PRODUCTION (9)
R – SPENT FUEL REPROCESSING (3)
W – LOW LEVEL SOLID WASTES (6)

NUCLEAR POWER PLANTS (WASH-1174 9/74):
■ OPERABLE (52)
▲ BEING BUILT (69)
● PLANNED (100)

Figure 5

Operating and Planned Nuclear Facilities

per year prior to shutdown for expansion.
Neither the G. E. Morris facility nor the
Allied Gulf Barnwell facility has operated
commercially as of 1 January 1976. Characteristics of these plants are shown in Table
IV and Figure 6. The data have been extracted
from referenced dockets filed in the NRC public document room, Washington, D.C.

3. <u>Plutonium Fabrication for Recycle</u>. While
various experimental facilities exist and
a first commercial scale facility has been
proposed, no commercial plutonium fabrication plants have been placed in operation
to date. Finalization for the schedule of
such facilities has been held up by the
resolution on GESMO, ERDA's Generic
Environmental Impact Statement on use of
Plutonium in Mixed Oxide Fuels (<i>11</i>). This is
now expected during 1977 (<i>12</i>). Characteristics of plutonium fabrication facilities
(<i>13</i>) are shown in Table V and Figure 7.

The significance of the above fuel facility status for
the U. S. nuclear industry is two-fold:

1. There is a critical need for the development
of the fuel cycle back-end capabilities if
healthy development of light water systems
is to continue.

2. The delay to date in installation of fuel
facilities for the back end or closing of
the fuel cycle provides a unique opportunity for study of optimum location of
such facilities. This is indeed a part of
the objectives of the NRC's present
nuclear energy center site survey.

Pending installation of new reprocessing facilities,
substantial light water reactor fuel storage is anticipated.
The forms of the spent fuel storage and relationship to proposed federal facilities are shown in Figure 8. A typical
estimate (<i>14</i>) of the buildup of spent fuel is shown in Figure
9 together with estimates of contained plutonium and the plutonium fabrication capability for which it would call.

TABLE IV

U. S. COMMERCIAL SPENT FUEL REPROCESSING PLANTS

	West Valley Reprocessing Plant	Barnwell Nuclear Fuel Plant	Midwest Fuel Recovery Plant
Plant Name:			
Docket No.:	50201	50332	50268
Owner: Site:	Nuclear Fuel Services,Inc. West Valley, New York, 30 miles south of Buffalo	Allied-Gulf Nuclear Serv. Barnwell, South Carolina, East side of Savannah river plant.	General Electric Company Morris, Illinois, South side of Dresden nuclear power station.
Status:	Processed 630 MTU 1966-1970. Expansion and hardening by about 1979.	Schedule startup 7/76. Spent fuel storage one year earlier.	Constructed 1967-74, discontinued for operability problems. Being modified for fuel storage only 1975.
Capacity:	300 MTU/yr 1966-1970 750 MTU/yr expansion.	5 MTU/D, 1500 MTU/yr (less for higher enrichment)	Reprocess 1 MTU/D, 300 MTU/yr. Store 90 MTU as is.
Fuels:	Low enrichment LWR UO_2 or PuO_2-UO_2; high enrichment equipment being removed; thorium-uranium oxide potential.	Low enrichment LWR, 3.5% U-235 or 29 KG fissle Pu per MTU, 40,000 MWD/T average 160 days decay.	Low enrichment (5% U-235) LWR, UO_2 or PuO_2-UO_2.
Process:	Purex (Tributyl phosphate in kerosene-like solvent).	Purex with proprietary features	Aqua-fluor (Purex and U fluorination).
Products:	Pu Nitrate, UF_6	PuO_2, UF_6, NP nitrate.	UF_6, Pu and NP nitrates.
Waste Forms:	Offgas stack, hardware vault, liquid tanks.	Offgas stack, hardware vault, low activity tanks, high activity tanks.	Offgas stack, hardware vault, low activity tanks, high activity solids in cans in pool, dry chemicals (fluorides) vault.

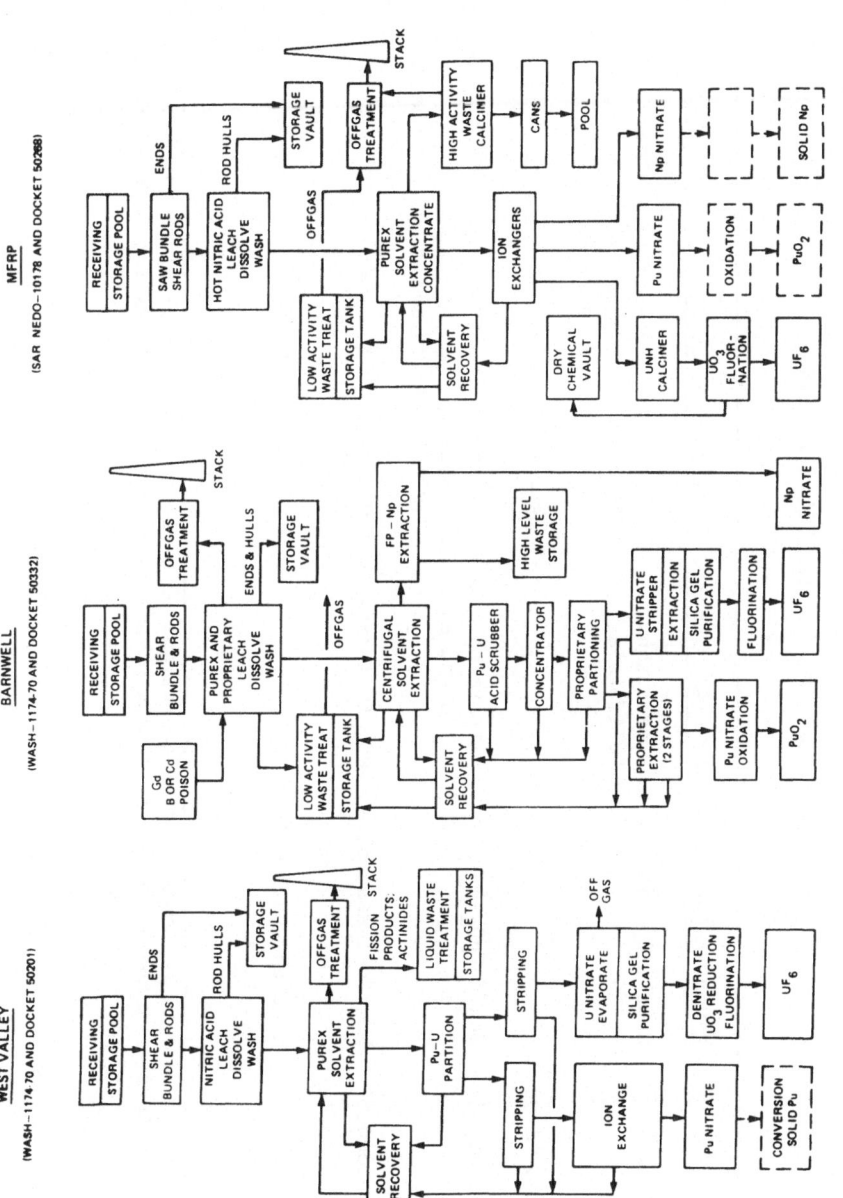

Figure 6

Fuel Reprocessing Plant Flowsheets

TABLE V

EXISTING MO FUEL FABRICATION FACILITIES

Licensee*	EXXON Nuclear	General Electric	Kerr-McGee	NUMEC	Westinghouse
Location:	Richland, Washington	Pleasanton, California	Crescent, Oklahoma	Parks Township, Pennsylvania	Cheswick, Pennsylvania
Capacity MT/yr	15	3	5 - 10	20	10 - 15
Feed Material	UO_2 + PuO_2	Nitrate solution (U and Pu)	Nitrate solution (U and Pu)	Nitrate solution (U and Pu)	Nitrate solution (U and Pu)
Plant Product:	MO_2 Fuel assemblies	MO_2 fuel rods	MO_2 fuel rods	MO fuel rods	MO_2 fuel rods
Plutonium Possession Limits (kg)	10 unencapsulated 100 total	15	360	2,000	120

* There are four other commercial organizations with licenses to process plutonium for fuel rod manufacture, but their present plutonium possession limits are too low for consideration as viable mixed-oxide fuel manufacturers.

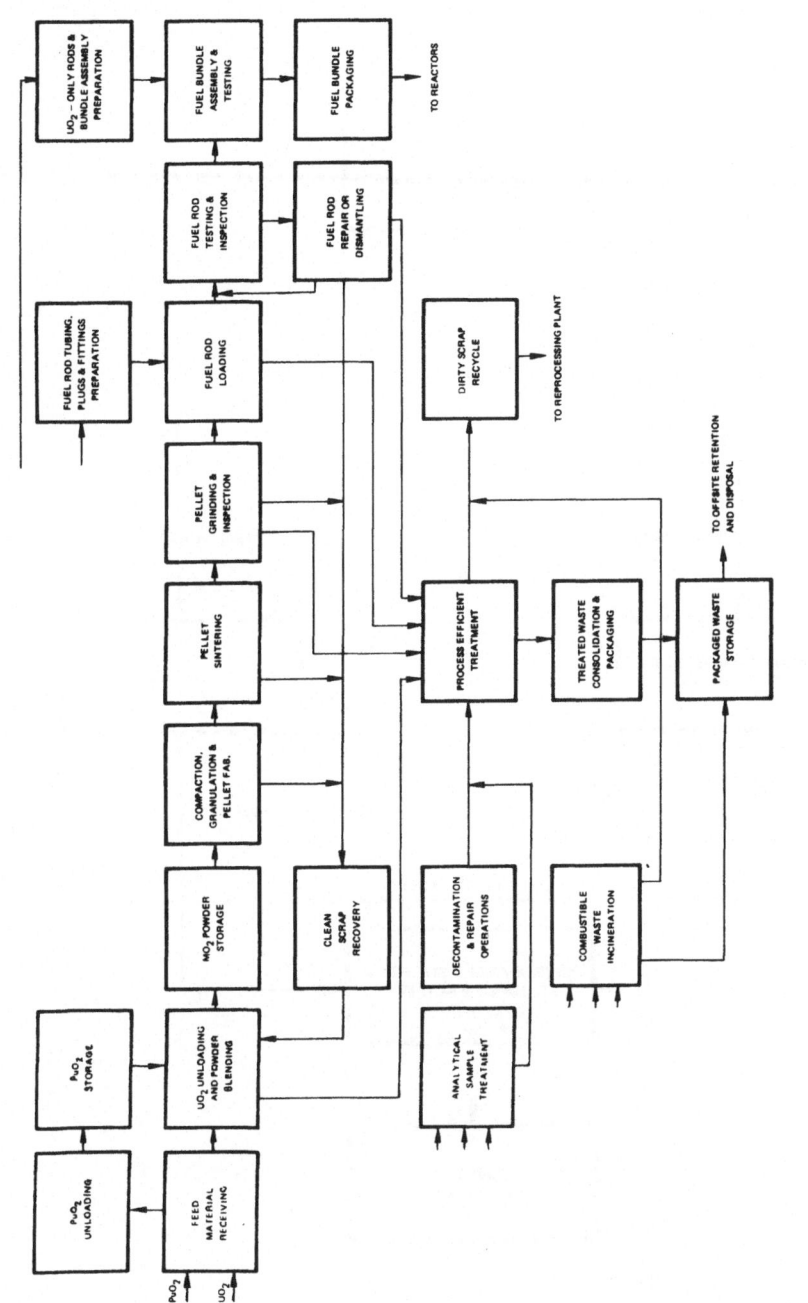

Figure 7

MO$_2$ Fuel Fabrication Plant Flowsheet

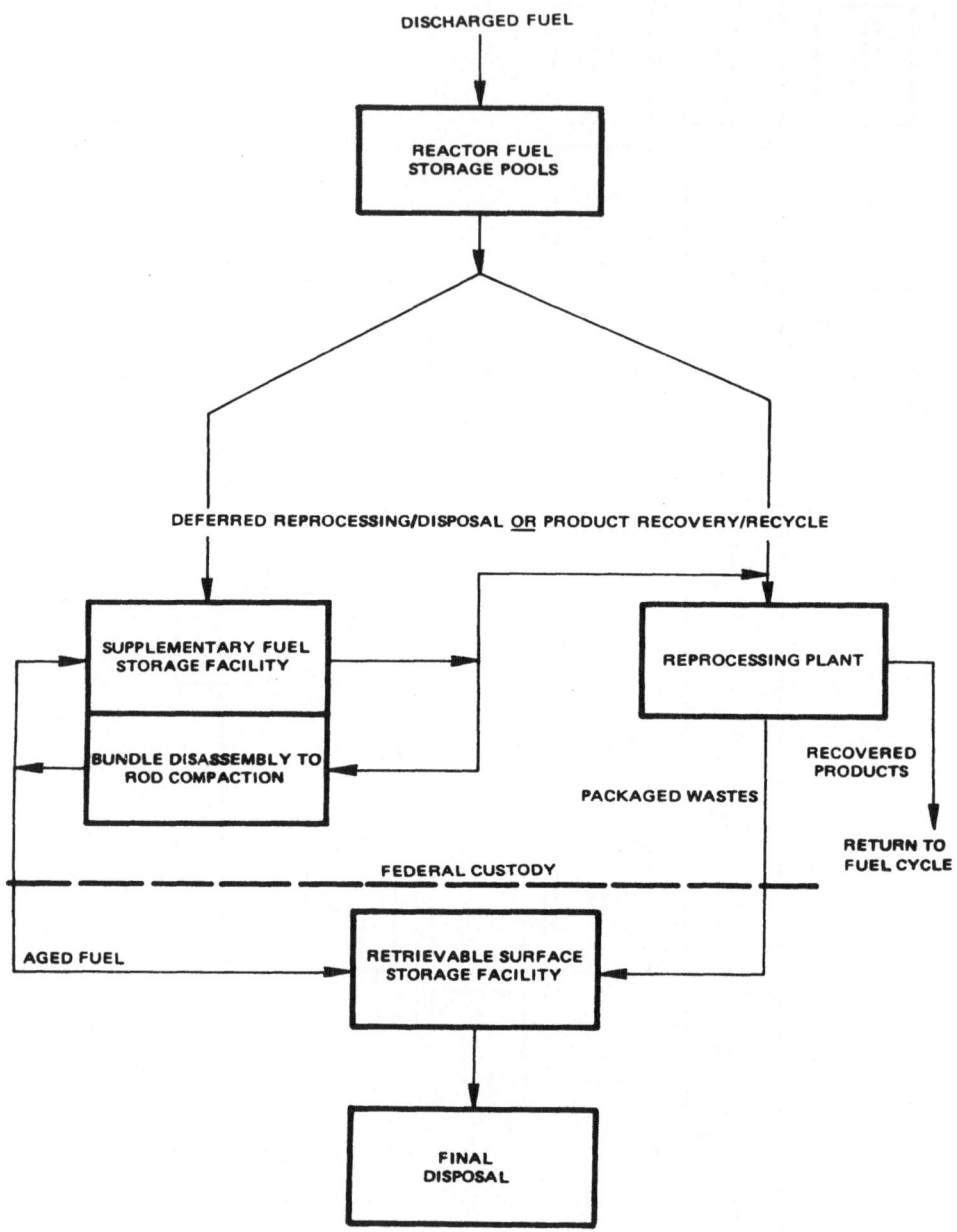

Figure 8
Alternatives for Disposition of Spent Fuel

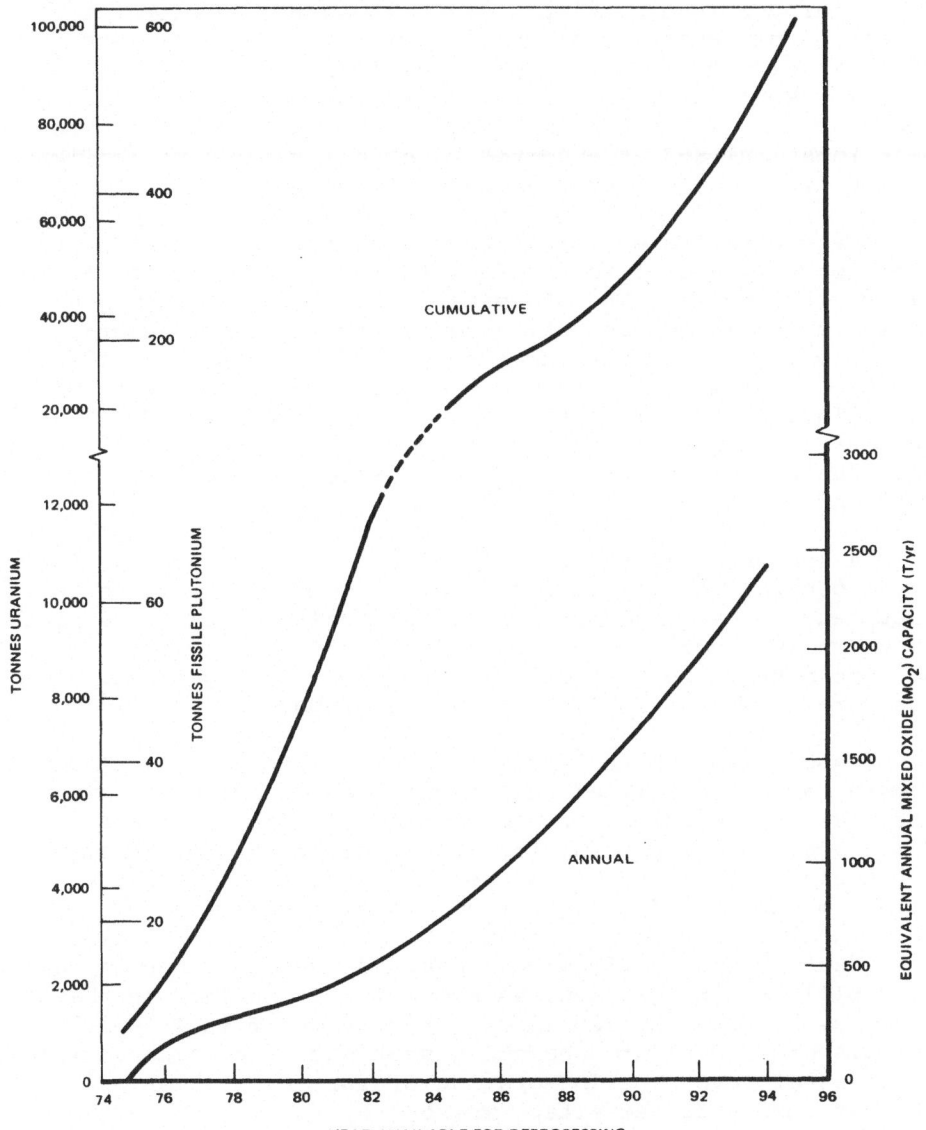

Figure 9

Projected Spent Fuel and Equivalent
MO₂ Fabrication Requirements

A similar published estimate of the possible development of reprocessing capabilities and interim spent fuel storage is shown in Figure 10. One additional fuel processing plant is referred to in this figure, over and above those already constructed. It is a plant proposed by EXXON (U. S. southeast location) and is expected to have a capability similar to the 1.5 x 10^6 kg per year Barnwell capability.

Recent estimates (5) of the unit processing costs for fuel recovery and plutonium fuel fabrication are shown in Figures 11 and 12. The steepness of the curves shows generally acknowledged incentive to progress toward larger-scale facilities. Reference is made on the charts to the volume of throughput that would correspond to a single LWR installation of four units, each of approximately 13 GW capacity. Thus, the logic of large centralized facilities and transportation of fuels to the processing centers is apparent.

A somewhat different perspective on this incentive for large centralized facilities results from replotting these cost data (5) as a variance on total nuclear fuel cycle costs as shown in Figure 13. This tempering of scale effects would be even more evident if one were to plot total generation cost trends versus fuel facility throughput.

Thus, although the nuclear industry has been developing toward large centralized fuel facilities, other cost components related to future transportation requirements, etc. could open up the possibility of differing arrangements. Of particular interest here is the concept of energy centers with some degree of integration of fuel facilities.

Recent studies (5) have appraised a whole range of levels of facilities integration, as shown in Figure 14. In these studies, a series of levels of integration of fuel facilities for dispersed nuclear generating stations were compared with a series of levels of integration of fuel facilities on a nuclear energy center site. These are represented by the left and right hand columns of Figure 14. In this figure, a gradual progression is shown from the separated nuclear generation and fuel facilities of the present, with a fuel cycle operating on storeaway basis (Case O), on through to full integration of fuel cycle and generating facilities (Cases C and D).

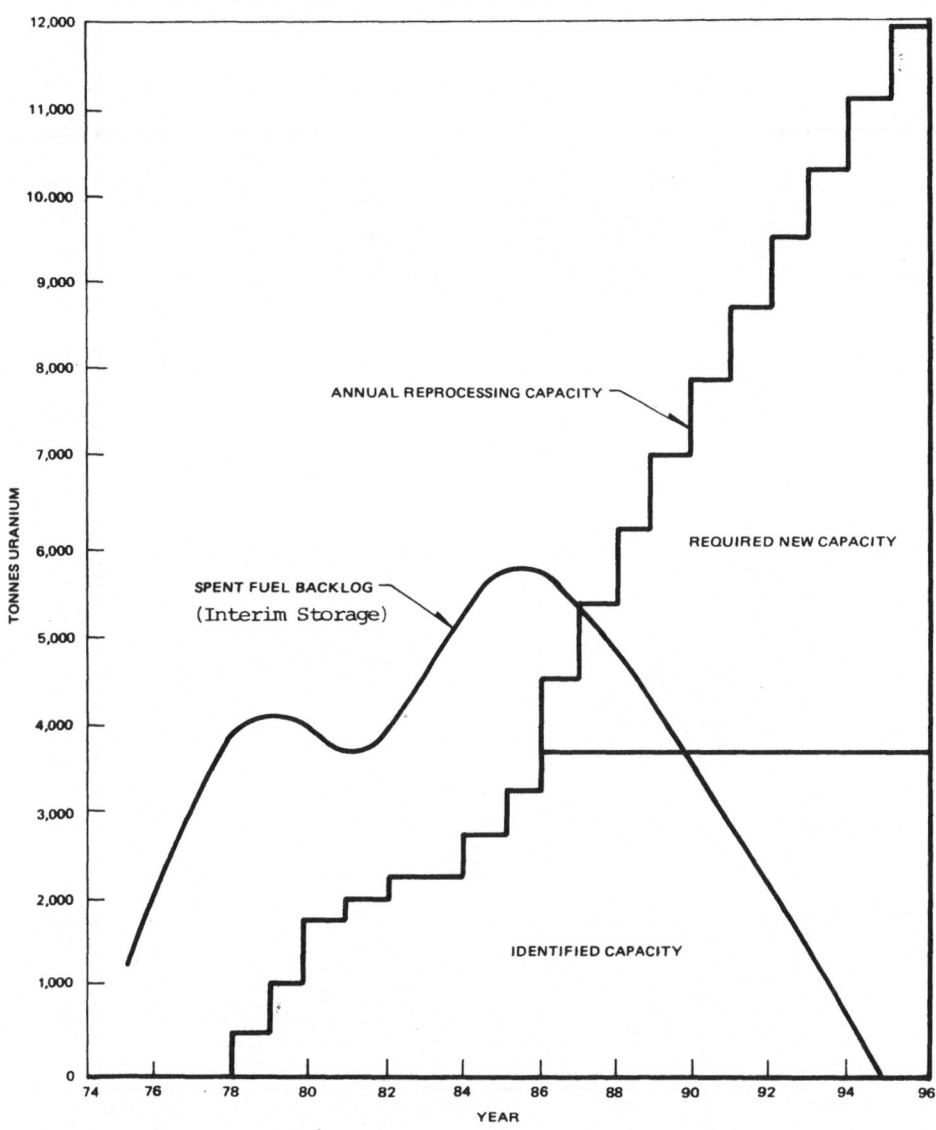

Figure 10
Reprocessing Capacity Required to Reduce Backlog by 1995

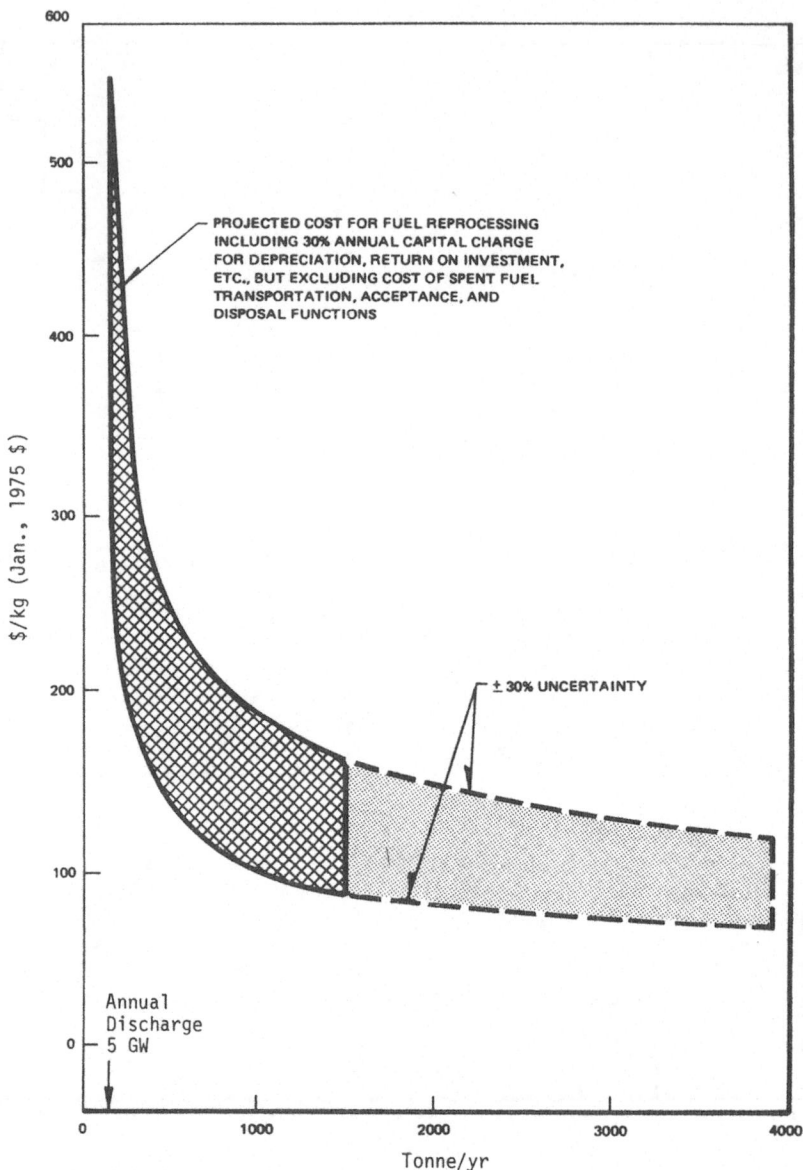

Figure 11

Unit Reprocessing Cost Versus Capacity (LWR Fuel)

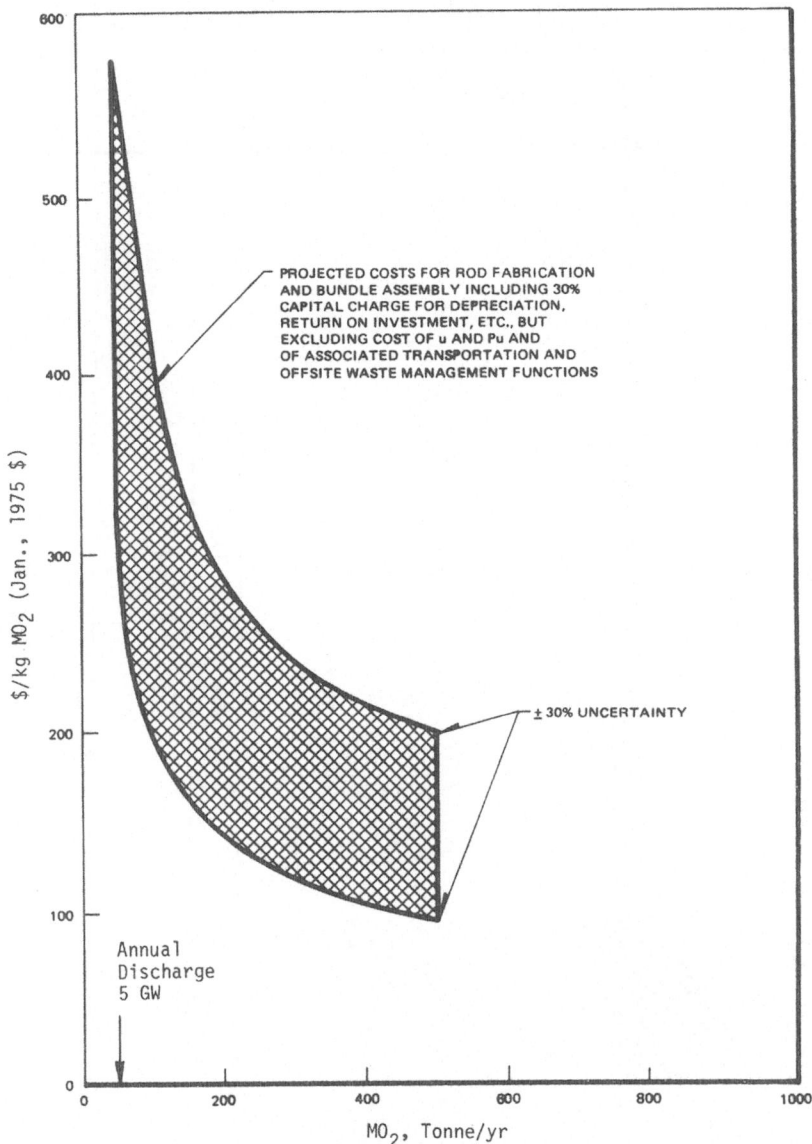

PROJECTED COSTS FOR ROD FABRICATION AND BUNDLE ASSEMBLY INCLUDING 30% CAPITAL CHARGE FOR DEPRECIATION, RETURN ON INVESTMENT, ETC., BUT EXCLUDING COST OF u AND Pu AND OF ASSOCIATED TRANSPORTATION AND OFFSITE WASTE MANAGEMENT FUNCTIONS

± 30% UNCERTAINTY

Annual
Discharge
5 GW

$/kg MO$_2$ (Jan., 1975 $)

MO$_2$, Tonne/yr

Figure 12

Unit MO$_2$ Fuel Fabrication Cost Versus Capacity

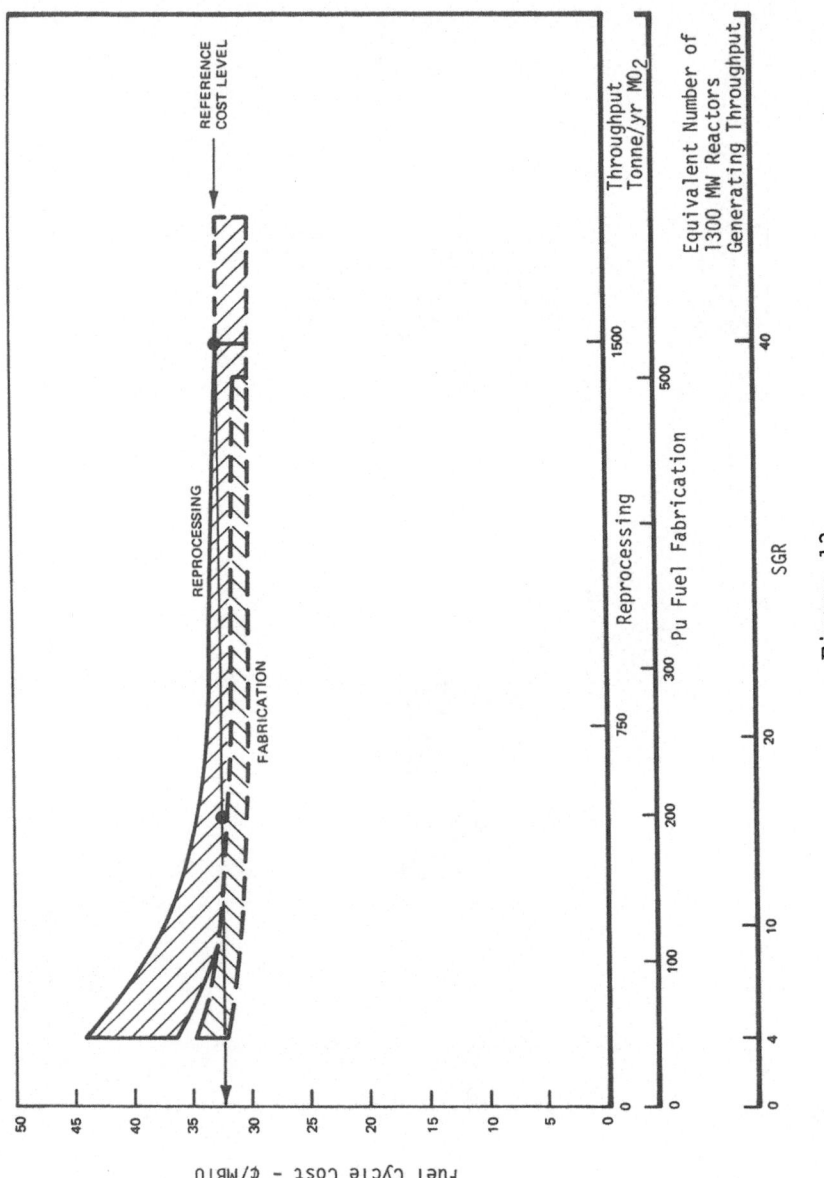

Figure 13

Impact on Fuel Cost of Recycle Facility Uncertainties and Throughput

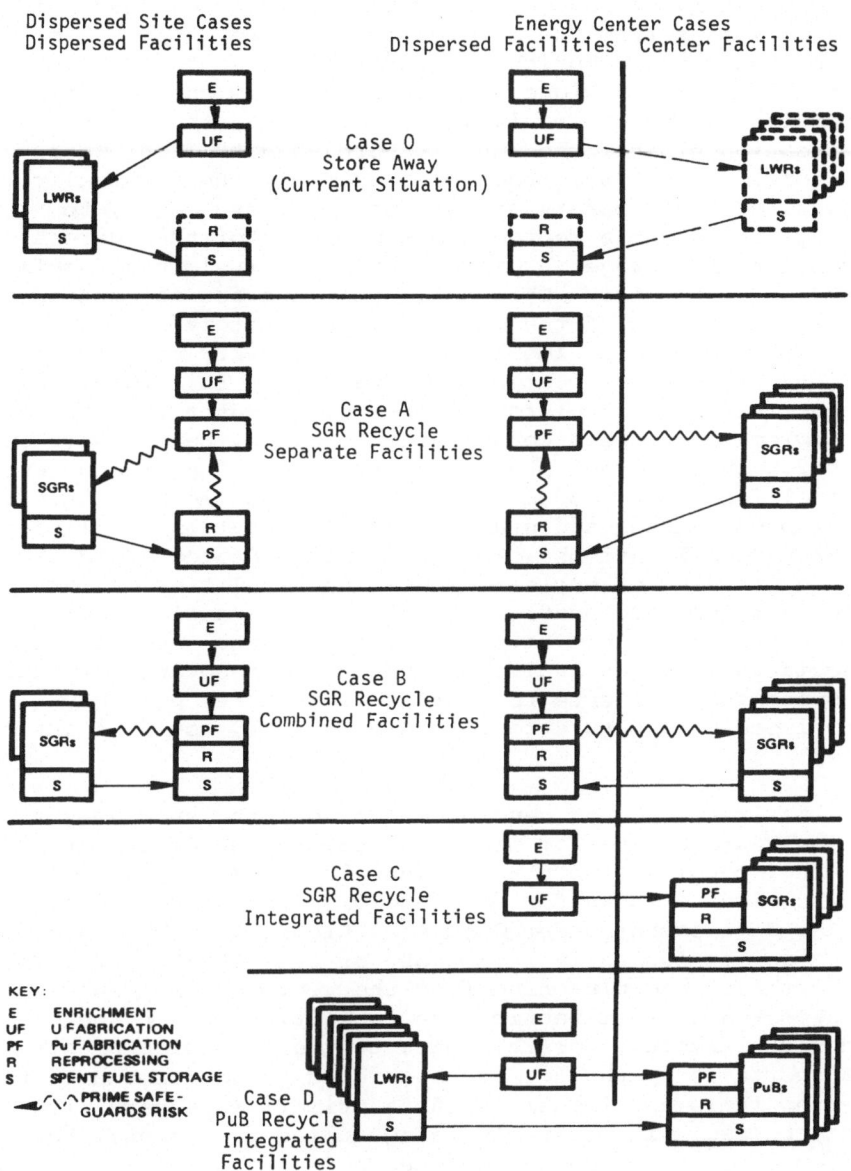

Figure 14

Fuel Cycle Integration Steps

Returning to Case O, the current situation, uranium
fuel materials from the mines are processed through a series
of federally-owned enrichment facilities located at Paducah,
Portsmouth and Oak Ridge. Enriched fuel is shipped from
these facilities to uranium fabrication plants. From the
fabrication plants, the fuel is furnished to light water
reactors. Spent fuel from those units is then typically
discharged to temporary storage on site, and hence, to
storage basins at a reprocessing plant. For the current
situation of storage, a backlog of spent fuel is accumulated
in these facilities. This could continue for as much as
10 years for much of the industry. Ultimately, however,
reprocessing will be implemented for recovery of uranium
and for recovery of plutonium, although at this point, it
is not certain that plutonium into other water reactors will
be implemented. When considering an energy center development for this current storeaway situation, the arrangement
of fuel facilities would be essentially unchanged. Existing
uranium enrichment and uranium fuel fabrication facilities
at dispersed locations across the United States would service a grouping of nuclear generating facilities making up
a nuclear energy center. This grouping of nuclear units at
the center would then have some fuel storage capability
associated with the individual units. Subsequent shipment
of spent fuel off-site to storage facilities, possibly
associated with future reprocessing plants, would be anticipated.

A first stage of development beyond the storeaway situation would be that of implementation of spent fuel reprocessing, coupled with uranium and plutonium recycle. This
progression is shown for both the dispersed site and energy
center development under Case A of Figure 14. The material
flows follow logically from those of Case O with the recovered uranium and plutonium from the reprocessing plant, being
transported to fuel fabrication facilities, or in the case of
recovered uranium, possibly back to the diffusion plant. The
transportation paths for the recovered plutonium and plutonium-
bearing fuels are identified as a primary safeguard risk;
nevertheless, with suitable safeguarding provision, there is
believed to be no reason why the separate facilities suggested for this so-called "SGR recycle" situation cannot be implemented. The term "SGR" refers to "self-generated recycle"
of plutonium; i.e., the reactor system is operated on a fuel
cycle in which it consumes, in subsequent fuel cycles, all

the plutonium present in discharged fuel from a given fuel
cycle. Again, under this SGR recycle, Case A mode of operation, the configuration for an energy center is relatively
similar. An energy center development using SGR recycle
and separate facilities will face the same plutonium transportation safeguards demands of the dispersed facilities
development. Any benefits from such a center development
would relate solely to improvement gained in construction
and operation of a large group of generating units at the
same site.

While the Case A type SGR recycle through system facilities was, in the past, the generally-accepted plan for fuel
facility development, there is now general recognition of the
attributes of combining plutonium fabrication facilities with
reprocessing facilities, either at a single site or at immediately adjacent sites. This situation is given in Case B.
This combined facilities configuration substantially lessens
the safeguarding risk, and should offer some other lesser
economies relating to shared site services, lesser transportation and packaging costs and time interval.

A further stage of development in this progression
toward increased integration is shown in Case C. It is
suggested that it is uniquely an energy center option, and
it is one in which the plutonium transportation burdens
are further minimized by locating the key fuel cycle components, reprocessing and plutonium fuel fabrication at the
energy center, along with the grouping of generating units.
Under this integrated facilities arrangement, these then
become dedicated fuel facilities, and plutonium, once having
been generated on site, is then recycled on this site, is
then recycled on this site in self-generation cycle mode;
and in principle, no regular schedule of off-site plutonium
shipments are required. It is suggested in Case C that
such an energy center, with integrated fuel facilities,
would still utilize the existing federally-owned uranium
enrichment facilities as a fuel source, and that this enriched fuel material still would be processed in the industry's existing dispersed uranium fabrication facilities,
thereby benefiting from the economies of scale of these
existing very large facilities. By comparison, the suggested integration of the plutonium fabrication facilities on
the energy center site under Case B implies that this fabrication will be on a small scale, as indeed, will be the

reprocessing. This represents an economic burden for the
energy center site unless a very large aggregation of generating units on that site is achieved; for instance, the
Barnwell 1.5×10^6 kg per year design capability corresponds to the discharge from some 40 light water reactor
units, each of approximately 13 GW generating capability.
While energy center concepts of this scale have been suggested (4,6), it is believed that the feasibility and economics of such large centers is questionable, and thus the
fuel cycle economies may not be achievable.

A further variant of Case C is one in which the volume
throuput for the integration fuel cycle facilities at an
energy center are boosted, without actually calling for the
large block of on-site nuclear generation referred to in
Case C above. This somewhat unique arrangement depends on
the coupling of dispersed generating facilities or other
energy center generating facilities with those at a PuB
center. The form of the coupling is through spent fuel
shipments and the term "PuB" refers to the fact that all
the plutonium recovered from spent fuel from off-site and
on-site nuclear units is then recycled through the on-site
PuB's or plutonium burner units. Thus, the plutonium burner concept represents an alternative to the SGR mode of
plutonium recycle. It implies a special grouping of nuclear
units at the energy center site that are dedicated to this
plutonium burner role, and which would, in consequence, have
a much higher loading of plutonium in their reload fuel.
In the limit, and in preference, this plutonium loading
would be the source of essentially all the fissile material
in the PuB reload fuel. Under this circumstance, the on-
site fuel cycle facilities could have a production capability for three to four times the volume of the spent fuel
generated by the on-site PuB units themselves. This suggests that, under a PuB mode of operation at an energy center, a Barnwell-sized facility might be coupled to 10 to 15
PuB's on site, as opposed to the 40 LWR's mentioned above.
The remaining 25 to 30 LWR's would then be off-site and
located either at other energy centers or at dispersed site
facilities. Indeed, this PuB energy center concept permits
a coupling between the existing industry dispersed generating units and a future energy center mode of generation.
It would also eliminate any need for regularly scheduled
plutonium shipments outside the site boundaries of the energy
center. · It would call for the design of plutonium burner

reactors able to accept this much higher level of plutonium
loading in the core. Ultimately, it is anticipated that
such plutonium burners will be the breeder reactors, but in
the interim, it is believed that the light water reactors,
and possibly other systems, can fulfill this role. With
some modest sacrifice in fuel cycle economy, it is believed
that the plutonium burner energy center concept can be reduced in scale to as few as four plutonium burner units at
the energy center site, coupled with possible a 500-ton-
per-year reprocessing plant and a 200-ton-per-year mixed
oxide fuel fabrication facility. This range of options has
been developed quite rigorously in a series of studies of
alternate nuclear energy center configurations and related
dispersed site studies. The coverage of this prior study
work is shown in Figure 15. The various modes of fuel cycle
operation considered are shown, together with the various
facility arrangements. Most significantly, a series of four
cases for a plutonium burner energy center concept are called
out, together with their coupling, to dispersed units.
Their case numbers are, specifically, "N4PD", meaning
"Nuclear, 4 units, Park case, D". "Park" is used synony-
mously with energy center in this series of case desig-
nations. The other cases are: N6PD, and two alternatives
for N20PD. The material flows, under equilibrium fuel cycle
implied by this array of cases, are shown in Table VI.

B.   Candidate Fuel Cycles

The U. S. nuclear program, at this point, has become
almost entirely focused on the LWR's as the near-term nuclear
steam supply systems. Thus, the LWR's become the primary
candidates for at least initial energy center application.
The Liquid Metal Fast Breeder Reactor (LMFBR) remains a
longer-term development. Discussions continue within the
industry on other options, particularly alternate or improved fuel cycle possibilities for the period prior to
LMFBR commercialization. Up until recently, the HTGR was
considered by some to offer that improved fuel utilization
potential. The light water breeder reactor, yet another
possibility of advancing beyond present LWR fuel cycle performance, depends upon introduction of the thorium fuel
cycle for an enhanced neutron economy much as the HTGR does.
The enhanced neutron economy of this thorium cycle only
really shows up after an initial supply of U-233 has been
generated. Furthermore, the production of uranium-233 from

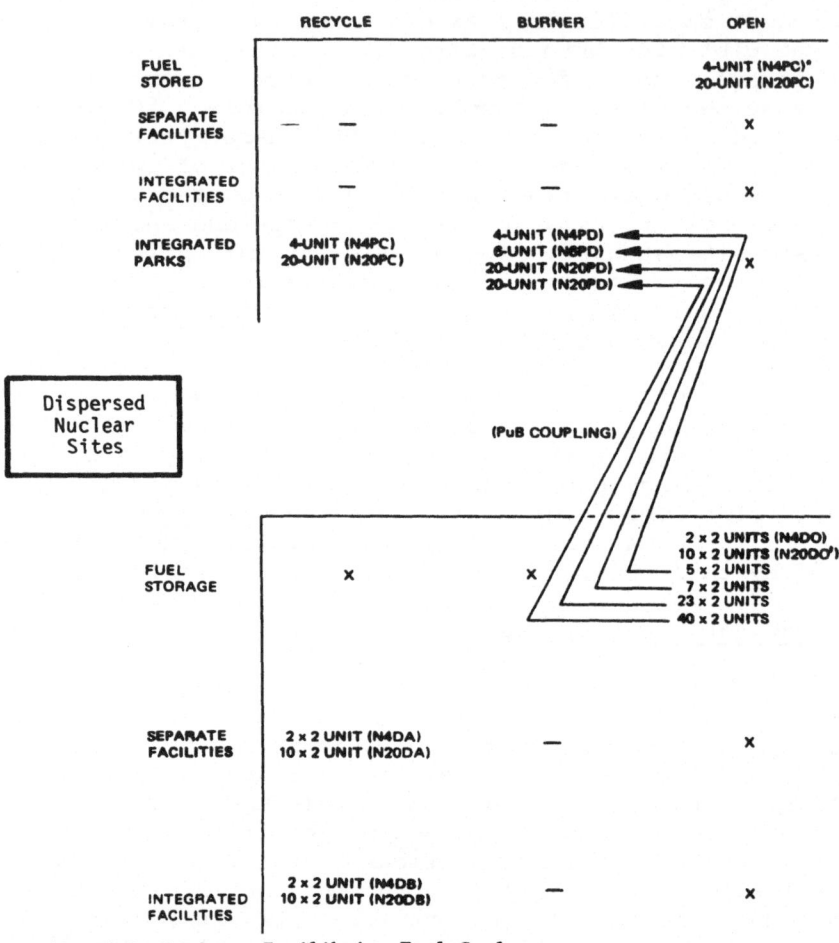

Figure 15

Parameter Coverage - Centers Versus Dispersed Sites

TABLE VI
ANNUAL LOGISTICS; LWR EQUILIBRIUM FUEL CYCLES

Logo Type	O	O'	A	B	C	A	B	C	D	D	D	D'
Power Generation	Dispersed Sites or Energy Parks		2 Dispersed Sites		Energy Park	10 Dispersed Sites		Energy Park	PuB Energy Park Sites or Satellite Parks	PuB Energy Park + Dispersed Parks		
Open Cycle Reactors MW	5200	26000	---			---			12100	18200	60500	104000
SGR Pu-Recycle Reactors MW	---	---	5200			26000			---	---	---	---
PuB Reactors MW	---	---	---			---			5200	7800	26000	26000
Total Power MW	5200	26000	5200			26000			17300	26000	86500	130000
Fuel Reprocessing and Recovery:												
Heavy Isotopes MT	---	---	146			728			484	728	2422	3640
Fissile U MT	---	---	1.0			4.9			2.9	4.4	14.6	25
Fissile Pu MT	---	---	1.9			9.4			6.2	9.4	31	39
Mixed Oxide Fuel Fab.:												
Heavy Isotopes MT	---	---	44			218			146	218	728	728
Fissile Pu MT	---	---	1.9			9.4			6.2	9.4	31	31
Discharged Fuel Storage:												
Spent Fuel MT	146	728	---			---			---	---	---	---
Fissile Pu MT	1.0	4.9	---			---			---	---	---	---
SNM Fissile Pu MT	---	---	---			---			---	---	---	8.1
Required Off-Site Support: Uranium Fabrication												
Heavy Isotopes MT	145	728	102			510			338	510	1694	2912
Fissile U MT	4.7	23.5	3.3			16.4			10.9	16.5	54.7	94
U₃O₈ Ore ST	970	4860	650			3250			2260	3400	11300	19400
Separative Work MT	730	3640	470			2340			1700	2550	8470	14600
Fuel Reprocessing and MO₂ Fabrication	None	None	Separate	Inte-grated	Inte-grated	Separate	Inte-grated	Inte-grated	in PuB	in PuB	in PuB	in PuB
Transportation:												
Heavy Isotopes MT	322	1612	530	530	332	2650	2650	1660	1730	2600	8650	11650
Fissile U MT	6.0	29.8	8.5	8.5	7.5	42.6	42.6	37.7	27.7	41.6	138	238
Fissile Pu MT	1.0	4.9	5.6	3.8	---	28.2	18.8	---	2.3	3.4	11.3	19.6
Case Designation	N4DO N4PO	N2OOO N2OPO	N4DA	N4DB	N4PC	NQDA	N2ODB	N2OPC	N4PD	N6PD	N2OPD	N2OPD'

Cases: O - Open cycle (store-away); A - Dispersed facilities; B - Integrated facilities;
C - Park facilities; D - PuB facilities

thorium depends upon the same initial fissile uranium-235
fuel supply necessary for the current LWR's. Thus, all
candidate fuel cycles, plutonium based or U-233 thorium
based, evolve around burning U-235 initially. This fuel
cycle relationship is illustrated in Figure 16. Uranium
ore from the mines, the starting point for the entire nuclear industry, is at the center of the chart. The initial
U. S. fissile material endowment in terms of uranium-235 in
uranium ore in the various U. S. mineral deposits was of
the order of 20,000 t of uranium-235 in ores costing
under $100 per t to mine. This 20,000 t of material
represents the starting point and the life-blood of the
entire U. S. nuclear industry to date. This consideration
and the future options that may be derived therefrom will
be discussed in some detail later, in Section IIC. The
mined uranium ore flow is traced through milling, conversion and enrichment, yielding the enriched fuel material
for both the light water reactor fuel cycle and potentially
for a mixture with thorium-bearing fuels for the LWBR or
HTGR fuel cycles. The fuel cycle flow paths are categorized
in Figure 17 according to their current industry implementation status and their vulnerability from the viewpoint of
safeguards for SNM,* a potentially significant factor bearing on the configuration of energy centers. The low enriched materials for the light water reactor fuel cycles
are not classified as a source of concern from the viewpoint
of safeguards. The more highly enriched materials are those
of interest for a thorium cycle. Perhaps more significant
are the plutonium fuel materials recovered from the processing of spent light water reactor fuel, and for use as a replacement for uranium-235 enrichment in either the light
water reactor fuel cycles or the thorium fuel cycles, and
for use in the LMFBR systems. At the present time, essentially all these SNM flow paths, together with the entire
thorium cycle and LMFBR fuel cycle, have not reached the
stage of commercial implementation. The vast majority of
industry fuel activity is on the initial flow path from the
mines to the milling, conversion, enrichment, and on through
to uranium fabrication and loading into light water reactors.

---

*   "Strategic SNM" is plutonium, uranium-233 and
        uranium enriched to greater that 20 percent in
        the uranium-235 isotope (materials from which
        a nuclear explosive can be fabricated).

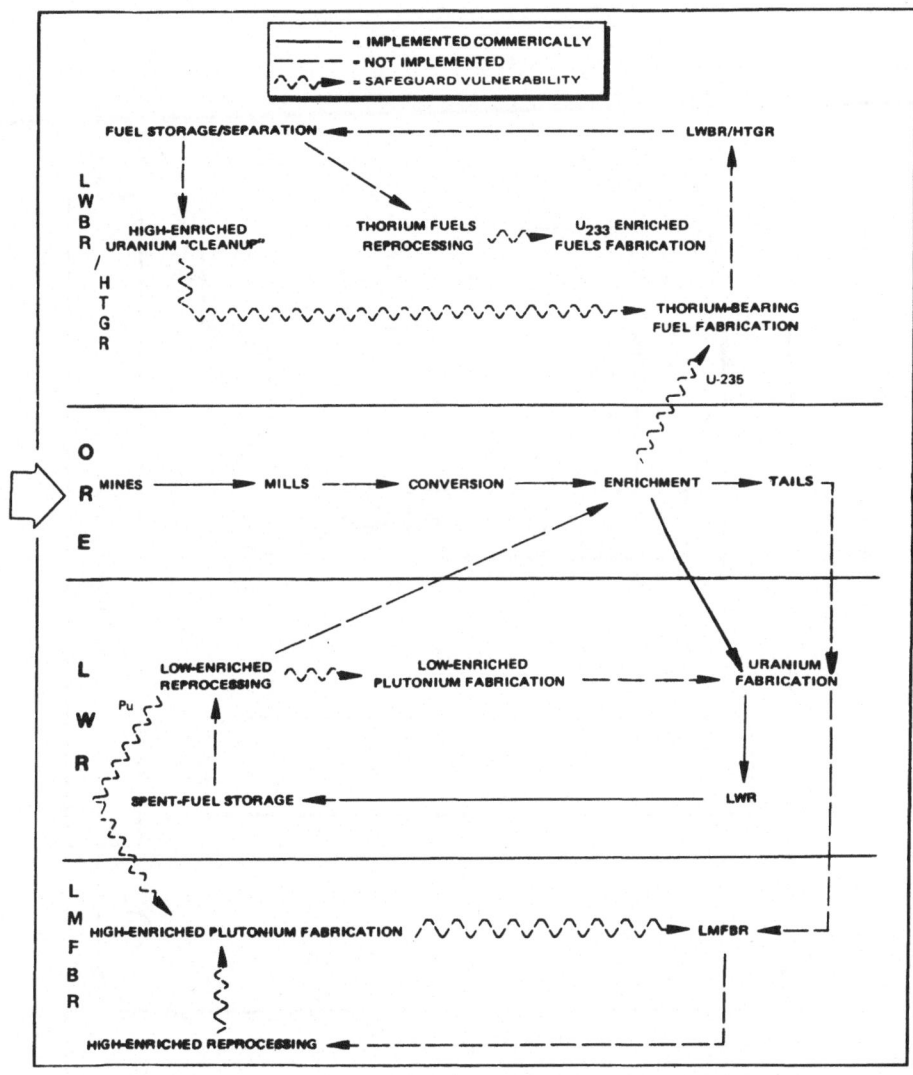

Figure 16

Anticipated Nuclear Fuel Cycle Relationships

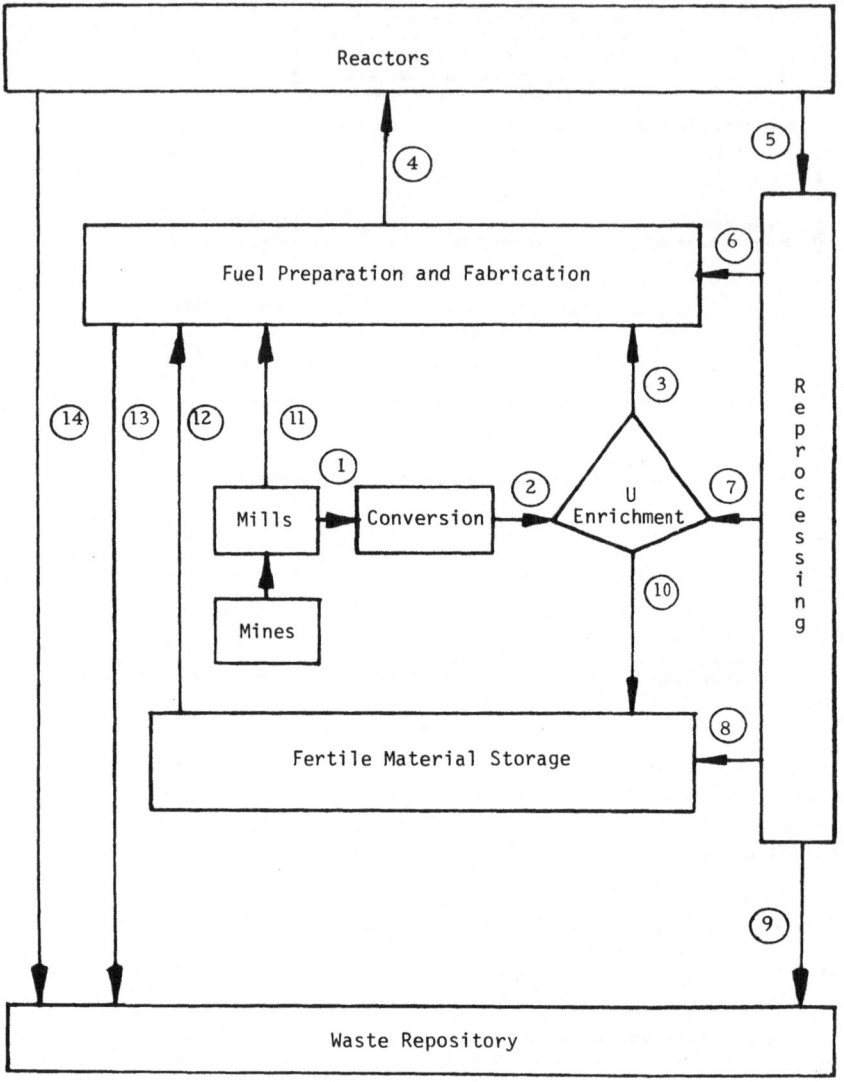

Figure 17

The Nuclear Fuel Cycle

Spent fuel coming from the LWR's continues to go into storage, either in fuel storage basins on site, or other temporary storage facilities primarily associated with existing
fuel-processing facilities. Indeed, while the nuclear industry has been in a period of rapid growth for the past
10 years or more, the back end of the fuel cycle that characteristically involves:

      a.   spent fuel shipment
      b.   spent fuel storage
      c.   spent fuel reprocessing
      d.   recovered uranium recycle to the
           enrichment plant
      e.   recovered plutonium fabrication
           and recycle

and all are as yet in a relatively early stage of commercial
development. While facilities have been constructed, few
have been operated, and their total capacity is less than
10% of that, which may be required by the industry over the
next decade or two.

In order to analyze the requirements of the various
fuel cycles, the generic fuel cycle flow paths and associated
nodes as shown in Figure 18 are identified. This, then can be
simplified for the individual cycle configurations, as shown
in Figure 18. Specific data consistent with these configurations, using the available published data on individual
reactor characteristics, are listed in Table VII. The BWR
and PWR systems are representative of currently-offered
commercial light water reactor systems. Typically, data
have been inferred from industrial sources such as fuel track
and from the LMFBR and LWBR environmental impact statements.
The information is shown for both once-through and for recycle conditions. Two levels of recycle are considered; the
first is recycle of uranium with storage of recovered plutonium; the second is recycle of both uranium and plutonium. A
further variant of the LWR systems is the consideration of
full plutonium-loading for the recycle cores. This case is
termed "the PuB, or plutonium burner case". It contrasts
with the more commonly-considered self-generated recycle
(SGR). This self-generated recycle implies a level of fuel
at equilibrium. As shown in Table VII, this represents
approximately one-third of the total fissile loading of a
light water reactor. It has been the basis for most of the

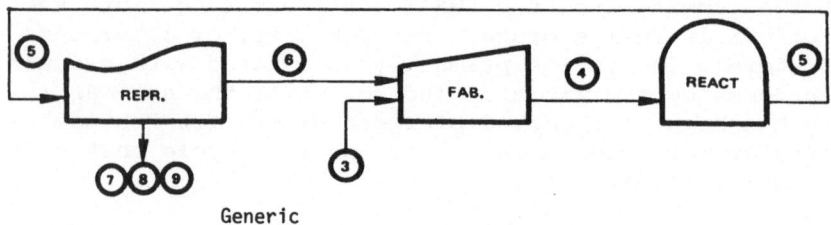

Generic

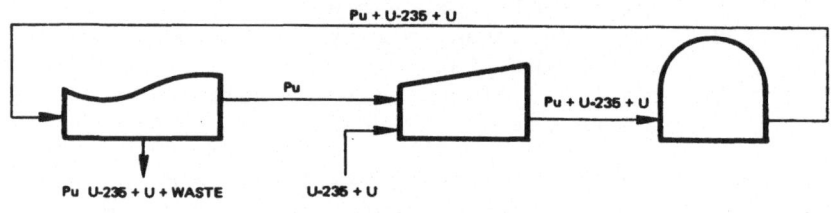

LWR with Pu Recycle

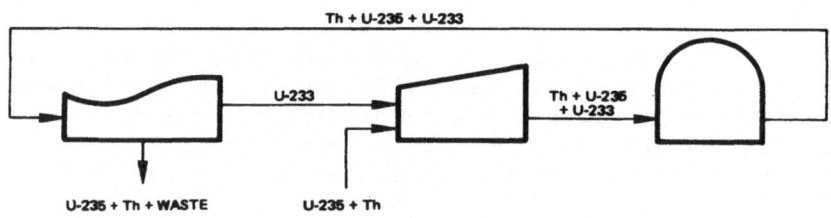

LWBR/HTGR with U-233 Recycle

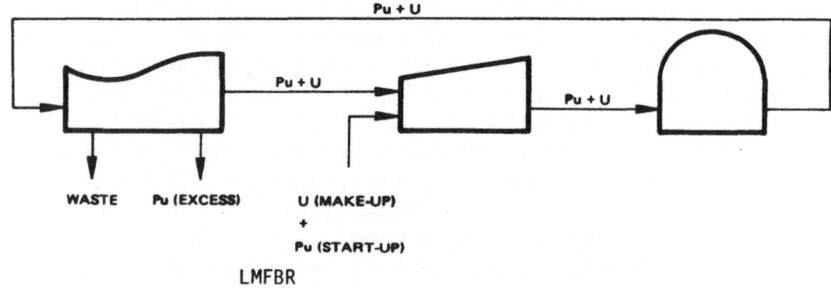

LMFBR

Figure 18

Nuclear Fuel Cycles

TABLE VII

TYPICAL FUEL-CYCLE DATA PER 1000 MWe OUPUT

(80% Capacity Factor; 0.25 Percent U-tails)

Reactor Coolant	Light Water						Inert Gas		Liquid Metal	Light Water
Reactor Type	BWR			PWR			HTGR		FBR	LWBR
Initial Loading Requirements:										
Heavy Isotopes MT	111 (U)			80 (U)		80 (U)	32 (TH)		~50 (U)	186. (TH)
Fissile Isotope kg	2246 (2.03 W/O U-235)		2884 Pu-f	1694 (2.11 W/O U-235)		2240 Pu-f	1387 (93 W/O U-235)		~2500 (Pu-f)	5.4 (U)
U3O8 ST	554		--	420		--	389		0	4300 U-238
Separative Work Units MT	219		--	170		--	322		0	0
Equilibrium Fuel Cycle	Open Cycle	SGR	PuB	Open Cycle	SGR	PuB	Open Cycle	Recycle U-233	Breed Pu	Breed U-233
Annual Requirements:										
Heavy Isotopes MT	31	31	31	28	28	28	7.6	7.6	20	106 (Th)
Fissile Makeup kg	834 (U-235)	559 (U-235)	349 (Pu-f)	904 (U-235)	632 (U-235)	437 (Pu-f)	578 (U-235)	334 (U-235)	None	None
Fissile Re-cycled kg	None	353 (Pu-f)	722 (Pu-f)	None	360 (Pu-f)	760 (Pu-f)	None	192 (U-233)	1100 (Pu-f)	2450
Fissile Sold kg	240 (U-235) 171 (Pu-f)	161 (U-235)	None	242 (U-235) 188 (Pu-f)	189 (U-235)	None	34 (U-235) 172 (U-233)	26 (U-235)	60 (Pu-f)	0
U O Ore (No U Recovery) ST	175 (223)	117*	None	184 (235)	123*	None	153	86*	None	None
Separative Work Units MT	102 (105)	68*	None	119 (124)	80*	None	127	72*	None	None

* No adjustment for reactivity reduction effect of U-236 in recycled U

+ Mo. first cycle

plutonium recycle evaluations by industry up to the present
time. Recently, however, with the greater interest in nuclear parks, integrated fuel cycle facilities and potential
for reduction of SNM risks, more detailed consideration has
been given to the plutonium burner concept, in which recovered plutonium from LWR discharged fuel would be retained
on the fuel reprocessing facility site, and preferentially
fed into some dedicated on-site plutonium burner reactors.
The PuB cases thus correspond to the maximum plutonium loading for LWR's. These data represent an extrapolation from
the SGR cases. The comparative off-site fissile material
shipments for the case of LWR PuB's versus other recycle
and storeaway fuel cycles, as previously defined in Figure
14, are shown in Table VIII.

C.   Plutonium Burner Considerations

     Some significant studies have been conducted over the
past few years of the ability to operate LWR's with a full
plutonium loading (15); that is, with fuel material consisting of natural or depleted uranium (tails) and enriched
with plutonium to the required level of 3.5%. These studies
indicate such an LWR to be practical, and that existing nuclear systems, with some modification of the fuel design
and control clusters, can accommodate a full plutonium
core loading. Some questions remain regarding transition
cores in which a gradual changeover from uranium to plutonium enriched fuel bundles is envisaged, but this does not
need to be a constraint on the proposed plutonium burner
energy centers. The uncertainties relate primarily to a
possible power distribution mismatch between plutonium and
uranium enriched zones of the core. For a plutonium burner
in an energy center, however, there well can be an excess
of plutonium available from the very start of plant operations under what is termed the preferred or reversed priorities mode of development.

     The sequence for the reversed priorities mode of devel-
opment of a PuB energy center is:

          Phase 1   fuel storage facility construction
          Phase 2   fuel reprocessing plant construction
          Phase 3   plutonium burner reactor construction,
                     together with associated plutonium
                     fuel fabrication

TABLE VIII

CHARACTERIZATION OF OFF-SITE FUEL SHIPMENTS PER YEAR FOR 26,000 MWe INSTALLED

Path on Fig. 17	Nature of Shipment	Case O Heavy Isotopes MT	Fissile U MT	Fissile Pu MT	Case A Heavy Isotopes MT	Fissile U MT	Fissile Pu MT	Case B Heavy Isotopes MT	Fissile U MT	Fissile Pu MT	Case C Heavy Isotopes MT	Fissile U MT	Fissile Pu MT	Case D Heavy Isotopes MT	Fissile U MT	Fissile Pu MT
3	Enriched UF$_6$	728	23.5	---	728	16.4	---	728	16.4	---	728	16.4	---	649	16.4	---
4	Fabricated Mixed Oxide Fuel	---	---	---	262	0	9.4	262	0	9.4	---	---	---	---	---	---
4	Fabricated Uranium Fuel	728	23.5	---	466	16.4	---	466	16.4	---	466	16.4	---	649	16.4	---
5	Spent Fuel	728	6.3	4.9	728	4.9	9.4	728	4.9	9.4	---	---	---	649	4.4	3.4
6	Recovered Plutonium (SNM)	---	---	---	---	---	9.4	---	---	---	---	---	---	---	---	---
7	Recovered Uranium	---	---	---	466	4.9	---	466	4.9	---	466	4.9	---	649	4.4	---

A number of studies of this PuB energy center construction and startup transient have been made. They show the
feasibility of achieving levelized loadings for the various
on-site facilities through use of spent fuel and recovered
plutonium storage options. The general characteristics of
the startup transients for such PuB energy centers are traced
in the analyses of four-unit and twenty-unit centers illustrated in Figures 19 and 20. The variation of several key
energy center parameters, with time, is traced. Time zero
is defined as the startup of the first PuB reactor on the
energy site. For the preceding 10-15 years, LWR's are shown
to be installed and operating at dispersed sites and shipping spent fuel to the storage facilities at the energy
center.

This shipment of spent fuel to storage at the energy
center site is shown to continue for nearly 10 years prior
to startup of the on-site reprocessing facility. Thus,
spent fuel inventory goes through a well-defined peak or
double peak for the larger two-state facility startup. At
the start of reprocessing, the fuel is open-cycle discharged
fuel from off-site LWR's; only after more than five years of
operation does the facility start to process PuB fuels with
their higher plutonium concentrations.

The recovered plutonium fuel inventory at the energy
center site itself goes through similar maxima a few years
after start of reprocessing at the point where the plutonium
fuel fabrication facility is coming on line to provide startup and reload cases for the PuB reactors at the energy center
site. Again, due to the large initial core loading demands
for the PuB units, assumed here to start up at one-year intervals, there will be need for some advance production of
plutonium fuels and inventory thereof. These fabricated
fuel inventory transients are relatively modest and are not
shown in the figures.

The resultant total U. S. circumstance for a PuB energy
center program in which the total LWR installation amounts
to approximately 400 GW by the year 2000, some 60-80 GW of
which are at PuB energy center sites, is shown in Figure 21.

Two rates of PuB system installation at the energy centers are shown in Figure 21. The first (dashed line) case
corresponds to a steady plutonium-limited rate of introduction of PuB units starting in 1985 (power plants on line).

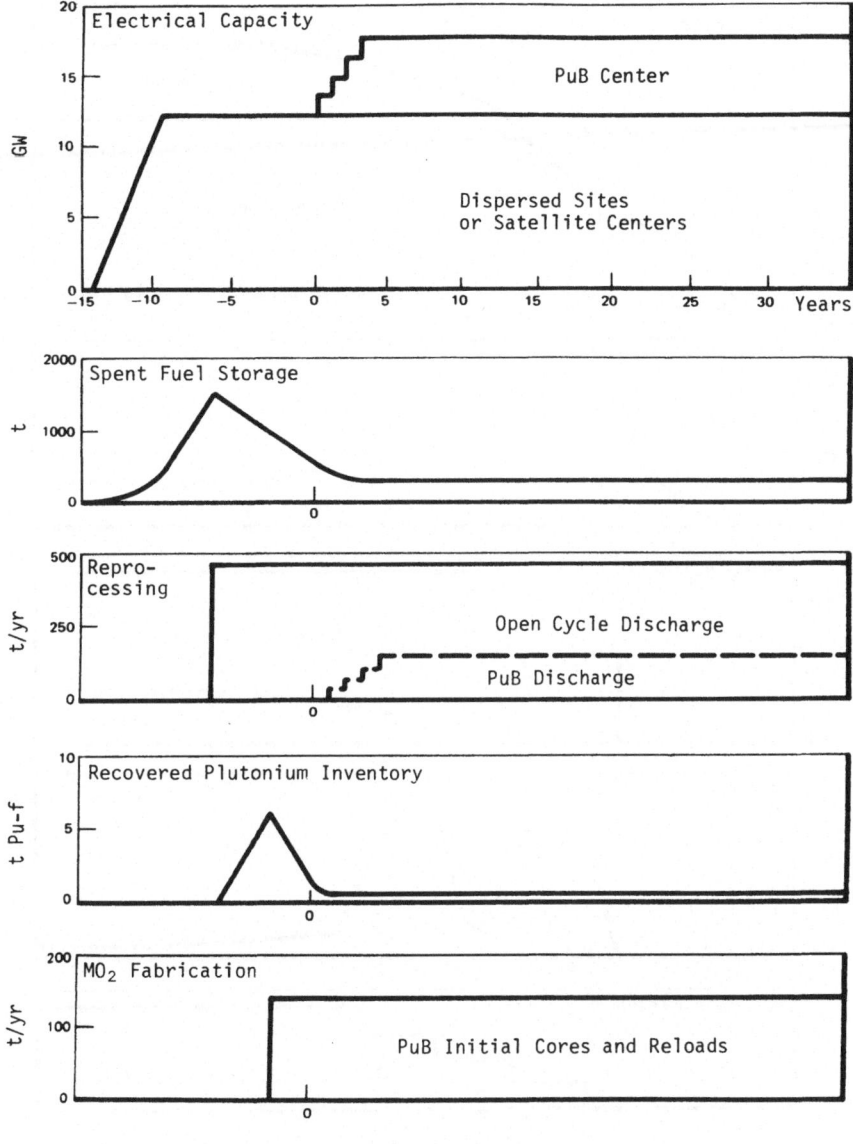

Figure 19

Fuel Cycle Transients in a Four-Unit PuB Center

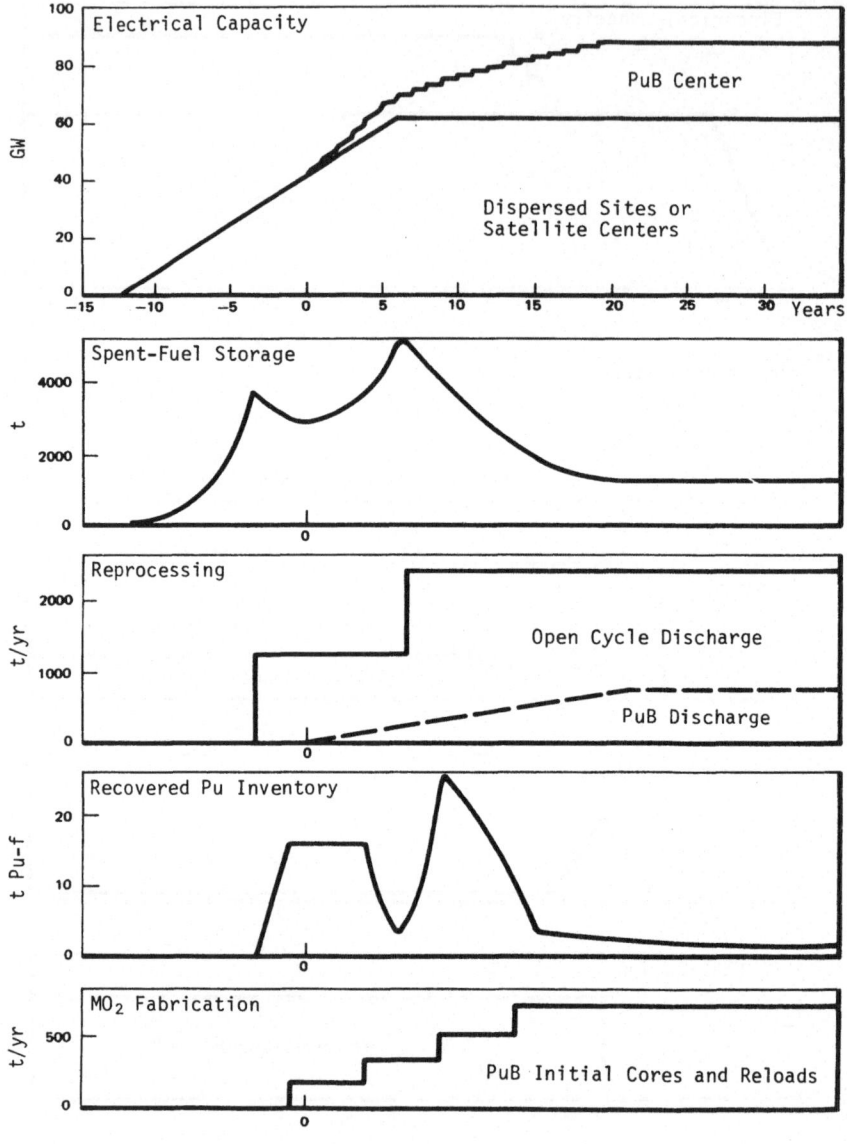

Figure 20

Fuel Cycle Transients in a Twenty-Unit PuB Center

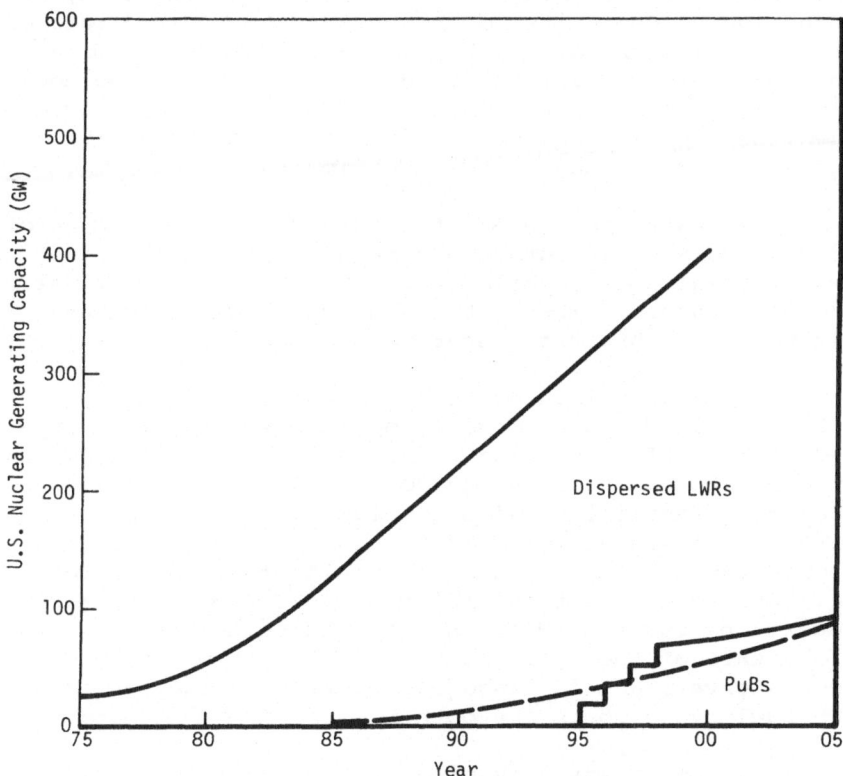

1999 Cumulated Flows and Inventories, tonnes Fissile Plutonium

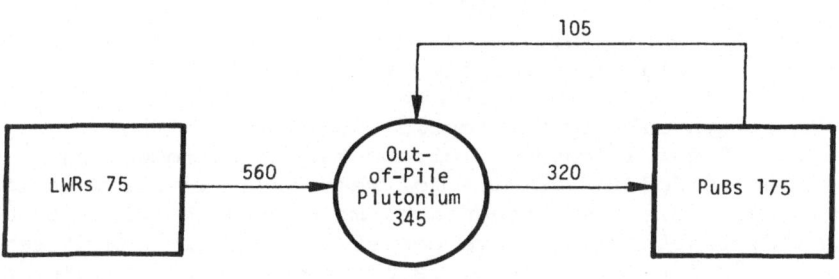

Figure 21

Plutonium Burner (PuB) Scenario

The second case corresponds to a further ten-year delay of
introduction of PuB units to 1995, and a more rapid rate of
introduction (essentially all new LWR commitments) for a
period immediately thereafter. Both cases then trend toward
an installed capacity of PuB units of approximately 100 GW
by the year 2005 if no other major demand for plutonium
(LMFBR's) has burgeoned in the meantime.

In the lower flow sheet of Figure 21, by 1999, much of
the recovered Pu (640 out of 690 t) has gone toward creating
the PuB initial cores, while only 640-560 = 80 t of fissile
Pu has been burned. Thereafter, the PuB's will increase
slowly until the breeders enter the picture.

One question raised by this concept of PuB energy
centers is that of the economic impact resulting from prolonged storage of discharged light water reactor fuel, pending the construction of energy centers with their suggested
reprocessing facilities and plutonium recycle capability.
Originally, it was anticipated that significant fuel cycle
economies would be achieved as a result of plutonium recycle. Fuel cycle costs of 13¢ to 15¢/TJ (10^9 joules) were
forecast (16) for the LWR's on this plutonium recycle basis.
Many elements of the fuel cycle cost have, however, changed
in the interval, and in consequence, recent study estimates
of LWR fuel cycle costs are both:

1. substantially higher, i.e., 30¢ to 40¢/TJ

and

2. essentially equal for storeaway and for
    plutonium recycle situations.

These typical LWR fuel cycle cost data are shown in
Table IX. Under a storeaway circumstance, reprocessing
costs are avoided and credits for recovered fissile material
are not available; in consequence, more uranium ore must be
purchased, together with more conversion and enrichment services. Fabrication, however, is not burdened by the presence
of plutonium, but storage provisions have to be made for
relatively long terms for spent fuel. A further consideration is that of plutonium recovery for sales to other reactor systems. The computed plutonium value, based on this
computation, is $6 per gram of fissile content, and by defin-

ition under that circumstance, the light water reactor fuel
costs again turn out to be 32¢/TJ. Reference to the aforementioned fuel cycle cost data of some years back indicates
that a plutonium value in water reactors of that time, of
approximately $7 a gram. The cost data on which these fuel
cycle computations are based are shown in Table X; they are
derived in part from the reprocessing and mixed-oxide fabrication cost data of Figures 11 and 12, which correspond to
present scale facility operations. One further remaining
question with regard to the concept of plutonium-burning
energy centers is that of the wisdom of the nation's embark-
ing on a program whereby the fissile material resource -
plutonium - generated in current light water reactors, is
progressively burned up over the next several decades, as
opposed to the originally-planned concept of feeding it
directly to liquid metal breeder reactors, in which it would
have potentially a higher fissile value. The reason for the
suggested introduction of plutonium burner energy centers is
the delay in the program for the LMFBR commercialization. The
fact remains, however, that fissile material represents a
depletable resource in the absence of an LMFBR. Too long a
continuance of a program of plutonium burners without breeder
reactors could result in a difficult period ahead for the
nuclear industry due to insufficient availability of fissile
material. This, of course, depends in part on the performance characteristics of the breeder reactors and their
ability both to generate new fissile material for their
growth needs, and if necessary, to supply additional fissile
material back to the LWR's during the period of wind-down of
their operations, potentially somewhere through the period
2025-2050.

Some typical data for LMFBR doubling time characteristics are shown in Figure 22. The data were prepared using
the following "typical" parameters of a 1.2 GW LMFBR:

   1.   core average burnup ∽ 90 MWd/kg
   2.   core average specific power = 155 kW/kg U + Pu
   3.   reactor average fissile specific power
        ∽ 1100 KW/kg Pu-239+241
   4.   core fuel residence time = 24 months
   5.   nuclear power plant capacity factor = 0.80
   6.   net efficiency power generation cycle = 0.41

TABLE IX
LWR FUEL CYCLE COSTS WITH AND WITHOUT Pu RECYCLE

Cost Component	CONTRIBUTION TO COST ($¢/1.055 \times 10^9$ J)*			
	Store Away	Plutonium Recycle	Plutonium Sales	
Fresh Fuel:				
Uranium Ore	12	8	12	6**
Conversion &				
Enrichment	10	7	10	6**
Plutonium	--	2	--	-
Fabrication	6	11	6	2**
Spent Fuel:				
Storage	4	-	--	-
Shipment	-			
Reprocessing	-	6	6	1**
Waste Storage				
Recovered Fissile				
Credits	-	(2)	(2)	(2)**
Totals:	32	32	32	13**

*   1975 dollars, no escalation of costs.
    Other assumptions: Equilibrium core operation at 80%
    capacity factor, 0.25 W/O tails, 8% interest and
    present worth factor, 16% fixed charge rate, unescalated.

** Using 1971 estimated costs[16], Pu value of $7/g versus
    reference 5 assumed $6/g fissile.

TABLE X
LWR FUEL CYCLE COST ASSUMPTIONS

Cost Component		Values Used (1975$)	
		NSF Study	1971 Estimate
Fresh Fuel:			
Uranium Ore	$/lbU$_3O_8$	13	7
Conversion	$/kgU	3	2
Enrichment	S/kgSWU	42.5	26
Plutonium	$/g	3 to 6*	7
	$/kgUO$_2$	112	
Fabrication			42
	$/kgMO$_2$	350	
Spent Fuel:			
Storage	$/kgU/YR	10	–
Transportation	$/kgU	10 ⎫	
Reprocessing	S/kgU	160 ⎬	45
Waste Handling	$/kgU	40 ⎭	

*   Indifference values.

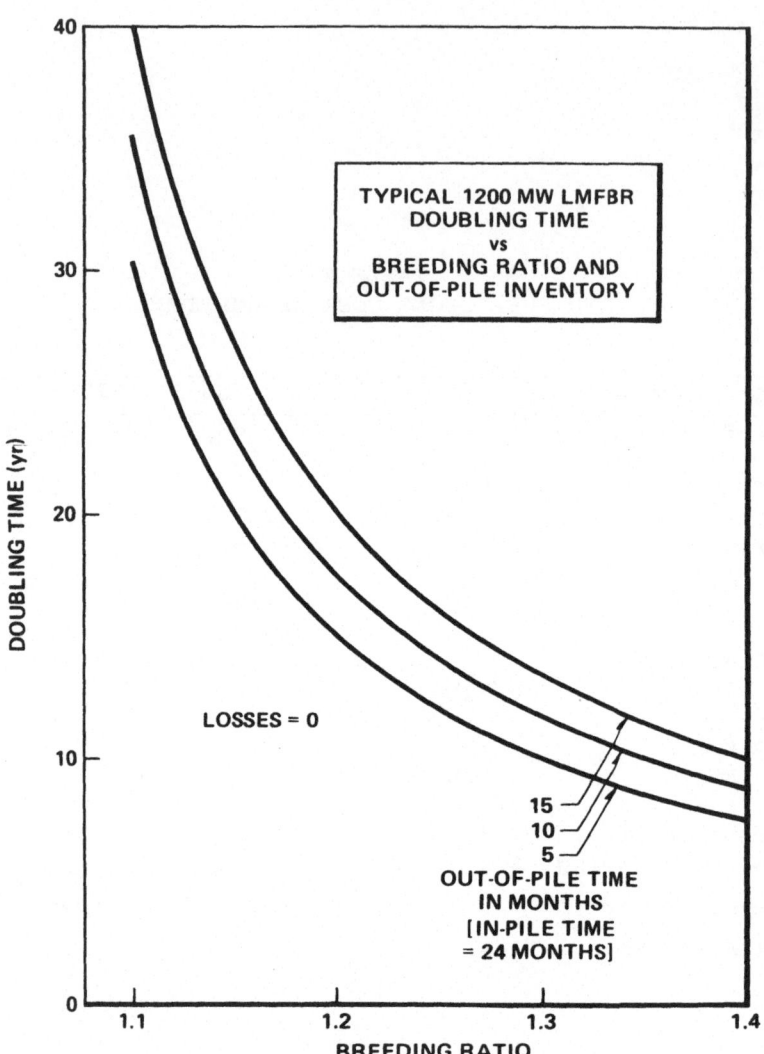

Figure 22

LMFBR Doubling Time Versus Breeding Ratio
and Out-of-Pile Inventory

The significance of out-of-pile time to the overall breeding performance of an LMFBR is clearly shown. Colocation of LMFBR's and fuel facilities at an energy center can help this performance aspect.

D. The Plutonium Resource Utilization Question

The basic concept of plutonium burning, while potentially appealing from the viewpoint of avoiding accumulation in storage of large quantities of actinide materials should be considered with caution relative to overall national interest of optimum fissile material resource utilization.* This point has come into clearer perspective within recent months. The updated perspective relates, as noted earlier, to three trends:

1. the revision of the LMFBR schedule for commercialization

2. the increased level of uncertainties on available U. S. uranium resources

3. uncertainties in the development of improved converter systems

To this point, an analysis has been conducted of uranium utilization in the U. S. nuclear economy with and without plutonium burner implementation. In fact, implementation of plutonium burners is believed little different from plutonium recycle as assumed in GESMO, at least from a fissile material resources conservation viewpoint. The basis for the analysis outlined here is a forecast derived from data in the recent ERDA planning document (ERDA 48). The recommended scenario within that document (Scenario V) has been used. Even so, the analysis based on this forecast implies a significant problem ahead on fissile material resource availability, particularly if plutonium burning in LWR's

---

* Also, more recent analyses are suggesting that multiple recycle of plutonium in a thermal spectrum may result in higher levels of buildup of less desirable, long-lived actinides than those present in first cycle plutonium discharged from present LWR's.

were to be implemented as suggested in the above-described
LWR plutonium burner concept.

The nuclear installation rate comparable with ERDA 48,
Scenario V is shown in the center box of Figure 23.

This forecast has then been extended for the purpose
of the present analyses of energy centers to something more
comparable with the expected lifetime of an energy center;
i.e., out beyond 2000, to the year 2025. The extrapolation
is at a linear rate, as shown by the dashed portion of the
curve. It is noteworthy that even so, the suggested installed nuclear generation by 2025 is no higher than that
for the year 2000 as assumed in many pre-1975 industry forecasts. The specific nuclear installation rate assumed for
the ERDA draft, Generic Environmental Statement for Mixed
Oxide Fuels (GESMO) analysis, for example, is also shown in
Figure 23.

Two variants of Scenario V are also shown, both with
similar total nuclear generation, with respectively no LWR
predominance through some improved fuel cycle such as suggested by ERDA's light water breeder reactor (LWBR) program.
In all cases, the rates of installation have been continued
beyond the year 2000 at a linear rate, i.e., no further
growth in the supporting industry beyond essentially present
levels, save for the breeders. In the case of the breeders,
their maximum growth rates are dictated by fissile material
availability in the '90's. These installation rates imply a
nuclear energy generating capacity of approximately 460 GW
by the year 2000. The fissile material resource situation
implied by these conservative alternates has been computed
and is shown in Figure 24. The ordinate of this figure is
in terms of metric tons of fissile material available either
in the U. S. uranium resource of minable ore bodies or as
fissile material in various locations within the nuclear
economy. The significant point is that, within the first
couple of decades of the 21st century, essentially all the
U-235 resource is consumed. Indeed, this is generally consistent with the suggestion of ERDA 48 which presumes plutonium burning in light water reactors. The interesting
point is how little fissile material (plutonium) stockpile
exists at any point while this rapid depletion of the total
resources is in process. There is apparently little flexibility, little margin for error. Indeed, it appears to be
a situation in which nuclear options could rapidly close out

for the industry. The accompanying figure for Scenario V*
is, in some respects, more alarming in that U-235 depletion,
as might be expected, occurs even more rapidly. There is,
however, under this circumstance, a significant stockpile
of fissile material (plutonium) building up. Indeed, the
total fissile material in the economy trends to a level in
excess of 7,000 t over this period in which recoverable
uranium ore in the ground trends toward exhaustion. An
illuminating overall mass and energy balance for this suggested 50 year span of the nuclear program might be approximated by:

<u>Pre-1975</u>   <u>Post-2025</u>

20,000 t U-235 $\rightarrow$ 10^{10} MWD + 7500 t   + 7000 t
(ore in ground) (elec. gen.)   (fissile material in
                                     system)   (U-235 in tails)

Under this Scenario V* situation, 50 years from now,
breeders presumably would be generating sufficient fissile
material to support the continuance of the suggested relatively modest (16 GW/y) linear rate of additions. Continuing fuel supplies for the ~50% installed capacity of thermal reactors is, from then on, less certain. Their fuel
demands would tend to compete with those for the additional
breeder reactor installations. This circumstance, a significant stockpile of fissile material (plutonium) in existence,
and indeed the total fissile material in the economy, trends
to in excess of 7,000 t. A further set of estimates for
Scenario V** is shown, and it suggests that given delayed
large-scale LMFBR introduction to beyond the year 2000, further deterioration of the fissile material resource picture
might be avoided if indeed improved performance LWR cores
can be achieved, such as suggested by ERDA's LWBR program.
Indeed, this might provide some additional options for future
direction for LWR industry development.

Many other variants for fuel cycles and parks can be
visualized beyond those described here. Further evaluation
is undoubtedly appropriate, and many questions remain as to
the real worth of stockpiling versus burning plutonium before the LMFBR arrives. Such questions bear directly on the
matter of a preferred set of nuclear fuel cycles and their
further development and integration in any proposed energy
parks program.

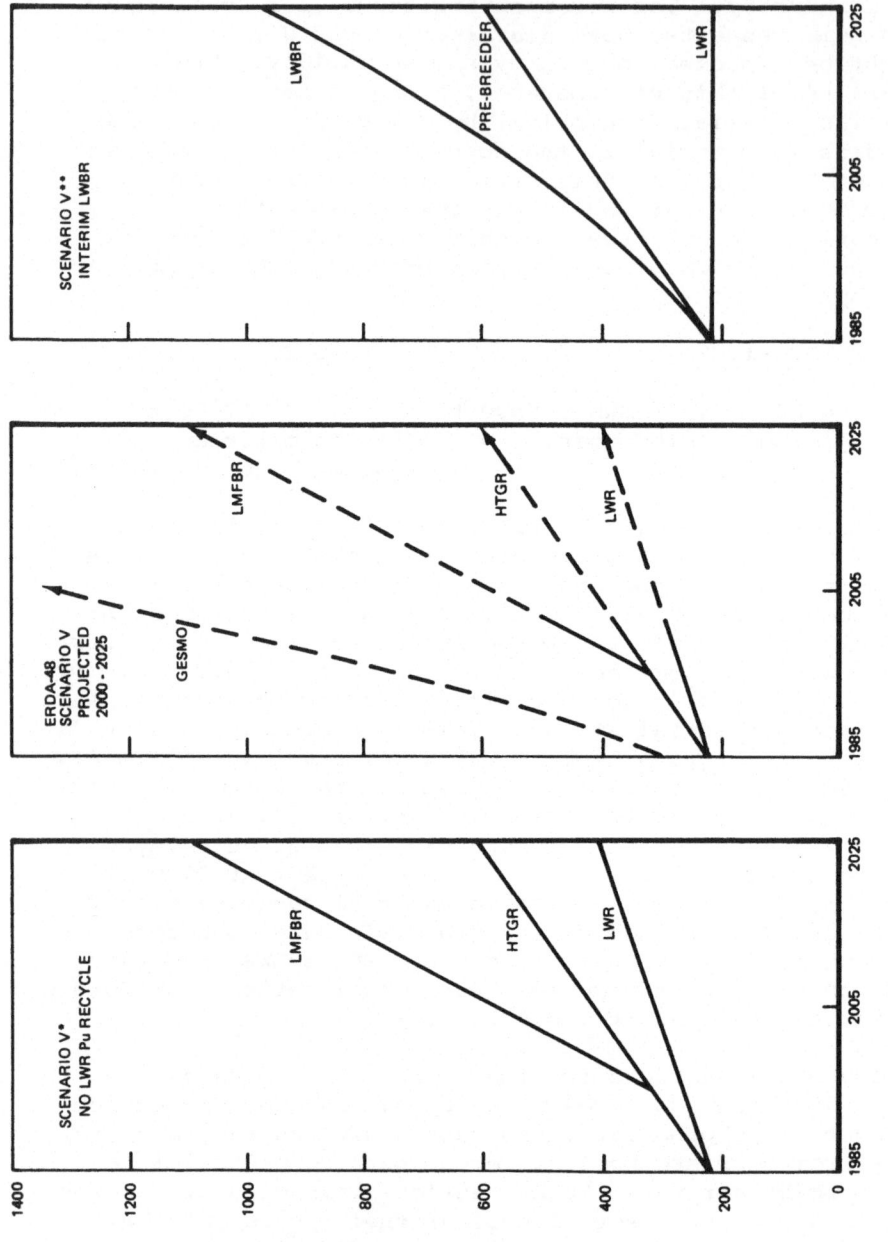

Figure 23

Assumptions of Nuclear Power Growth, 1985–2025

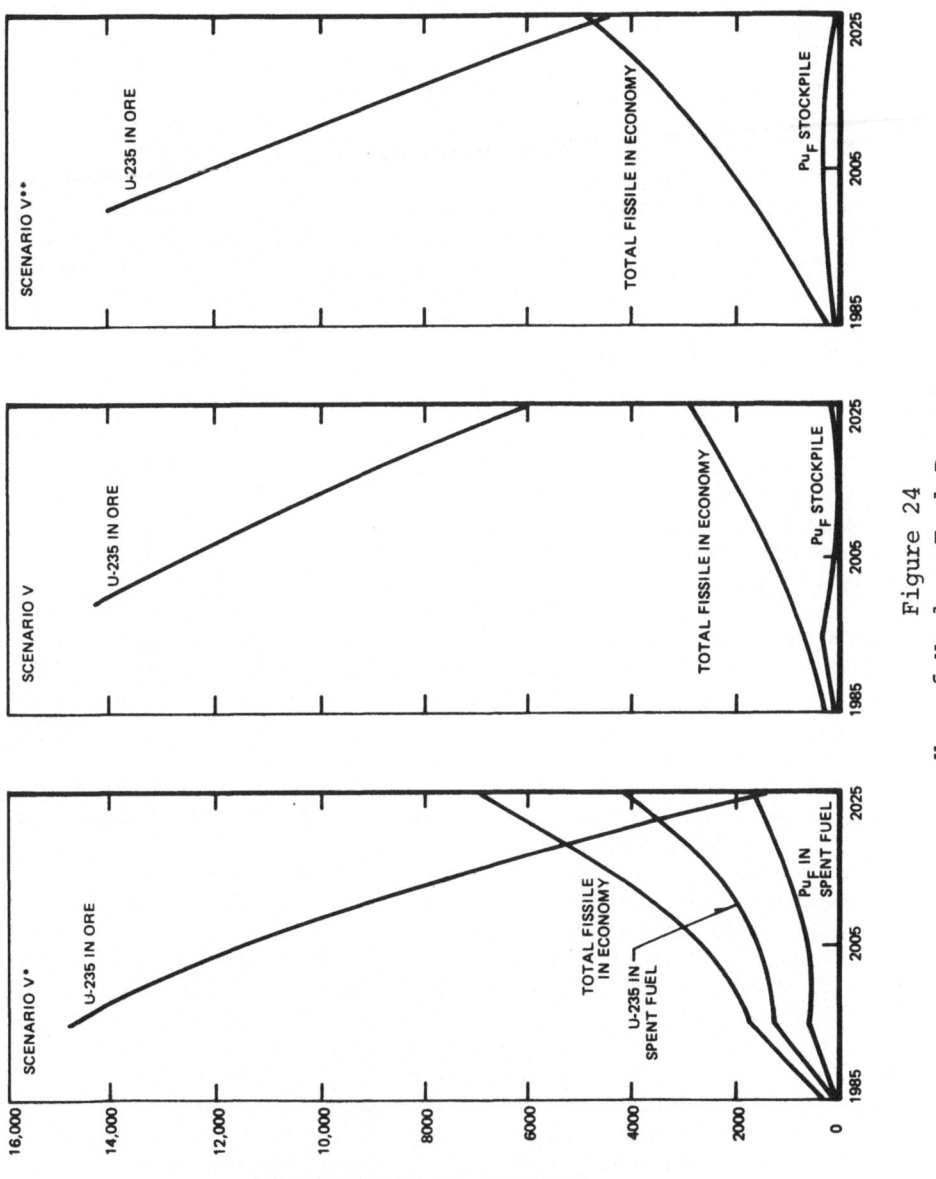

Figure 24
Use of Nuclear Fuel Resources

III. MODULAR CONSTRUCTION

A.   Conceptual

The construction cost experiences of the past decade,
for nuclear power plants, is typified by the trends of Figure 25. Progressive increases in direct construction cost,
added items of scope associated with safety systems, increased schedule length and escalation have all contributed
to this trend. Modular construction, which provides scope
for reducing congestion at the job site and for moving a
portion of the construction labor from the field to a shop
environment, has been investigated as a means of combating
this mounting cost trend.

Limited levels of modulatization are employed in present
LWR construction, but it still turns out that some 40% of the
total plant construction cost relates to on-site material and
labor. A further 45% of the cost is related typically to shop
manufactured components and modules shipped to the site. The
bulk of remaining cost before escalation and interest during
construction is for engineering. This illustrated in Figure
26. The potential opportunity presented by nuclear energy
center development is to extend substantially the use of
modular construction, thus moving more labor from the field
into a shop environment to improve productivity and to reduce schedules.

A nuclear energy center offers the potential for many
units to be constructed to the same stnadardized design. It
permits consideration of much larger-sized modules and of
substantial construction facility investment at the site.
This may take the form of necessary major lifting facilities
for large modules and an on-site factory in which to construct them. Various concepts are being explored (17,18),
based on these principles, and the schematic site layout for
one of these, based on the use of a 1,000 t gantry crane, is
shown in Figure 27. This gantry crane, capable of spanning
the 130 m width of the total power plant, sets the whole
character of the proposed modular design concept. Its 1,000 t
lifting capacity establishes an upper limit to module size
(weight). The sequencing of plant construction and the overall plant layouts are significantly effected by consideration
of the heavy lifts, their timing, duration, clearances, etc.
A proposed modular assembly sequence for one portion of a

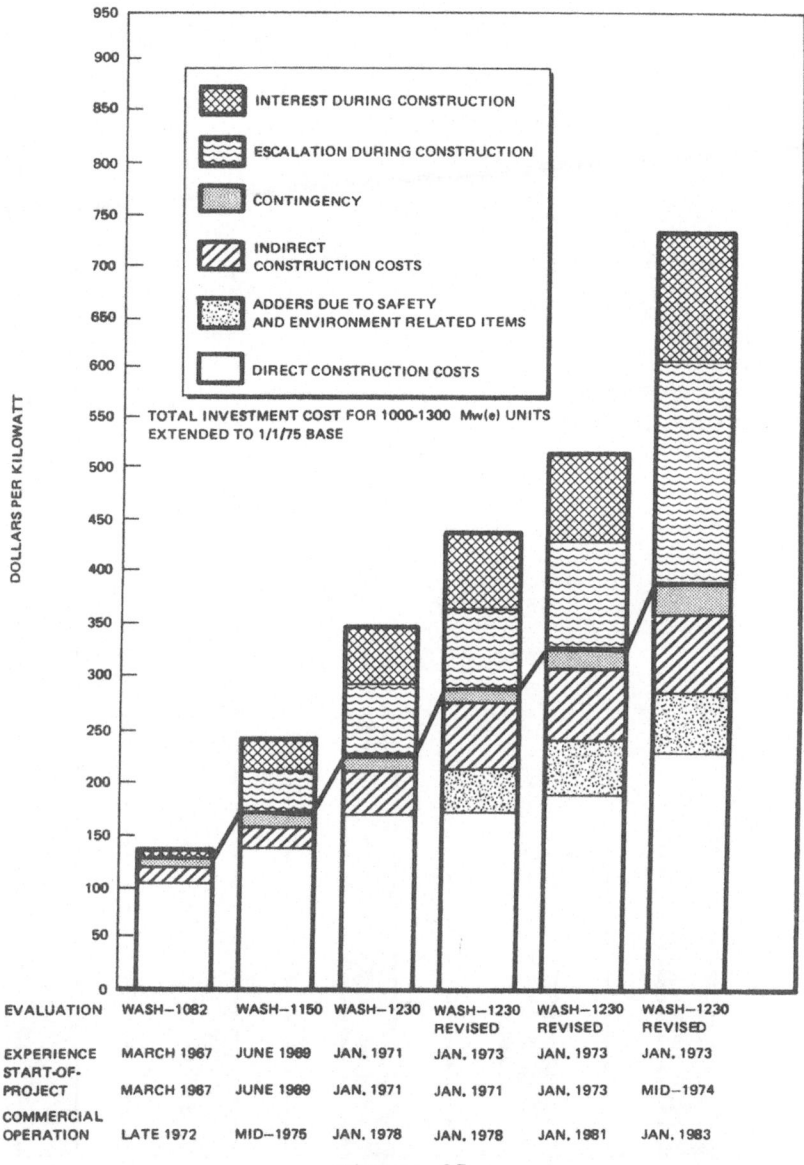

Figure 25

Relationship of Nuclear Plant Costs
To Trend of Prior Estimates

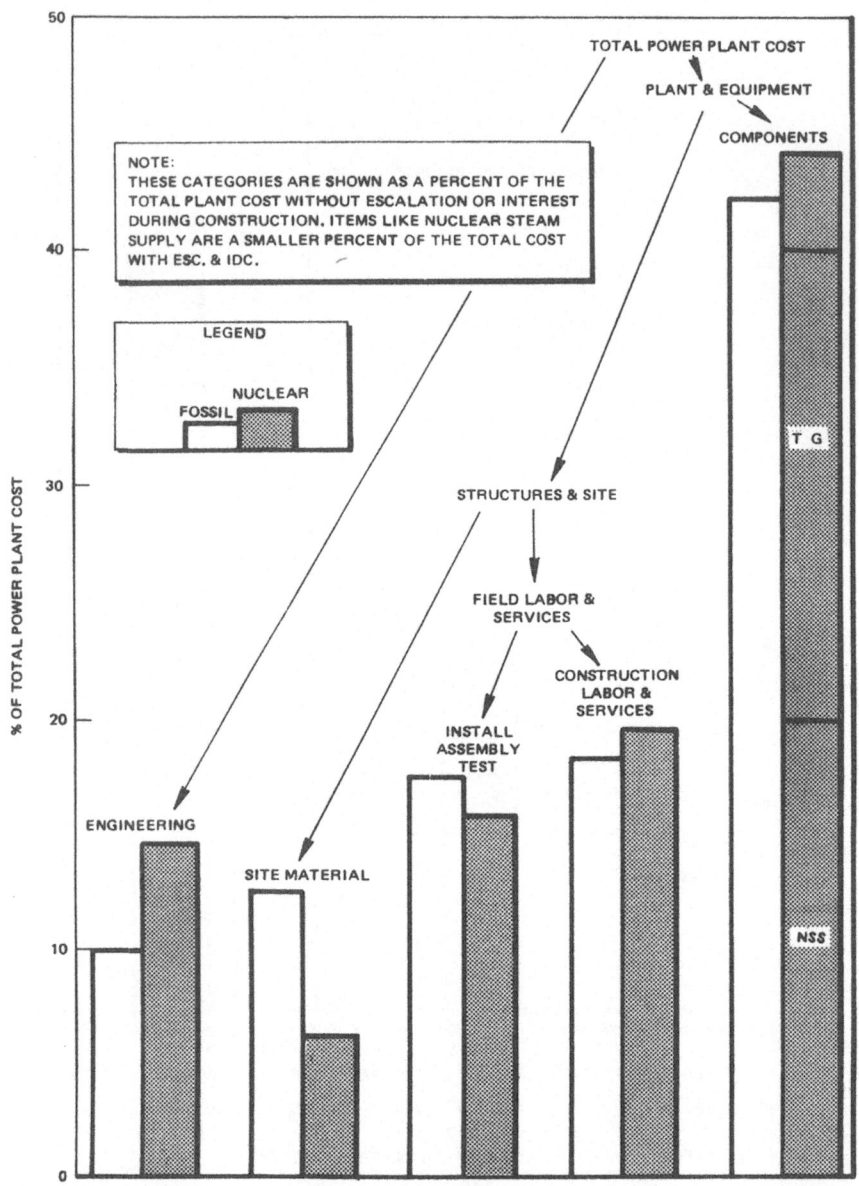

Figure 26

Cost of Typical Nuclear Power Plant

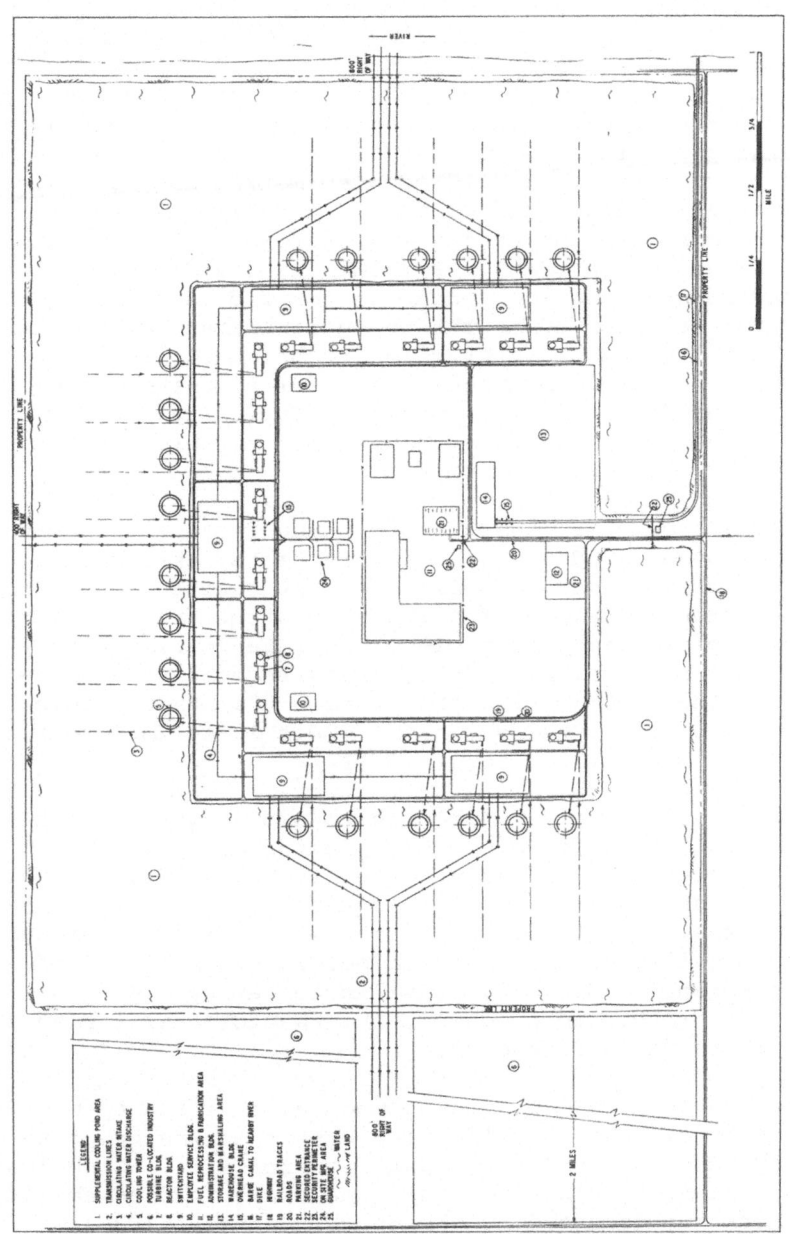

Figure 27

Energy Center Site Layout

reactor building construction is shown in Figure 28. In
this illustration, steel form work is proposed for the
major structural modules. These structural forms are
assembled as a module with rebar in place, and then the
module is lifted to its final location and the structural
and shielding concrete poured. This sequence is then continued for the several hundred modules making up the entire
power plant and equipment complex. Scheduling and module
sizing and configuration become key factors in the potential
for success of such a plant. Preliminary estimates suggest
that standard construction schedules for a dual unit plant
might be reduced from 60 months to approximately 44 months
through use of modular construction techniques. This schedule defines the period between receipt of a construction
permit and the start of fuel loading. Typically, an additional 24 months of licensing activities may precede this
receipt of a construction permit.

Alternate modular construction concepts also have been
explored in other studies. In many respects, the pouring of
complete concrete floor and wall slabs at a site factory location and transportation of a finished structural module to
its final location is preferable. Concrete pouring with the
plant under construction is drastically reduced, as is the
form work. The problem with this approach is that module
weight becomes large, and given a 1,000 t lift limit more,
smaller-sized modules are required (17).

A further alternative directed toward larger modules
with attendant scope for equipment pre-installation, together with pre-installed service interconnections, is that
based on marine construction techniques. In this case, steel
is used as the full structural member with internal compart-
mentation provided as necessary to permit water filling for
shielding purposes following final installation (18). An
illustration of this approach for the reactor auxiliaries
building is shown in Figure 29.

A yet further variant of the modular approach to energy
center construction is the limiting case that was proposed
by Offshore Nuclear Systems Corporation for floating nuclear
plant construction. In this case, the heavy lift capability
is, in part, supplemented by the concept of floating the
plant (barge) at various stages of construction to the new
work station. The concept of applying barge-type construction

Step 1. Install and anchor the reactor pedestal-weir-wall steel frame (Module 1). Erect formwork and install reinforcement for shield building wall (below grade).

STARTING POINT OF SEQUENCE

A. Reactor building foundation mat poured (450 m³)
B. Lowest course of containment in place and anchored (400 t)
C. Floor liner plates
D. Structure foundation plates } Cast in place (170 t)
E. Reinforcing bar dowels for shield building walls }

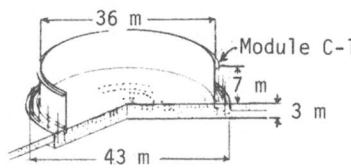

In Place Weight - 11970 t

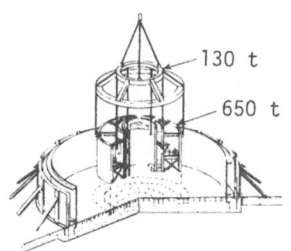

Step 2. Pump in and compact all concrete for the reactor pedestal-weir wall and drywell floors, and for shield building foundation wall. Prepare reactor vessel assembly (Module 2) for installation.

Step 3. Install and anchor reactor vessel assembly (Module 2) and prepare the reactor shield wall (Module 3) for installation.
Begin installation of substructure modules for adjacent buildings.

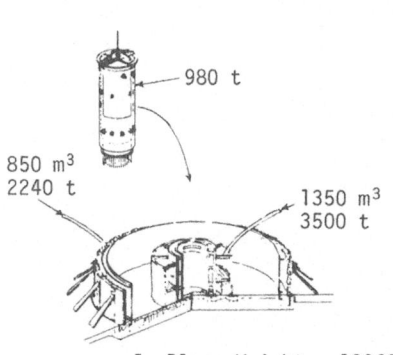

In Place Weight - 18360 t

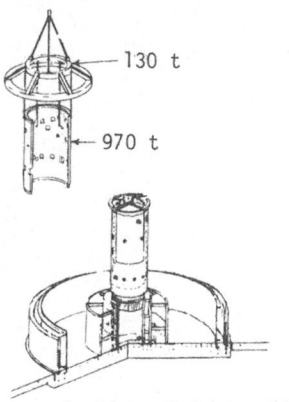

In Place Weight - 19340 t

Figure 28

Typical Modular Construction Sequence Start

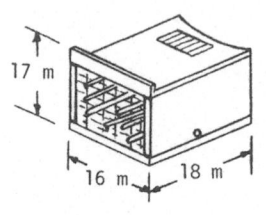

Structure	645 t
MSL Piping & Framing	90 t
Other Piping	100 t
Total	835 t
Water added at the site	2650 t

Module AB-3

Steam Tunnel and Piping

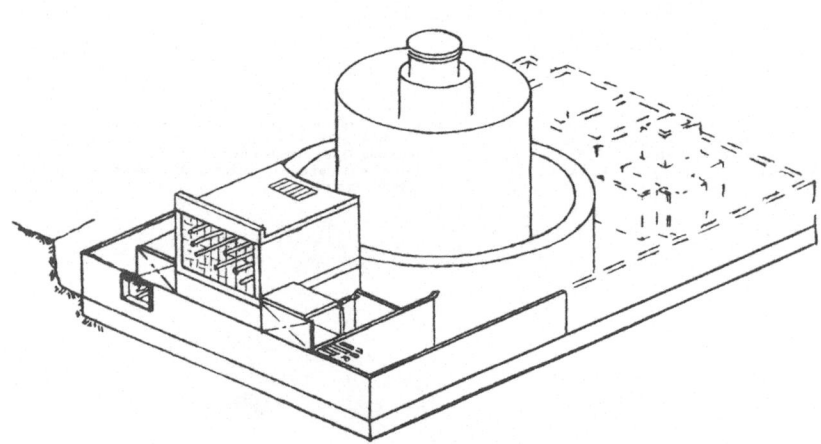

Figure 29

Auxiliary Building Construction Sequence

to estuary-type energy center locations has been proposed on
several occasions.

B.   Modular Components

     The specific concept of modular construction for an
energy center based on use of prefabricated structural concrete modules and separately installed equipment modules
has been studied in some depth by combined design teams from
General Electric and C. F. Braun.   Examples from these studies
are reviewed here.   They are representative of the design concept that is believed nearest to realization as a candidate
for energy center construction.   The design is based on the
existing BWR-6 and associated turbine island; no major configuration changes have been made to accommodate the modular
construction approach although cost savings should be achievable if a plant were to be redesigned from the ground up with
modularized construction economy as a primary objective.   A
partial listing of structural and equipment modules for the
reactor and turbine buildings, together with their schedule
for installation is shown in Figure 30.   Similar schedules
have been developed for all other components of the energy
center facilities; they are consistent with the suggested 44-
month installation schedule for individual units and a one-
per-year plant turnover basis, all with the use of a single
gantry crane for major lifts.

     The assumed general arrangement of the reactor and fuel
and auxiliary buildings and the identification of the individual structural modules is shown in Figure 31.   Similarly,
the turbine generator building structural modules are identified in Figure 32, and a partial description of these modules and of the additional mechanical piping and electrical
modules is shown in Tables XI and XII.

     The sequence of construction can be traced from founda-
tions on up by reference to the task numbering system.   In
concept, the guiding principles were:

          1.   Excavate and prepare foundations according to
               conventional construction.

          2.   Install precast structural and shield walls
               in largest feasible module sizes.

3. Make necessary mechanical fastenings with
    a minimum of on-location concrete pouring
    and grouting, and use permanent steel for
    form work wherever possible.

4. Install major equipment modules into
    cubicles formed by pre-installed structural
    walls.

5. Enclose cubicles with major equipment
    packages in place, using precast roof
    slabs.

6. Repeat sequence for ensuing floor levels.

C.   Underline On-Site Factory

The success of the modular construction approach to a
nuclear energy center project depends upon the ability to
set up a factory on site that can turn out modules at a com-
petitive cost level. Steady work load, standardized product
and factory-type working conditions all are identified as
items potentially in its favor.

The proclaimed purpose of modular construction capability in an on-site factory is to achieve increased productivity and more readily attainable adherence to quality standards as a result of improved working conditions. No change
from construction lablr unions and construction labor rates
is planned for the on-site factory work force, although the
suggestion has been made that special labor rates might be
negotiated. The potential short life span of the factory and
the importance of startup, shutdown and learning curve effects
all represent uncertainties for such an installation that need
further evaluation. Given an energy center containing at
least 10 units, a 15-year life for the facility might be expected, based on a one unit per year turnover rate. Such a
turnover rate is believed to be near optimum.

Typically, the major elements proposed for such on-site
modular components factories (one plant per year turnover
rate) in studies to date include:

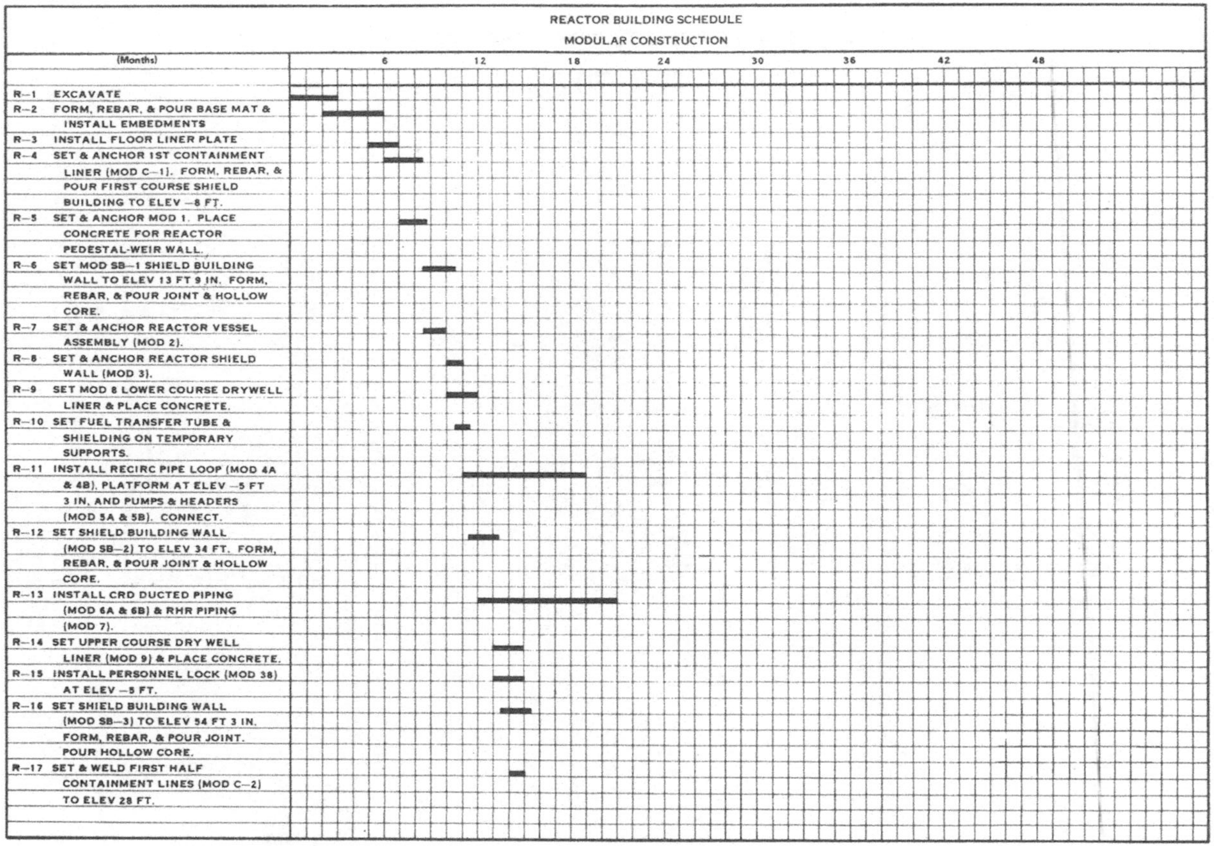

REACTOR BUILDING SCHEDULE
MODULAR CONSTRUCTION

(Months)		6	12	18	24	30	36	42	48

R—1   EXCAVATE

R—2   FORM, REBAR, & POUR BASE MAT &
        INSTALL EMBEDMENTS

R—3   INSTALL FLOOR LINER PLATE

R—4   SET & ANCHOR 1ST CONTAINMENT
        LINER (MOD C—1). FORM, REBAR. &
        POUR FIRST COURSE SHIELD
        BUILDING TO ELEV —8 FT.

R—5   SET & ANCHOR MOD 1. PLACE
        CONCRETE FOR REACTOR
        PEDESTAL-WEIR WALL.

R—6   SET MOD SB—1 SHIELD BUILDING
        WALL TO ELEV 13 FT 9 IN. FORM,
        REBAR. & POUR JOINT & HOLLOW
        CORE.

R—7   SET & ANCHOR REACTOR VESSEL,
        ASSEMBLY (MOD 2).

R—8   SET & ANCHOR REACTOR SHIELD
        WALL (MOD 3).

R—9   SET MOD 6 LOWER COURSE DRYWELL
        LINER & PLACE CONCRETE.

R—10 SET FUEL TRANSFER TUBE &
        SHIELDING ON TEMPORARY
        SUPPORTS.

R—11 INSTALL RECIRC PIPE LOOP (MOD 4A
        & 4B), PLATFORM AT ELEV —5 FT
        3 IN. AND PUMPS & HEADERS
        (MOD 5A & 5B). CONNECT.

R—12 SET SHIELD BUILDING WALL
        (MOD SB—2) TO ELEV 34 FT. FORM,
        REBAR. & POUR JOINT & HOLLOW
        CORE.

R—13 INSTALL CRD DUCTED PIPING
        (MOD 6A & 6B) & RHR PIPING
        (MOD 7).

R—14 SET UPPER COURSE DRY WELL
        LINER (MOD 9) & PLACE CONCRETE.

R—15 INSTALL PERSONNEL LOCK (MOD 38)
        AT ELEV —5 FT.

R—16 SET SHIELD BUILDING WALL
        (MOD SB—3) TO ELEV 54 FT 3 IN.
        FORM, REBAR. & POUR JOINT.
        POUR HOLLOW CORE.

R—17 SET & WELD FIRST HALF
        CONTAINMENT LINES (MOD C—2)
        TO ELEV 28 FT.

Figure 30

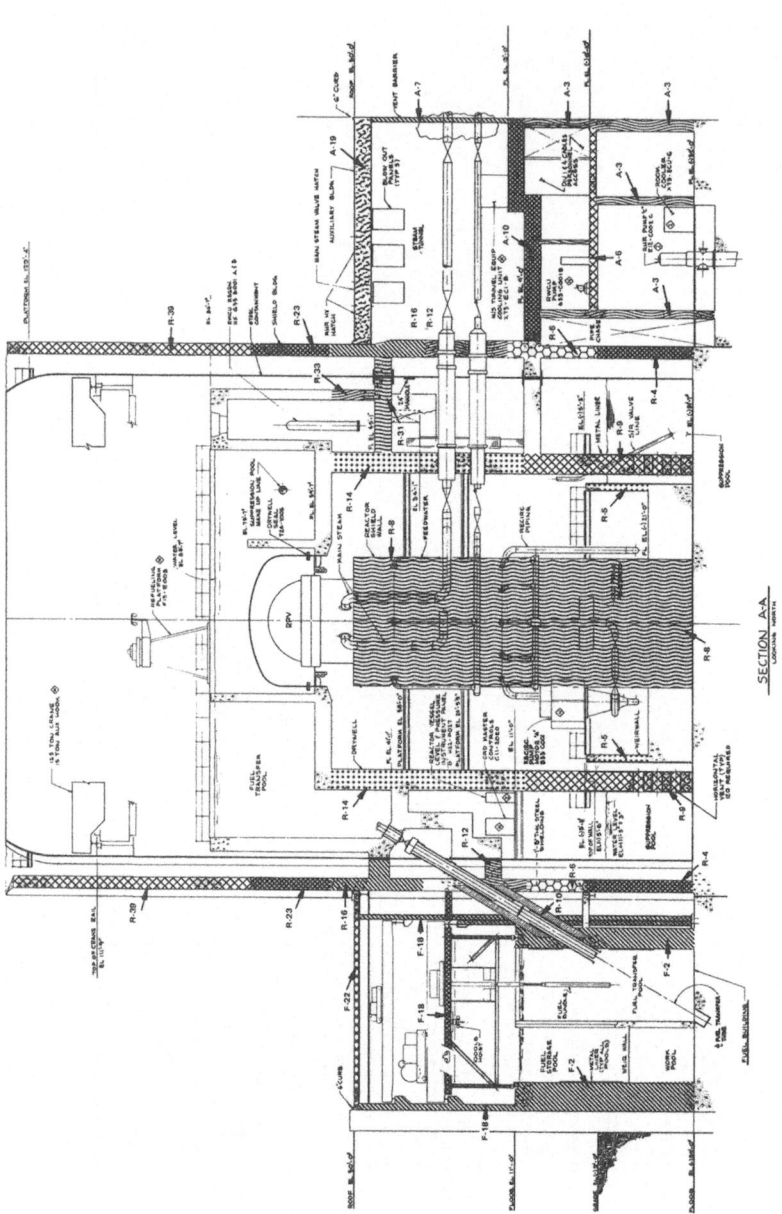

Figure 31

Reactor, Fuel, and Auxiliary Buildings

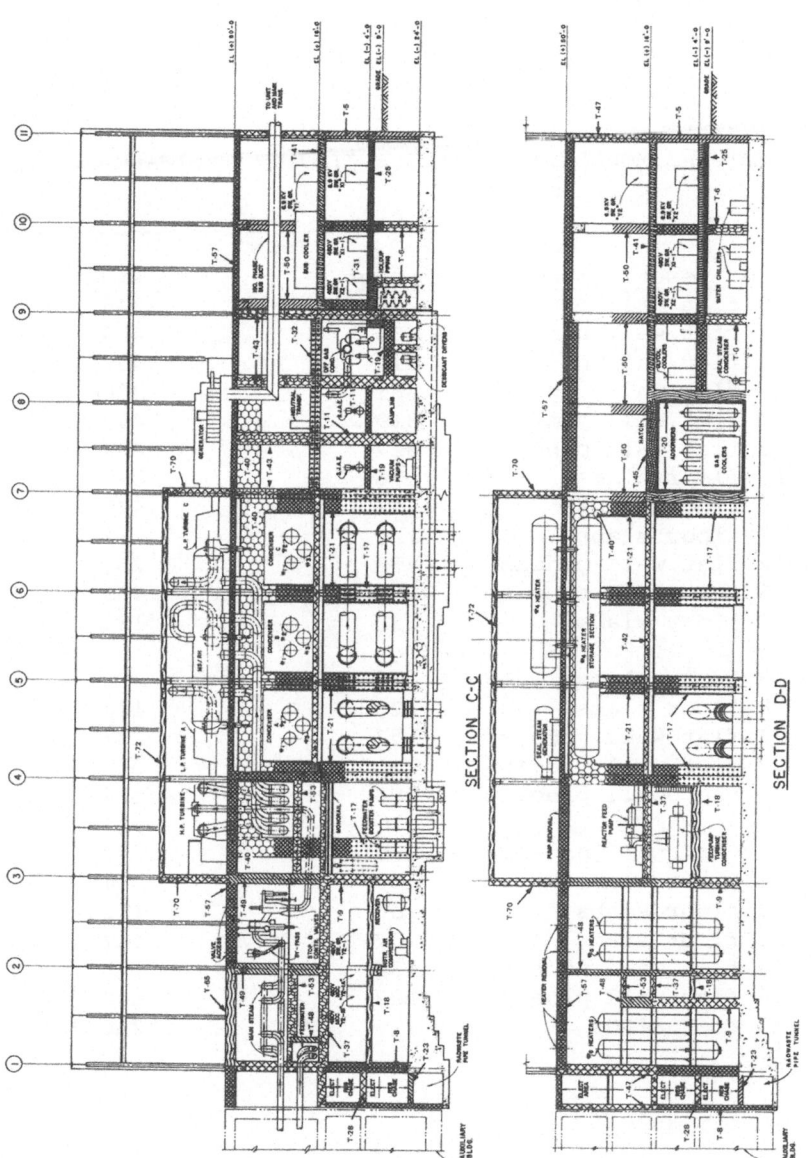

Figure 32

Figure 32
Turbine Building

TABLE XI
STRUCTURAL MODULES: TURBINE BUILDING: TYPICAL

		Tons	
Schedule	Description	Average	Maximum
T-5	8 flat wall panels	248	345
T-6	3 flat wall panels	364	458
T-8	17 flat wall panels	310	458
T-9	9 flat wall panels	361	386
T-11	11 flat wall panels	319	483
T-13	10 flat wall panels	227	245
T-18	12 floor slabs	176	230
T-25	18 floor slabs	242	420
T-28	4 floor slabs	132	132
T-29	2 flat wall panels	252	252
T-30	6 flat wall panels	267	405
T-32) - 12 floor slabs		132	196
T-37)			
T-41	18 floor slabs	156	205
T-42	8 floor slabs	126	144
T-43	9 flat wall panels	157	227
T-47	27 flat wall panels	181	380
T-48) - 6 flat wall panels		287	427
T-49)			
T-50	6 flat wall panels	200	273
T-53	8 floor slabs	323	487
T-54	3 floor slabs	186	192
T-57	38 floor slabs	207	447
T-65	4 flat wall panels	132	132
T-70	14 flat wall panels	122	140
T-72	12 roof slabs	167	230

TABLE XII
TYPICAL MECHANICAL, PIPING, AND ELECTRICAL MODULES
TURBINE BUILDING

Schedule	Description
T-4	Maximum length prefabricated circulating water pipe runs.
T-15	a HVAC chillers and pumps b Seal steam condenser c Vacuum pumps d Sampling system e Off-gas dryers f Dryer regenerators g Condensate pumps h Condensate transfer pumps i Closed cooling water exchangers j Closed cooling water pumps k Service and instrument air compressors and equipment l Closed cooling water treatment tank and surge tank m Electric auxiliary boiler n Feedwater booster pumps
T-16	Radwaste pipe modules
T-22	Off-gas holdup piping
T-24	First level regulated electric chase cable trays
T-26	a Steam jet air ejectors b Off-gas coolers
T-27	a Feed pump turbine condensers b 480 volt switch gear c 480 volt motor control centers d 6.9 kV switch gear e Off-gas sampling and analyzing

Concrete Module Facilities: $\sim 10{,}000$ m^2

- batch plant
- rebar assembly yard
- precast yard

Mechanical Module and Support Facilities
    5,000 m^2 covered

- pipe shop   - electrical shop
- steel fab shop   - sheet metal shop
- nuclear components   - rigging shop
- T-G assembly   - paint shop
- mechanical module   - warehouses
- project offices   - motor pool

The effectiveness of these facilities in assembling large modules depends on the ready availability of a heavy lift capability. A typical layout proposed for the on-site factory is thus one in which there is near direct service from a gantry crane that spans the power plant job sites. The on-site factory layout shown in Figure 33 is a possible compromise in which a transporter is used to have assemblies from the various factory sites to the gantry crane service area. A possible improvement over this arrangement is one in which dedicated and unlimited clearance railroad shipping within the energy center site is employed.

The gantry crane service area and scheduling of lifts turns out to dictate the whole on-site factory concept:

1. Lifts should not travel over largely
    completed/operating plant; i.e.,
    construction sequence is:

    ① , ⑥ , ② , ⑤ , ③ and finally ④ .

2. Underlying area evacuation may be desirable
    for certain heavy lifts.

3. Traverse time and effort for crane travel
    distances $\gg 1$ km may prove unreasonable.

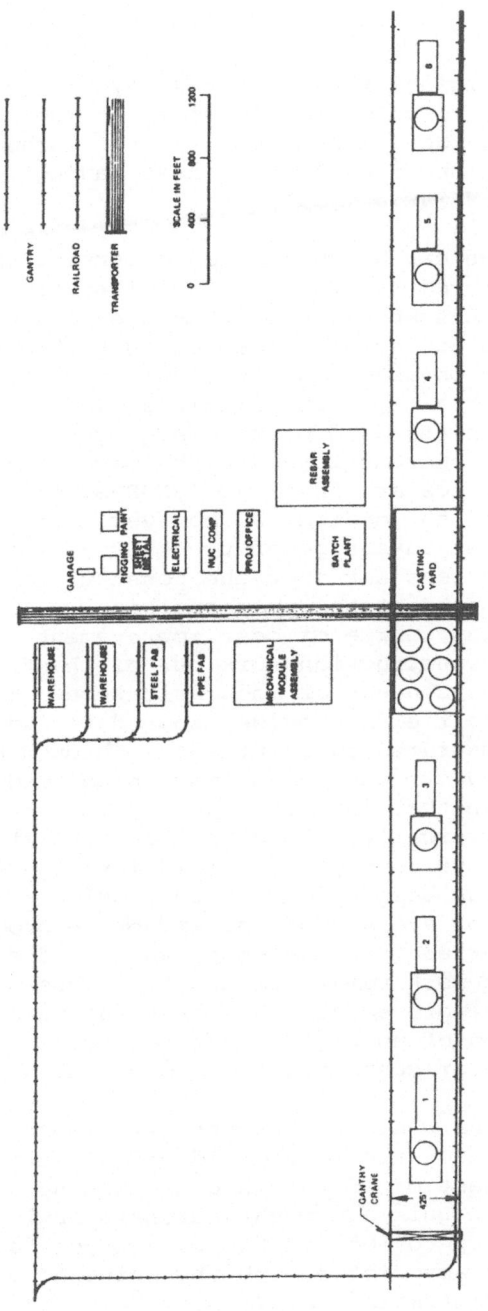

Figure 33

Possible On-Site Factory Layout

D.   Uncertainties

     Much of the original impetus for study of energy center
development related to potential for cost savings. Cost savings through shared facilities, cost savings through standardization or replication of design and cost savings through on-
site construction of large modules.

     Some confidence can be expressed concerning the ability
to save costs on shared facilities, administrative buildings,
maintenance, etc. Elsewhere, the subject of shared fuel cycle
facilities has been addressed. Cost savings through standardization and through on-site construction characteristically
are subject to uncertainty. In principle, many of the savings associated with standard plant design can be achieved
from multiple unit construction over the same time period,
even if differing sites are involved (SNUPPS, for example).
A certain portion of the design is, however, site unique and
thus a potential cost advantage of the energy center, provided
designs can be held constant over the construction period of
a large number of units at the proposed energy center site.
The biggest potential source of cost improvement relates,
however, to moving construction work out of the field into
the shop through the concept of modularization. Experience
in other engineering fields provides some data and the basis
for the trends illustrated in Figure 34. Through the use of
large, pre-assembled structural and mechanical modules, estimates have been made that just over 50% of the field construction effort (typically, 10-12 man hours/kW at $18/hour
in 1975) might be moved out of the field into the shop. A
two-way advantage has been suggested to result from this, a
potential doubling on average of the effective productivity
of the work transferred into the shop, and a modest improvement in the on-site work remaining due to a lesser level of
congestion. These improvements have been estimated to yield
an overall reduction of up to 30% in construction labor over
that for an equivalent conventionally constructed plant.

     In addition, the concept of repetitive construction by
the same labor force on a number of identical units on the
same site has been estimated to yield a learning curve effect
of several percent. Bulk equipment purchase savings are
estimated to range up to 10% and to average possibly 5%,
while indirect cost associated with the original plant design, project administration, etc., have been estimated to
be reduced by up to 30% or 40%.

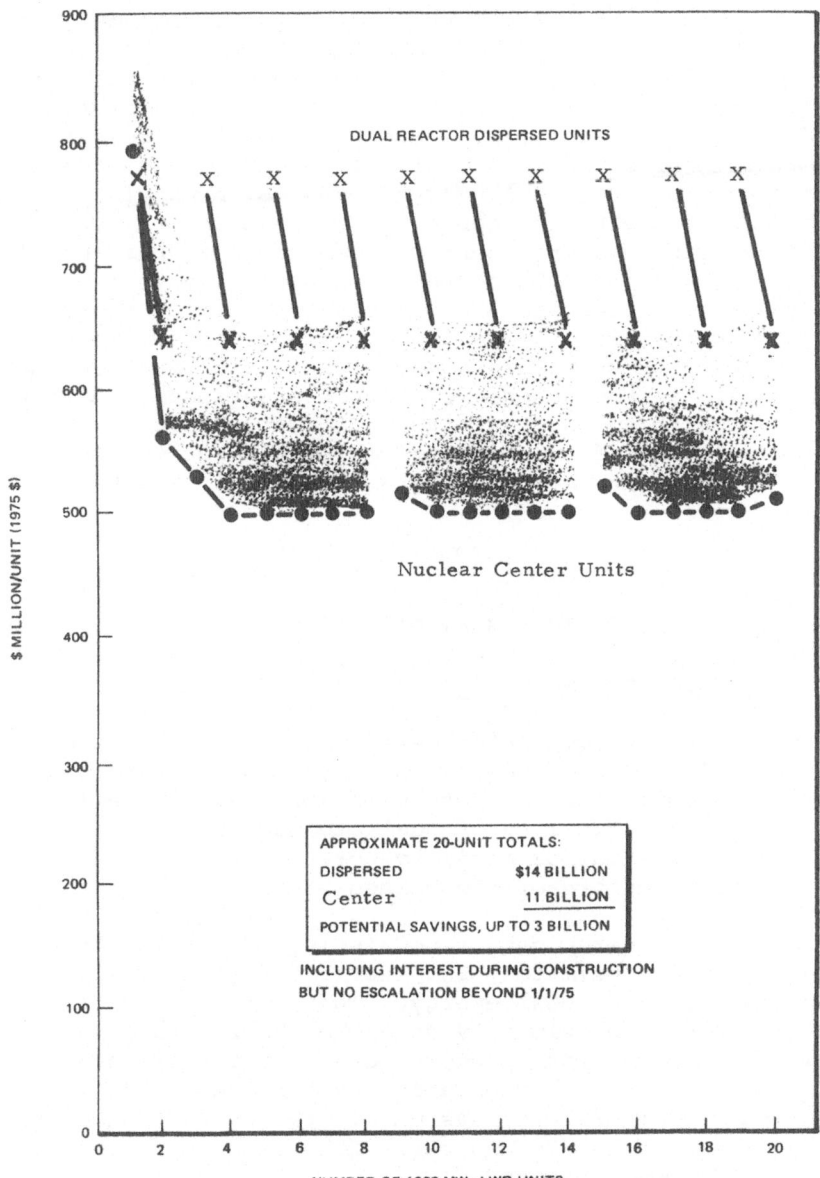

Figure 34

Construction Cost Comparison
Center Versus Dispersed Generation

The culmination of all these factors, plus some judgments concerning uncertainties suggests the major potential for cost savings of Figure 34. In this figure, typical dual reactor plant costs are shown at an average of approximately $550/kW (1.3 GW units at $767 and $647 million, respectively) including interest during construction at 8%, but no escalation provisions beyond January, 1975. By comparison, a first unit at the energy center is shown to be somewhat more expensive, due to energy center front-end cost burdens of as much as $70 million. Subsequent units, however, start to show major savings up to the point at which some factory or crane relocation is proposed (after eight units). A substantial band of uncertainty must remain with such estimates at this time in the absence of direct experience. Nevertheless, even with an uncertainty band of + 30%, it appears that there may be significant incentive for a large module approach to construction at energy centers of possibly six to eight units or more.

IV.   HEAT REJECTION

A.   Center Siting Implications

Potential energy center developments are addressed to a point in time (post-1980) when once-through fresh water cooling for their heat rejection is not a viable design option and the availability of adequate water supply for large-scale (20 GW thermal or more) evaporative cooling may be in doubt in many sectors of the United States.

A 20-unit energy center consisting of 1.3 GW net units operating at approximately 33% thermal efficiency will reject approximately 52 GW heat energy. This corresponds to the consumption of approximately 45,420 m^3/hour total usage including provisions for blowdown, etc., of approximately 17.0 m^3/s of water if evaporative towers are used. There are a limited number of river systems in the United States that are large enough to furnish this without major perturbation of their minimum flow condition; i.e.:

	Minimum Flow (m³/s)
Mississippi	3566
Ohio	719
Missouri	115
Tennessee	557
Red	41
Arkansas	56
White	147
Wabash	48
Cumberland	99
Mississippi (above Missouri)	452
Non-simultaneous sum	2264
St. Lawrence	4500
Columbia	2200
Snake	334
Willamette	102
Susquehanna	42
Mobile	
Alabama	212
Tombigbee	14
Hudson	29

NOTE: Minimum flow is based on 1961-1965
        Department of Interior, Geological
        Survey data.

Normally, evaporative towers require additional (comparable amount) water for blowdown, and still more water for boiler makeup. The above observations suggest:

1. Large energy centers would pose a potential
    water consumption problem in many sectors
    of the United States.

2. Local storage to carry through minimum flow
    period and thus better utilize available
    water runoff appears logical.

3.   Oceanside locations for energy centers,
      utilizing a sea water heat sink offer
      some attraction.

4.   Smaller water demands from more modestly-
      sized energy centers and/or use of
      combined wet and dry cooling towers may
      prove more appropriate for certain water-
      short areas of the United States.

The possibility of utilizing reject heat for process
industry, space heating, etc. has been investigated extensively. Unfortunately, the scale of industry needs at any
one process plant location typically is very small, relative
to the reject heat from even one 1.3 GW nuclear unit, let
along from an energy center in total. Waste heat utilization will not normally reduce the net water requirements
of a complex below those of the power installation alone.

A further significant consideration regarding large
quantities of waste heat rejection is the potential impact
on climatic conditions. While understanding of the meteor-
olical effects of heat islands, such as large cities or
groups of power plants at an energy center is advancing,
there remains some major questions concerning the point at
which significant weather disturbances can appear downwind.
Heat reject levels for a 26 GW energy center will be comparable to heat rejection from a city with a population of
five million. A city of five million population may extend
over an area of several hundred square miles; by comparison,
the energy center typically would occupy approximately ten
square miles. Thus, the reject heat energy density may be
as much as ten times as high at such an energy center. Statistically, significant downwind weather effects have been
identified for a number of the larger U. S. cities in recent year (19).

Further studies are in progress on this point. In the
absence of definitive findings, there is some logic to:

1.   preferential consideration of smaller energy
      center complexes (∿ 10 GW thermal) that do not
      represent such a major extension beyond
      current levels of heat rejection to the
      atmosphere by evaporative cooling.

2.   alternately consider preferential location
     of energy centers at sites where heat re-
     jection to the ocean may be used, and thus,
     dispersion of the reject heat over a much
     larger area is attainable.

B.   Water Availability

While a rigorous nationwide analysis of water availability versus potential energy center requirements for cooling represents a major undertaking, there are sufficient
data in the literature to suggest that, in most areas, total
consumptive water requirements for all power generation are
a small or even negligible fraction of total runoff. Data
from the Water Act Group Report by Espey Houston & Associates (20) are indicative of this (see Table XII).

Some potentially water-limited areas as defined by
Water Resource Region are:

     Texas Gulf
     California
     Arkansas, White, Red
     Upper and Lower Colorado
     Great Basin

The first two of these areas coincides with major power
generation regions of the United States, and not entirely by
coincidence, they also are bordered by a significant length
of seacoast. Sea water cooling is thus potentially avail-
abile if coastal or near-coastal energy center locations
were to be selected. The remaining water-deficient regions,
while not bounded by seacoast, also are well-removed from
major energy growth areas, and thus are unlikely sites for
energy center development.

It is therefore conceivable that at least modest size
(10 GW-20 GW) energy centers requiring on the order of
1.89 x 10^6 m^3/day could be located in most regions, of substantial anticipated power growth. Consideration of temporary storage provisions (makeup ponds) of possibly a
modest investment in water transportation for supply to the
evaporative towers, and of saline water sourcing for Gulf
and California coast locations appears relevant.

TABLE XIII

20-YEAR DROUGHT ANNUAL WATER SHORTAGES (-) FOR YEAR 2000, ENERGY PREDICTION (PRE-1974) AND RETROFIT OF ONCE-THROUGH COOOLING TO WET COOLING TOWERS

Principal FPC Regions	Specific Water Resource Council Regional Location	Annual Flow Available 95% of Years (10^9 gal/day)	Water Consumption Use Excluding Steam-Electric (10^9 gal/day)	Retrofit Consumption Use by Steam-Electric Generation in 2000 (10^9 gal/day)	Water Shortage (-) From Total Water Consumption (10^9 gal/day)
I	New England	112.0	1.90	0.81	96.5
I	Middle Atlantic	116.0	6.89	1.75	104.3
III	S. Atlantic-Gulf	42.4	4.25	4.44	35.3
I,II,IV	Great Lakes	67.5	3.64	1.32	62.0
II	Ohio	24.4	2.14	2.26	23.2
III	Tennessee	28.5	0.40	0.46	24.7
IV	Upper Mississippi	24.6	1.73	1.13	19.8
V	Lower Mississippi		2.52	1.06	
VI	Souris-Red-Rainy	1.91	⎱14.24	0.016	⎱8.8
VI	Missouri	23.9	⎰	0.61	⎰
V	Arkansas-White-Red	33.4	8.44	1.19	22.8
V	Texas-Gulf	11.4	10.83	1.84	-2.4
V	Rio Grande	2.1	4.82	0.11	-2.6
VI	Upper Colorado	7.5	⎱12.09	0.12	-3.9
VIII	Lower Colorado	0.85	⎰	0.33	-2.4
VII	Great Basin	2.46	5.42	0.005	
VII	Pacific NW	138.0	13.71	0.31	122.6
VIII	California	25.6	31.25	1.03	-11.0
Total Contiguous United States:		663.0	124.3	18.8	497.7
		Table 5.3 Column 1	Table 5.7 2 Medium	Table 6.2 2000 assumed Retrofit of wet	Table 7.4 95% 2000 Retrofit

Reference: Espey, Huston & Associates, Section IX in Utility Water Act Group Report (June 1974)

C.   Possible Trends

Two major constraints have been identified for energy centers relating to heat rejection: the supply of water and the meteorological impact of exhaust plumes from evaporative towers given water availability. The question might well be asked concerning the appropriateness in the future of dry cooling towers for energy centers, given these circumstances. Brief review suggests that, with dry towers, the plume buoyancy and thus the potential for meteorological effects is, if anything, greater, and thus no help.

With regard to the minimization of water usage, such as towers or a variant of them, peak shaved dry/wet tower combined installations could have a future role, particularly if coastal or near-coastal siting in California or Texas Gulf regions were not available. The situation then becomes one of economic balance between:

> pumping water the required distance from and back to the coast for once-through seawater cooling, versus
>
> pumping water for evaporative cooling (plus handling blowdown), versus
>
> utilizing waste water for evaporative cooling (Sun Desert Plant, San Diego Gas & Electric), versus
>
> peak shaved dry/wet tower if water costs exceed approximately $0.26/m^3, but not more than $2.60/m^3, versus
>
> all dry towers if water costs were to exceed $2.60/m^3.

Thus, many cooling system design options exist for envisages energy centers, particularly if those sized are not so large as to threaten potentially significant changes in meteorology of the region.

V.   CONCLUSIONS

The concept of nuclear energy centers now is attracting significant interest in both industry and regulatory sectors of the electric utility business.

In the minds of some, it represents a next logical step in nuclear development, particularly as a response to certain environmental, nuclear materials safeguards and construction cost escalation problems of the industry. In the minds of others, the nuclear energy center may represent the exchange of one set of known constraints for an alternative that includes institutional and social uncertainties.

Both viewpoints are significant. The challenge is believed to be one of accommodating the valid considerations of each, namely:

1.   There is a basic societal value to a certain level of preplanning and some centralization of energy facility development, potentially including certain fuel cycle facilities.

2.   There is potential for improved construction techniques utilizing, where possible, upgraded working conditions.

3.   The importance of economic balance in siting and water utilization considerations is noteworthy.

4.   Avoidance of risk of significant meteorogical effects from very local heat release rates to the atmosphere is an important caution until greater understanding of the phenomenon is achieved.

It would appear that if energy center developments are to be encouraged, a series of evolutionary steps with modestly-sized installations rather than some more revolutionary advance in one step to very large centers, offers greater promise of long-term success. The evolutionary steps well may involve initial consideration of small nuclear energy centers of less than optimum size, from a full modular con-

struction standpoint. Such development may be within the
range of capability of both suppliers and individual utilities or small groupings of utilities; thus, in the near
term, tempering the question of need for some industry restructuring. While such small (5 GW-10 GW) energy centers
may preclude opportunities for major advances regarding
equipment installation techniques, other improvements such
as the potential for on-site fuel facilities still may be
practical, given adoption of the PuB concept. The water
requirements and atmospheric heat release burden locally
would not represent order of magnitude changes from prior
experience.

The concept of what has been termed a reversed priorities mode of energy center development is consistent with
the suggested evolutionary introduction of energy center
development into the electric utility industry. It could
consist of the following steps:

> Step 1 - Select site for potential development
> of an energy center, but initially develop it
> for fuel storage from existing LWR's at sur-
> rounding dispersed sites.
>
> Step 2 - When sufficient spent fuel inventory
> has built up and the technology/regulatory
> uncertainties are resolved, construct a fuel
> processing facility and commence uranium
> recycle.
>
> Step 3 - When sufficient plutonium inventory
> on site has built up and technology/regulatory
> uncertainties are resolved sufficiently to
> permit economic plutonium recycle, commence
> building dedicated plutonium burning reactors
> on the energy center site, together with the
> plutonium fuel fabrication facilities to
> support them.

The above sequential program, even for a 5 GW PuB energy
center site, well could extend over 20 years. It might progress with similarly-sized (5 GW-10 GW) centers, in parallel
at as many as 10 separate sites across the United States
during the 1980's and 1990's. It could provide a valuable
base for nuclear development after the year 2000, by which

time many of the other uncertain issues of the present would
be better resolved including:

1.   uranium resource and logic of prolonged
      plutonium recycle in LWR's

2.   the commercialization and energy center
      role for the LMFBR

3.   construction economies to be derived on
      site manufacture of supermodules

4.   meteorological effects of large, localized
      heat rejection to the atmosphere

5.   advanced in electrical transmission be-
      tween large energy centers and the load
      centers

REFERENCES

1.   Energy Independence Plan submitted by Chairman Ray
      to the President December 1, 1973.

2.   Energy Reorganization Act of October 19, 1974.

3.   Puerto Rico Energy Center Study TID 25602, July, 1970.

4.   Evaluation of Nuclear Energy Centers, WASH 1288,
      January, 1974.

5.   Assessment of Energy Parks versus Dispersed Generating
      Facilities, NSF 75-500, May 30, 1975.

6.   NRC's Nuclear Energy Center Site Survey, ANS Winter
      Meeting, San Francisco, November, 1975.

7.   State/Regional Perspectives on Designated Sites and
      Energy Parks, Baroff, et al. (National Geologic Conf),
      ANS Winter Meeting, San Francisco, November, 1975;
      also, NUREG 75/018 of March, 1975.

9. A National Plan for Energy Research, Development and
    Demonstration: Creating Energy Choices for the
    Future, ERDA 48, Scenario V, June 28, 1975.

10. 26th Annual Electrical Industry Forecast, Electrical
    World, September 15, 1975.

11. "GESMO - The Generic Environmental Impact Statement
    on the Use of Plutonium in Mixed Oxide Fuels in Light
    Water Reactors" (Draft) WASH 1327, August, 1974.

12. NRC Announces Procedures for Considering Wide-scale
    Use of Mixed-oxide Reactor Fuel, NRC Announcement
    No. 75-720, November 12, 1975.

13. Environmental Survey of Uranium Fuel Cycle, WASH 1248,
    April, 1974.

14. "The Back End of the Fuel Cycle," AIF Fuel Cycle Conf.
    Paper, B. Wolfe, March 20, 1975.

15. "Plutonium Fuel Management Options in Large PWR's,"
    Hellens & Shapiro, Combustion Engineering, November,
    1973.

16. "Reactor Fuel Cycle Costs for Nuclear Power Evaluation,"
    WASH 1099, December, 1971.

17. "Modular Construction of Nuclear Power Centers," C. F.
    Braun & Co., March 1, 1975 (work in support of ref. 5).

18. "Modular Construction Using Steel-Water Modules for
    Nuclear Power Centers," Nuclear Services Corporation
    (work in support of ref. 5).

19. "On Possible Undesirable Atmospheric Effects of Heat
    Rejection from Large Electric Power Centers," Rand
    Corporation, R-1628-RC, L. R. Koenig, December, 1974.

20. "Consumptive Water Use Implications of the Proposed
    EPA Effluent Guidelines for Steam Electric Power
    Generation," Espey Houston & Associates, Austin,
    Texas, May 31, 1974.